# Surface Reactive Peptides and Polymers

# Surface Reactive Peptides and Polymers

## Discovery and Commercialization

C. Steven Sikes, EDITOR
*University of South Alabama*

A. P. Wheeler, EDITOR
*Clemson University*

Developed from a symposium sponsored
by the Division of Industrial and Engineering Chemistry
at the 197th National Meeting
of the American Chemical Society,
Dallas, Texas,
April 12–13, 1989

American Chemical Society, Washington, DC 1991

Scplae
CHEM

**Library of Congress Cataloging-in-Publication Data**

Surface Reactive Peptides and Polymers: Discovery and
Commercialization: Developed from a symposium sponsored by
the Division of Industrial and Engineering Chemistry at the 197th
National Meeting of the American Chemical Society, Dallas,
Texas, April 12–13, 1989 / C. Steven Sikes, A. P. Wheeler, editors.

    p.    cm.—(ACS Symposium Series; 444)
Includes bibliographical references and index.
ISBN 0–8412–1886–2

  1. Proteins—Surfaces—Congresses. 2. Peptides—Surfaces—
Congresses. 3. Phosphoproteins—Surfaces—Congresses.
4. Surface chemistry—Congresses. 5. Biomineralization—
Congresses.    I. Sikes, C. Steven. II. Wheeler, A. P.
III. American Chemical Society. Division of Industrial and
Engineering Chemistry. IV. American Chemical Society.
Meeting (197th : 1989 : Dallas, Tex.). V. Series.

QP551.S915   1991
574.19′245—dc20                      90–19384
                                     CIP

The paper used in this publication meets the minimum requirements of American
National Standard for Information Sciences—Permanence of Paper for Printed Library
Materials,              ANSI              Z39.48–1984.           ∞

PRINTED IN THE UNITED STATES OF AMERICA

# Foreword

THE ACS SYMPOSIUM SERIES was founded in 1974 to provide a medium for publishing symposia quickly in book form. The format of the Series parallels that of the continuing ADVANCES IN CHEMISTRY SERIES except that, in order to save time, the papers are not typeset, but are reproduced as they are submitted by the authors in camera-ready form. Papers are reviewed under the supervision of the editors with the assistance of the Advisory Board and are selected to maintain the integrity of the symposia. Both reviews and reports of research are acceptable, because symposia may embrace both types of presentation. However, verbatim reproductions of previously published papers are not accepted.

# Contents

INDEXES

# Preface

PEPTIDES, WITHOUT QUESTION, are among the most fascinating chemical materials, for they contribute to most life processes. For example, brain peptides like the endogenous opioids help to control such fundamental human responses as love, hate, pleasure, fear, cold, pain, sleepiness, and possibly even the entire range of human emotions.

Some peptides are recognized for their impact on the body's neural or hormonal systems, but other peptides are best-known for their tastes. Perhaps one of the most studied molecules ever, and the only peptide commercialized on a large scale, is the synthetic sweetener aspartame, a dipeptide of $N$-aspartyl-phenylalanine methyl ester. An estimated 30 million pounds of aspartame are produced annually. Given the range of chemical activities of peptides and their potential usefulness, it is no wonder that peptide research has become a huge enterprise with a bright future. Biomedical or other commercialization of new peptides for human use no doubt will be accompanied by much necessary study, debate, and caution, and thus introduction of these materials may be delayed. However, industrial and agricultural applications can be found for many peptides and polypeptides—including anticrystallization agents, dispersants, and other polyelectrolyte materials; peptide growth substances for animals; and peptide antibiotics—and these may be more readily available for commercialization.

The 25 chapters of this book cover these and other novel peptides, as well as the more traditional polymers, and the authors come from both academia and industry. A unifying theme is that the interaction of a peptide or other material with a solid surface is a fundamental component of the action mechanism of these chemicals. Such interactions range from the binding of a peptide onto an inorganic crystal for regulation of its growth to the binding of a peptide growth substance to a cellular surface and the resulting impact on cellular growth and differentiation.

Chapters 1 through 13 draw examples from the mineralization sciences, defined broadly to include the regulatory interactions of soluble organic materials with any crystalline or solid phase. The emphasis is on peptides that regulate the formation of calcium carbonates and phosphates. Primary structures, mechanisms of action, methods of synthesis, and derivatization are all considered. Potential applications of the peptides include scale inhibition, particle dispersion, promotion of crystallization, corrosion control, and oral health care products.

Chapters 14 through 19 focus on the power and sophistication of biotechnology as applied to peptides and include discussions of growth factors, regulation of calcium-enriched tumors, peptide antibiotics, and peptide antifreezes. Chapter 14 describes some technical hurdles encountered in preparing for commercialization of bovine somatotropin, a growth substance now scheduled to be produced at the level of hundreds of millions of grams per year.

Chapters 20 through 22 take a more industrial tone, looking at problems of adhesion, corrosion, and dispersion. Some real-world issues of synthesis are addressed, along with some simulations of industrial operating conditions.

The final three chapters offer a challenge. Chapters 23 and 24 describe how organisms regulate silica deposition, and Chapter 25 discusses how silica deposits are dealt with in industrial settings. Because silica polymerization is amorphous at the molecular level, one commonly held view is that it may be impossible to design a polymer that can interact with silica to prevent silica deposition. Yet organisms seem to accomplish both promotion and inhibition of silica formation with amazing precision. Development of a polyamino acid (or another polymer) that could regulate silica deposition would be a useful discovery.

Overall, the book promotes the use of peptides and polyamino acids in a variety of settings. An advantage of these materials is that they generally are nontoxic and biodegradable. Consequently, the book is intended not only as a source of information for scientists and engineers in industry, academia, the military, but also as evidence for government officers and policy-makers that environmentally compatible alternative materials are being developed for a number of applications.

## Acknowledgments

We are grateful to the Industrial and Engineering Chemistry Division of the American Chemical Society for providing the forum for this work and to the Graduate School and Coastal Research and Development Institute of the University of South Alabama; the Mississippi–Alabama Sea Grant Consortium; the South Carolina Sea Grant; and the Alabama Department of Economic and Community Affairs for providing partial funding.

Our thanks to Jerelyn Cox, Denise Weisbach, Joy Earp, and Carolyn Wheeler for administrative assistance at the Mineralization Center. Caroline Byrns, Mitzi Brett, and Mary Dean of the Office of Publication Services at the University of South Alabama provided production assistance. (We are also grateful to these people for being patient and understanding in dealing with us.) Our editor at ACS was Robin Giroux. Her publishing experience and insight into our problems were most useful. She was both professional and supportive.

Finally, we appreciate and admire our authors' research and contributions to this book. We wish them well and hope that in some way this book will help to promote their endeavors.

C. STEVEN SIKES
University of South Alabama
Mobile, AL 36688

A. P. WHEELER
Clemson University
Clemson, SC 29634–1903

September 18, 1990

# Chapter 1

# Phosphoproteins as Mediators of Biomineralization

**Arthur Veis, Boris Sabsay, and Chou Bing Wu**

**Division of Oral Biology, Northwestern University, Chicago, IL 60611**

Acidic proteins, and phosphorylated acidic proteins in particular, appear to play several important roles in the mechanism of mineralization in biological systems. They may direct the placement of the mineral crystals within and upon the organic matrix of tissues such as bone and dentin. They may also regulate the crystal growth rate. Further they may limit the size of the formed crystals by surface adsorption. Dentin has been selected as a model for the study of these problems because it is a rich source of a unique phosphoprotein, phosphophoryn, which may serve all of these regulatory functions. As a mineralization mediator the phosphophoryn appears to bind to the collagen matrix and to act to complex inorganic ions to initiate calcium hydroxyapatite crystallization in an ordered fashion on the matrix. Studies on the native and dephosphorylated protein show the profound effect of the phosphate groups on the calcium binding properties of the phosphoprotein. The phosphophoryn molecule has distinct domains which may relate to the different regulatory activities postulated for this one molecule.

In this paper, a general background on biomineralization is provided, defining concepts and hypotheses in their broadest terms. The dentin system is then used as one model for examination of these concepts and hypotheses.

## Biomineralization

The presence of deposits of inorganic crystals in living organisms is very widespread and the processes of formation of the crystalline phases are grouped together under the name "biomineralization", the "bio" in the term suggesting the active involvement of the cells of the host organism in the process. The relationships between mineral and organic phases are very diverse, but a few years ago Lowenstam and Weiner (1) identified two broad categories: biologically induced mineralization and matrix-mediated mineralization.

0097–6156/91/0444–0001$06.00/0

In biologically induced mineralization, macromolecules of biological origin interact with the microions, usually cations, to initiate a mineral phase. The mineral phase crystals, however, adopt crystal habits similar to those they would have formed from saturated solutions of the microions in the absence of the organic macromolecules. Furthermore, the crystal deposits are essentially random within the organic matrix. That is, the crystal orientations do not appear related to the structure of the underlying matrix. There is no regulation of crystal size.

In matrix-mediated mineralization one again has macromolecules of the organic matrix interacting with mineral microions (cations) to initiate the mineral phase. This is the same statement given above, but in this case the mineral phase grows within the preformed organic matrix and, most importantly, the minerals adopt some unique crystal habits with respect to either the nature of the crystal (possibly different from that produced by spontaneous crystallization from a saturated solution of the microions) or the orientation of the crystals relative to the underlying organic matrix. Matrix-mediated mineralization is further characterized by the crystals having a fairly narrow size range. All aspects of matrix-mediated mineralization appear to be highly regulated.

In our own studies (2) we had also recognized the distinctions between these two types of biomineralization processes. However, because of our interest in vertebrate systems, and bone mineralization in particular, we focused our efforts on the matrix-mediated mineralization process. We postulated that the matrix within which the mineral crystals grow is itself a two component system. The first part of this system is a structural protein. The organization of this protein defines both the internal architecture of the mineralized tissue and the overall shape of the tissue. The second part of the two component system is another protein, or another macromolecule, which can interact both with the microions to be incorporated into the crystals and with the structural protein. These dual interactions direct the placement and orientation of the initial mineral crystals within the matrix. Further crystallization may develop by growth on the initial crystalline deposits, so that all of the crystallization is not confined to this first phase interaction, but it is this first interaction which establishes the order and crystal habit.

Finally, it has to be made explicitly clear that the organic matrix is not homogeneous in composition and that in a mineralizing tissue there are well defined zones with different properties. All of the structure is generated by the cells of the particular tissue. In the immediate vicinity, or territorial matrix, of the cells there is generally a non-mineralized zone, or zone of crystallization inhibition (in bone this is the "osteoid", in dentin the "predentin"). Adjacent to this is a zone where crystal nucleation and growth occurs (in bone and dentin this is the "mineralization front"). Deeper within the mineralized tissue there is a final zone of crystal stabilization. A key tenet of our basic working hypothesis (2) is that specific, interactive matrix macromolecules regulate all stages of the mineralized phase initiation and maturation.

## Dentin: Type I Collagen and Phosphophoryn

We selected dentin as a relatively simple model for bone mineralization because only a single population of cells, the odontoblasts, is engaged in forming the matrix. No resorptive cells are present and the tissue does not remodel.

Like bone, the dentin matrix has Type I collagen as the structural component. The collagen fibers which constitute the matrix are fairly uniform in size. The initial mineralization at the mineralization front, at the boundary between predentin and dentin, takes place in association with the collagen fibers. Several studies directed explicitly to the question (3) have shown that there is very little to distinguish the type I collagen of dentin from the predominant type I collagen of the soft tissues.

A large number of noncollagenous proteins (NCP) are present in the dentin matrix, as illustrated in the gel electrophoretic pattern shown in Figure 1. One of these proteins, a highly phosphorylated protein which we have named phosphophoryn (4), is the most abundant of the NCP. The phosphophoryn (PP) is unique to dentin, none has been found in other tissues. However, we believe that there are functional analogs of this molecule in other mineralizing tissues. Bovine PP (bPP) is the best characterized of the phosphophoryns (5). It is a large molecule, $M_r \sim 150,000$, containing about 1130 amino acid residues, Table I. Aspartic acid accounts for about 450 residues and serine another 550. Of these serines about 90% are phosphorylated. Thus, about 950 of the 1130 total residues (84%) are potentially anionic. Moreover, since the phosphate groups have a second ionizable group with a pK of 6.8 (6), a substantial number of the phosphoserines must be doubly ionized at physiological pH and ionic strength. Thus, the net charge on the molecule *in vivo* is on the order of -1300, making it an extremely anionic macromolecule with a net charge per backbone residue > 1. In our very earliest experiments where free flow electrophoresis was used, this protein had the highest electrophoretic mobility of any component in the system and it was initially called the "fast" component.

Molecules equivalent to bPP have been found in the dentin of every species thus far examined. The unique anionic character is well preserved, but the apparent molecular weights vary from one species to the next. In the rat incisor riPP has a $M_r \sim 70,000$; in human dentin the highest weight hPP has $M_r \sim 110,000$. In both of these tissues, in contrast to the bovine case, there may be more than one class of PP, varying in both amino acid composition and the degree of phosphorylation. In the human, the variation in molecular weights may be related to an age-dependent *in vivo* degradation.

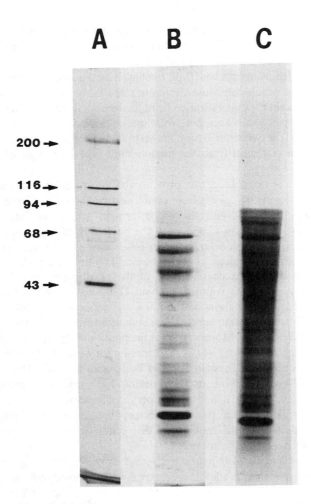

Figure 1. Gel electrophoresis of total 0.6 M HCl extract of rat incisor dentin, using a 5 to 15% acrylamide gradient, 0.1% SDS, reduced with mercaptoethanol. Lane A. Molecular weight standards. Lane B. Silver stain of the proteins. Lane C. The same as lane B but stained with Stains All after silver staining. Note that different bands are stained. Although there are obviously many protein bands, the HCl extract does not solubilize all of the matrix proteins. The band in lane C marked with the arrow is phosphophoryn. It is colored blue with the Stains All but not stained by silver.

Table I. The Amino Acid Composition of Bovine Dentin Phosphophoryn

| Residue | Number of residues/molecule |
|---------|:---------------------------:|
| Lys | 45 |
| His | 6 |
| Arg | 3 |
| **Asp** | **452** |
| Thr | 8 |
| **Ser** | **27** |
| **PSer** | **518** |
| Glu | 14 |
| Pro | 6 |
| Gly | 27 |
| Ala | 7 |
| 1/2 Cys | 1 |
| Val | 3 |
| Met | 1 |
| Ile | 3 |
| Leu | 4 |
| Tyr | 2 |
| Phe | 2 |
| Total | 1130 |
| GluNH$_2$ | 1 |

Calculated on the basis of a molecular weight of 155,000.
Data of Stetler-Stevenson and Veis (5)

## Phosphophoryn is a Domain Structure Protein

Attempts to sequence any of the PP have not met with much success. The overriding content of just two amino acids, plus the high content of phosphoserine, makes the problem very difficult from the perspective of direct sequencing. Thus, the sequencing will probably require the cloning of the PP gene and analysis with the use of the techniques of molecular biology. Such studies are under way in our laboratory, as well as in a few others but not much information is available as yet.

We have gained some useful information, however, by examining the amino-terminal sequences from peptides produced by limited trypsin digestion and by limited acid hydrolysis with weak acids. In this latter procedure the acid hydrolysis cleaves out the aspartic acid residues in peptide sequences. The peptides inserted between the Asp residues are stable to mild acid hydrolysis and can be collected and examined for amino acid composition and sequence. This has turned out to be a formidable task since in addition to free Asp, other free amino acids and some very small peptides (di, tri) were released. However, these data clearly suggest the presence of domains with blocks of sequences such as:

$$\{Asp\}_x$$
$$\{Asp-(P)Ser\}_y$$
$$\{(P)Ser\}_z$$

The symbol (P)Ser is used to indicate that the residue may be either serine or phosphoserine. Acid hydrolysis does not cleave out every Asp residue and some larger peptides can be recovered. Work on these peptides is in progress.

The tryptic digestion studies, also still in progress, have allowed two preliminary conclusions. First, some atypical, low Ser and Asp sequences containing some of the more hydrophobic residues are in the end regions of the molecule, and one peptide with the amino-terminal sequence

$$\{Ser-PSer-PSer-Ser-PSer-PSer-Ser-Ser-Ser-\}$$

has been isolated. The presence of this peptide confirms the presence of blocks of $\{(P)Ser\}_z$ noted above. the details of these experiments will be presented elsewhere. However, the PP molecule can apparently be modeled as comprised of:

$$\{H-domain\ 1\}\ \{H-domain\ 2\}\ \{Asp\ domains\}_x\ \{Asp-Ser\ domains\}_y$$
$$\{(P)Ser\ domains\}_z$$

In this model "H-domains" indicate relatively hydrophobic regions with a lower content of Asp and (P)Ser. These studies are very exciting and should ultimately permit a detailed consideration of the PP properties and mechanisms of action. In the meantime these data do suggest that the molecule may be multifunctional in the sense that different domains may be involved in different aspects of the mineralization process.

## Phosphophoryn is a Calcium Ion Binding Protein

Neither phosphoserine nor aspartic acid side chains in polypeptides have a particularly high binding affinity for calcium ions, yet, as shown in Figure 2, one can distinguish two clear classes of calcium ion binding affinities. The binding constants shown in Table II for the high affinity class, are really quite modest when compared to the binding affinities of intracellular proteins such as calmodulin. However, the bPP makes up for that by its enormous binding capacity. We assume that the weaker second class of binding sites reflects a typical colloidal reversal of charge phenomenon, in which the high positive surface charge brings in counter anions and a second layer of diffusely bound divalent cations. Note that in the experiments of Figure 2 and Table II the calcium ion binding was measured in the presence of 0.5 M KCl, a supporting electrolyte concentration that would swamp out nonspecific ionic interactions. The addition of even millimolar concentrations of $Ca^{2+}$ to bPP in 0.5 M KCl is sufficient to convert the bPP in dilute solution from a random chain to a more ordered structure, probably to a $\beta$-sheet like conformation. At physiological ionic strength the calcium

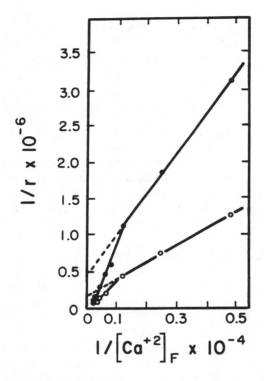

Figure 2. Double reciprocal plots of the binding of calcium ion to bPP in the presence of 0.5 M KCl, 0.01 M Tris-HCl, pH 8.3. The upper plot is at a bPP concentration of 0.05 mg/ml, lower plot at 0.5 mg/ml. These differences clearly show the non-ideal, bPP concentration dependence of the binding, as well as the biphasic nature of the calcium ion-bPP interaction. Reprinted with permission from ref. 11. Copyright 1987 Springer-Verlag.

binding coefficient for bPP is higher than in 0.5 M KCl, on the order of 3.6 x $10^{-4}$ mol$^{-1}$ (6).

Table II. Binding of Calcium Ions to Bovine Dentin Phosphoryn at the High Affinity Sites

| bPP, mg/ml | moles Ca/mg bPP | moles Ca/mole bPP | $K_i$, moles$^{-1}$ |
|---|---|---|---|
| 0.05 | $2.28 \times 10^{-6}$ | 353 | 771 |
| 0.50 | $5.65 \times 10^{-6}$ | 876 | 780 |

Determined in 0.5 M KCl, 0.01 M Tris.HCl, pH 8.3.
Data of Stetler-Stevenson and Veis (11).

Studies with model peptides suggest that repetitive carboxyl-phosphate side chain sequences strongly promote calcium binding and enhance the apparent binding constant for the carboxyl group calcium binding by at least an order of magnitude. Thus repetitive {Asp-(P)Ser}$_y$ sequence domains may have a particularly strong role in the calcium ion binding behavior of PP (7).

### Phosphophoryn is a Collagen Binding Protein

Phosphophoryn binds directly to both monomeric collagen in solution or to the surfaces of preformed fibrils. In the *in vivo* situation the odontoblast secretes the collagen matrix and forms it into fibrils. The PP is secreted separately and then binds to the surfaces of the already formed fibrils (8,9).

The binding of bPP to collagen is essentially electrostatic and can be reduced at high ionic strength, but it is so strong as to have many elements of specificity. In experiments in which [125]I-bPP was added to collagen, cold bPP could displace the labeled bPP. However, phosvitin, an unrelated but equally highly phosphorylated protein required six-fold higher concentrations to displace an equivalent amount of [125]I-bPP from collagen (10). Serum albumin and other matrix proteins were unable to displace significant quantities of bPP from collagen.

### Phosphophoryn Bound to Collagen Fibrils Enhances Calcium Ion Binding of the Fibrils

The uptake of calcium ion by bPP-conjugated to collagen fibril surfaces was studied in an effort to determine if the electrostatic interaction between collagen and bPP diminished the ability of the bPP to bind calcium ions in solution (9). The data on the uptake of [45]Ca onto cold bPP-collagen fibers was unequivocal in showing that collagen fibers with associated bPP interacted with calcium as avidly as did free bPP. Figure 3 compares the calcium binding to collagen fibers

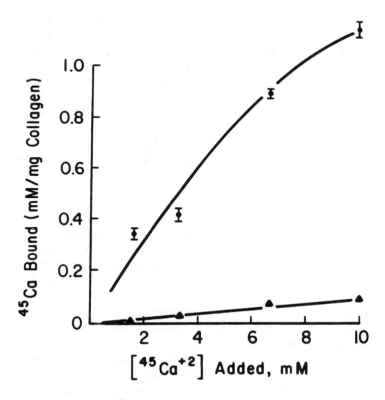

Figure 3. The uptake of [45]Ca by collagen in the presence of bPP, upper plot [●], and in the presence of osteonectin from bone, lower plot [▲]. The osteonectin data is essentially identical to that of collagen alone. The collagen concentrations, and the bPP and osteonectin concentrations were constant in every analysis in each assay. Reprinted with permission from ref. 10. Copyright 1986 Springer-Verlag.

in the presence and absence of surface associated bPP. The phosphate groups on the bPP were very important. Dephosphorylation markedly diminished both the calcium binding affinity and total binding, in accord with the peptide analog studies (7). In the studies described above in which $^{125}$I-bPP was added to preformed collagen fibers, the binding of the bPP to the collagen was on the fiber surface. In the bPP concentration ranges studied, binding never reached saturation and calcium binding to collagen fibers has not been studied under bPP saturation conditions.

In many of the early discussions of the potential role of PP in mineralization, a prime argument raised by critics of the idea was that the molar ratio of PP to collagen was less than one to one. Since the PP binding is to the fiber surfaces, and since one molecule of PP may sequester hundreds of calcium ions, it can function readily as a nucleating agent.

## Phosphorylation of Phosphophoryn

The phosphorylation of PP is an extremely interesting problem. At the moment we have no knowledge of either the mechanism of phosphorylation nor the kinases involved. The intracellular locus of phosphorylation is not known. The probable amino acid sequences within the Ser-rich domains of the PP are not known to be substrates for any of the known kinases. We have therefore begun a study of the protein kinases in order to determine what type of kinase might be involved. We selected ROS 17/2.8 cells, an osteoblast-like tumor cell line, as an appropriate source for kinases which might phosphorylate extracellular matrix phosphoproteins. An assay system was developed which permitted the specific identification, activation or inhibition of each of the known kinases. Native riPP, without any treatment to remove phosphate groups already present, was found to be a substrate for a ROS 17/2.8 kinase. However, application of the assay system to determine the kinase responsible showed that the kinase was unique. That is, none of the messenger dependent or messenger independent kinases thus far described in the literature was responsible.

The kinase, which we have named DPP-kinase, shares some features in common with casein kinase II. They are both cAMP-, cGMP-, and $Ca^{2+}$-independent; they can both utilize ATP and GTP as phosphate donors; heparin, spermine and sodium chloride inhibit both kinases; and, they both favor acidic substrates. However, the DPP-kinase is associated with the membrane bound, cell particulate fraction whereas the casein kinase II is found in the cytosol. Moreover, the two kinases have distinctly different pH optima for maximal activity. The DPP-kinase has a higher activity on the residual serines of non-dephosphorylated phosphophoryn than casein kinase II has on partially dephosphorylated casein.

Further work on this very important problem requires a better definition of the sequences which are the substrates for the DPP-kinase, and a detailed study of the rephosphorylation of the dephosphorylated PP. These studies are in progress.

## Generalizations on Biomineralization Based on Observations on the Dentin System

The development of dentin does appear to be an example of matrix-mediated mineralization. Although the role of each of the components has not been determined rigorously, we can consider the dentin system in terms of our basic postulates (2). The type I collagen fiber network surely serves as the structural framework which defines the shape of the tooth, the spaces within which the crystallization proceeds, and determines the orientations of the crystallites. For all of the reasons described above, we propose that the highly anionic phosphoprotein, phosphophoryn, is the principal bifunctionally interactive protein which, by interacting with the preformed collagen fibrils in a fairly specific manner, also determines, via its strong interaction with calcium ions, the locus of crystal nucleation within the collagen fiber network. The cellular control of this extracellular process of regulated crystallization residues in the repertoire of proteins produced for secretion and their delivery to the extracellular space. For example, the cellular sequestration of the collagen and phosphophoryn is such that the phosphophoryn is delivered directly to the mineralization front and deposited upon the collagen fiber network. A host of other macromolecular NCP components must also operate in regulation of the system. It is likely, for example, that one or more proteoglycans may inhibit induced mineralization within the predentin, that phosphatases present near the mineralization front may delay mineralization by dephosphorylating proteins which might otherwise induce premature mineral crystal initiation, and that other anionic components within the mineralization region might limit crystal growth by binding to growing crystal surfaces.

We believe that many biomineralization systems follow this same overall strategy for regulation of matrix mediated mineralization. The specific macromolecular components may vary depending upon the use and required metabolic stability of the mineral phase, but components of corresponding function will be found in quite different systems. Where proteins are involved in the process, phosphorylated species are likely to be involved because of the advantage such groups provide relative to carboxyl groups in enhancing calcium binding affinity. Where glycosaminoglycans are involved, the degree or nature of the sulfation may play a comparable regulatory role.

## Literature Cited

1. Lowenstam, H. A.; Weiner, S. In Biomineralization and Biological Metal Accumulation, Westbroek, P.; DeJong, E. W., Eds.; D. Reidel: Doredrecht, 1983; p 191.
2. Veis, A.; Sabsay, B. In Biomineralization and Biological Metal Accumulation, Westbroek, P.; DeJong, E. W., Eds.; D. Reidel: Doredrecht, 1983; p 273.
3. Volpin, D.; Veis, A. Biochemistry 1973, 12, 1452.

4. Dimuzio, M. T.; Veis, A. Calcif. Tissue Res. 1978, 25, 169.

5. Stetler-Stevenson, W. G.; Veis, A. Biochemistry 1983, 22, 4326.

6. Lee, S. L.; Veis, A.; Glonek, T. Biochemistry 1977, 16, 2971.

7. Lee, S. L.; Veis, A. J. Peptide Protein Res. 1980, 16, 231.

8. Weinstock, M.; Leblond, C. P. J. Cell Biol. 1974, 60, 92.

9. Maier, G. D.; Lechner, J. H.; Veis, A. J. Biol. Chem. 1983, 258, 1450.

10. Stetler-Stevenson, W. G.; Veis, A. Calcif. Tissue Int. 1986, 38, 135.

11. Stetler-Stevenson, W. G.; Veis, A. Calcif. Tissue Int. 1987, 40, 97.

RECEIVED August 27, 1990

# Chapter 2

# Macromolecule–Crystal Recognition in Biomineralization
## Studies Using Synthetic Polycarboxylate Analogs

L. Addadi[1], J. Moradian-Oldak[1], and S. Weiner[2]

[1]Department of Structural Chemistry and [2]Department of Isotope Research, Weizmann Institute of Science, Rehovot 76100, Israel

Control over crystal formation in biology is determined, to a great extent, by the specific recognition of a variety of acidic glycoproteins for crystal surfaces. Understanding the manner in which this occurs can be facilitated by growing crystals in the presence of simple synthetic analogs to these proteins, such as polyaspartic and polyglutamic acids, which can be either in solution or adsorbed on a solid substrate. These experiments demonstrate that the polymers are much more effective at influencing crystal nucleation and growth than the monomers, that the $\beta$-sheet conformation is essential for obtaining specific interactions, that rigid substrates are much better nucleators than fluid-like substrates, and that the different ligands associated with the macromolecules can cooperate to induce crystal nucleation or modulation.

Many mineralized tissues contain an assemblage of unique acidic glycoproteins. Their widespread distribution, their close association with the mineral phase and their characteristic acidic nature all support the notion that these macromolecules fulfill fundamental roles in biomineralization (1). A considerable amount of information exists on basic biochemical and some structural properties, as well as on calcium binding abilities of a variety of these macromolecules (2-4). One acidic protein, the aspartic acid and phosphoserine-rich phosphophoryn from dentin, has been particularly well characterized (5). It was also the first acidic protein from mineralized tissues to have been discovered (6). The *in vivo* locations of a few of these acidic macromolecules are known and there is very limited information, some of it indirect, on the spacial relations at a molecular level between the acidic macromolecules and the mineral phase (reviewed in 1). These macromolecules are particularly difficult to purify and characterize, and progress, to date, has been slow. Even if much more information on structure and localization were available, it would not necessarily provide answers to the manner in which

0097–6156/91/0444–0013$06.00/0

they function *in vivo*. Ideally what is needed in addition are *in vitro* assays for function, akin to those used for enzymes. As some of the primary activities of these proteins are presumably in regulating crystal nucleation and/or growth, the assays must assess the abilities of the macromolecules to influence crystallization processes under different conditions.

Two basically different types of "assay" experiments are used. The kinetic approach is by far the most common. In essence it evaluates the ability of the macromolecules to limit the nucleation process by changing the length of the induction period prior to crystal formation, or to affect the rates at which crystals grow. Monitoring is based on the fact that during crystallization protons are released with a concomitant change in pH (7). Alternatively the pH is maintained constant and the volume of titrant added reflects the status of the crystallization process (8). Using this approach with a wide variety of acidic glycoproteins from mineralized tissues in solution, it has generally been observed that they act as inhibitors of nucleation and/or growth (9-14). Macromolecules adsorbed onto solid substrates, however, have the tendency to induce crystal formation (15-17).

A different type of approach can be termed 'stereochemical', as it focuses on the manner in which the acidic macromolecules influence growing crystals by being adsorbed from solution onto some crystal faces and not others. Here too the macromolecule can first be adsorbed onto a solid substrate and its ability to induce nucleation off specific crystal planes is then assessed (18,19). Technically the key property monitored is the change in crystal morphology resulting from the fact that the protein inhibits growth of the faces onto which it is adsorbed. The strength of this assay is not, however, in the measure of the inhibition effect alone, but rather in differentiating specific from non-specific effects. Specific adsorption is based on the recognition by the protein of structurally and stereochemically defined molecular or ionic patterns on the crystal surfaces. This in turn may lead to control of crystal shape, size and in the case of nucleation, orientation of the induced crystals.

Studies using the stereochemical approach are, to date, limited to acidic glycoproteins extracted from mollusk shells (18,19) and sea urchin skeletal elements (20). The results (reviewed in 21,22) show that the macromolecules do indeed interact specifically with certain faces of a large variety of crystals. The affected faces all display a common stereochemical motif. Marked differences were observed in the mode of interaction of mollusk and sea urchin macromolecules with calcite crystals. Mollusk acidic glycoproteins, rich in aspartic acid and sulfated oligosaccharides, when adsorbed on a solid substrate, are able to nucleate calcite specifically off the 001 plane. This plane is composed entirely of calcium or carbonate layers arranged perpendicular to the $\underline{c}$ axis. The less acidic sea urchin proteins are not able to do this. They, however, when in solution are selectively adsorbed on the $\{1\bar{1}0\}$ faces of calcite aligned parallel to the $\underline{c}$ axis, and with further growth are occluded inside the crystal. The resulting calcite-protein composite material is still a very good single crystal, but has within it protein located at the boundaries of quasi-perfect mosaic blocks (Berman, A.; Weiner, S.; Addadi, L.; Leiserowitz, L.; Kvick, A., unpublished data).

The stereochemical approach provides some information on the interaction process by a comparison of the structure of the affected and non-affected crystal faces. It does not, unfortunately, provide complementary information on the structure of the interacting macromolecule. A partial solution to this problem is to use well defined synthetic analogs of the macromolecules with known structure. Here we review the information available to date on the study of a few simple synthetic analogs to acidic glycoproteins using the 'stereochemical' approach. Where possible, we compare the results with those obtained using synthetic analogs in the 'kinetic' approach.

## Choice of Synthetic Analogs and the Basic Questions Addressed

Very few proteins from mineralized tissues have been sequenced, and none of them appear to be the ones particularly rich in acidic residues. Thus there is no direct information with which to prepare synthetic analogs to the protein moiety. Some partial acid hydrolyzate studies of mollusk proteins (23) and dentin phosphophoryn (24) indicate that some of these proteins may contain repeating amino acid sequences in which Asp is present every second residue. Dentin phosphophoryn and probably the mollusk proteins as well, do contain stretches of polyaspartic acid (5,25). No systematic studies using aspartic acid-containing peptides of different sequences have been performed using the stereochemical approach. Sikes and Wheeler (7) have tested a variety of such peptides using the kinetic approach with interesting results, some of which will be discussed below. Our studies have focused on polyaspartic and polyglutamic acids with respect to the nucleation and growth of calcite, and of some model calcium dicarboxylate crystals.

We shall address the following questions: (i) Acidic proteins with many carboxylate containing residues interact *in vivo* with the crystals. We, therefore, first compare the effect of a carboxylate-rich polymer binding to the crystal surface to that of individual amino acid monomers. (ii) The secondary structure of different acidic matrix proteins has been observed to contain relatively large proportions of β-sheet conformation (reviewed in 1). We ask whether the presence of an ordered secondary structure in general, and of the β-sheet in particular, is essential for specifically influencing crystal nucleation and growth. (iii) Nucleating proteins probably act when adsorbed on a scaffolding matrix. It is therefore relevant to investigate the effect of both substrate and polymer rigidity versus fluidity on crystal nucleation by acidic glycoproteins, as well as to determine the properties necessary for oriented nucleation off a substrate to occur. We also address the effect of increasing amounts of immobilized charge on crystal nucleation. Here we refer specifically to the fact that some matrix proteins have been found to contain relatively large amounts of phosphoserine (5), or carbohydrates rich in carboxylate and sulfate groups (26). (iv) As many organisms control mineralization at the molecular level on a specific basis, we finally address the differences between specific and non-specific effects of polymeric additives on crystal formation.

**Effect of Monomeric and Polymeric Additives on Crystal Growth.** Additives of all kinds have long been known to influence crystal growth (27,28) by being adsorbed on crystal surfaces and consequently impairing further crystal growth. Charged polymers in particular are well known inhibitors of nucleation and growth of ionic crystals. For this reason the use of polyvinyl sulfonate and polyglutamic acid have been proposed as antiscaling agents (29), and charged polysaccharides are used in the industrial crystallization of sodium chloride (30).

A series of studies performed in the solid state group at the Weizmann Institute systematically analyzed, by means of induced morphological modifications, the stereochemical relations between the surface structure of the crystal faces selectively adsorbing an additive, and the molecular structure of the additive itself (reviewed in 31). This analytical approach was subsequently extended to the interactions between acidic glycoproteins from mineralized tissues and various calcium salts, including a series of calcium dicarboxylates; calcium fumarate, malonate, maleate and tartrate (18). The crystals were shown to adsorb proteins selectively on certain planes, thus demonstrating the stereochemical recognition capability of carboxylate-rich protein domains for certain calcium-counterion patterns present on different faces.

Crystals of calcium fumarate were grown in the presence of protein (5 $\mu$g/ml), and of aspartic or glutamic acid at concentrations equal to 10X, 100X and 1000X those of the acidic residues in the protein. Concentrations of amino acids up to 100X that of the protein did not result in any change either in the crystal growth kinetics or in crystal morphology. Glutamic acid at a concentration 1000X that of the protein decreased the crystallization rate. These crystals did not have the modified morphology typical of specific additive adsorption but rather that induced by a weak non-specific inhibitor. Sikes and Wheeler (32) also noted in a kinetic experiment that the free amino acids are inactive at a concentration equivalent to that at which polyaspartic or polyglutamic acid are very active inhibitors of crystal growth.

Marked increase in inhibition by polymers relative to their monomeric forms has also been observed in different types of crystal additive systems (33,34). In one case (34), the growth of crystals of glutamic acid hydrochloride was inhibited in the [001] direction by the presence in solution of various amino acids. The same amino acid residues were then linked to a polymeric backbone, such that their functional groups were still available for binding to the crystal surface. The resulting polymer was then used as an additive and the same inhibition effect was obtained as for the monomeric additive, but at a 100X lower concentration in solution per monomeric unit. In all instances, the effect of the monomeric inhibitors is lower by orders of magnitude than that of the corresponding polymer. There is, however, a fundamental difference between these experiments and the ones using poly-Asp or poly-Glu. In the former experiments the amino acid inhibitor either as a monomer or a polymer, binds to the (001) face of the glutamic acid hydrochloride crystals through its amino and carboxyl groups. The amino acid headgroup of the additive molecule occupies the exact lattice position of a regular molecule on the crystal surface. The binding site is defined in three dimensions and as a result the effect is thus stereochemically directed to one crystal face. The increased efficiency of the polymer is due to the concerted action of many groups, each one of which is active. In contrast, in polymers such as poly-Asp or poly-Glu the effect of single carboxylate groups on the calcium dicarboxylate crystals and on calcite, when present, is presumably due to electrostatic interactions which cannot be stereochemically defined unless various groups are organized in at least one or two-dimensional arrays. It is thus the organization of the carboxylate groups on the polymer surface that carries the stereochemical information. It can therefore be anticipated that the polycarboxylate secondary structure will be essential in determining the influence of the adsorbed polymer on crystal growth.

**The Effect of Polymer Secondary Structure on Crystal-Polypeptide Interactions.** The basic objective is to determine how the same polymer in different conformations or different polymers that are chemically equivalent but conformationally diverse, interact with growing crystals.

Polyaspartic acid, MW 20000, predominantly assumes the $\beta$-sheet conformation in solution at pH 7 to 9 and in the presence of calcium (Figure 1). This conformation is essentially not influenced by temperature in the range of 10 to 40 °C. Polyglutamic acid of the same molecular weight, on the other hand, mainly assumes the random coil conformation in solution in the presence of calcium. The circular dichroism spectra suggest, however, that at 40 °C the random coil conformation is in equilibrium with a minor amount of $\beta$-sheet. These observations are in agreement with the results of Keith (35,36), who showed that calcium salts of poly-Glu precipitated from solution at room temperature yield solid polypeptide salts in the $\alpha$-helix conformation and above room temperature, in the $\beta$-sheet conformation.

Calcium fumarate crystals grown in the presence of poly-Asp, both at 18° and at 35 °C are specifically inhibited from growing in the [010] direction (Figure 2). The effect is manifested at concentrations as low as 8 nmoles amino acid/ml. At 40 nmoles/ml total inhibition is observed even when the solutions are seeded with crystals. In the presence of poly-Glu at 18 °C and at concentrations up to 200 nmoles/ml, the growth of calcium fumarate is delayed, but the crystals develop with a regular morphology. At even higher concentrations they are only slightly affected in a non-specific manner. On the other hand, at 35 °C and at concentrations of 100-200 nmoles/ml, poly-Glu does specifically affect the [010] growth direction in a manner similar to that of poly-Asp. The concentrations required, however, are about ten-fold higher than for poly-Asp. These studies clearly show that the $\beta$-sheet conformation is important for allowing specific interactions to occur between polymer and crystal.

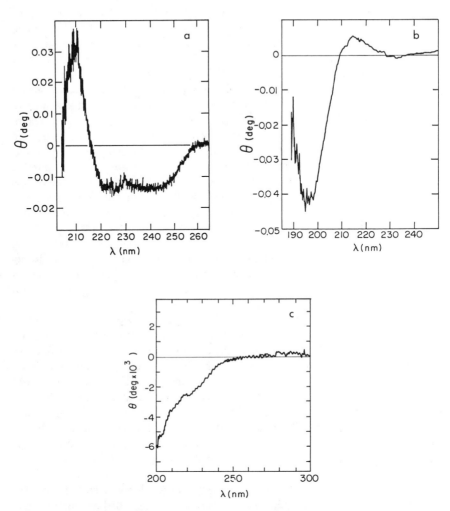

Figure 1. Circular dichroism spectra of polyaspartate and polyglutamate in the presence of 5 mM Ca++. The intensities on the ($\theta$) scale are arbitrary. (a) Polyaspartate (MW 20,000). The positive peak at 208 nm and the negative peak at 220 nm indicate that a major amount of the polymer is in the $\beta$-sheet conformation. (b) Polyglutamate (MW 20,000) measured at 10°C. The negative peak at 196 nm and the positive peak at 215 nm are typical of random coil conformation. (c) Circular dichroism spectra of polyaspartate (MW 6000) measured at room temperature. The spectrum matches that of a mixture of 60% random coil and 40% $\beta$-sheet, according to Greenfeld and Fasman (41).

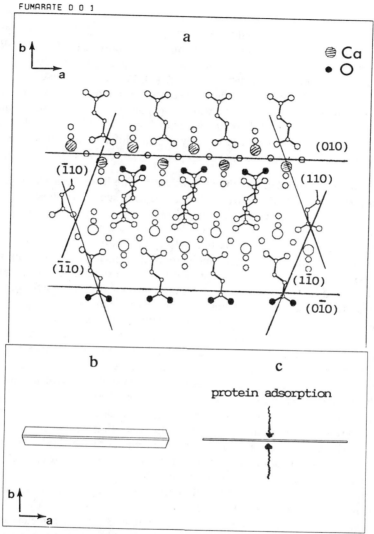

Figure 2. (a) Structure of calcium fumarate trihydrate viewed along the c̲ axis and delimited by the expressed faces {010} and {110}. The {010} faces, which are viewed edge-on, contain calcium-carboxylate miotifs emerging perpendicular to the faces (shadowed and bold atoms). (b), (c) Measured morphologies of typical crystals of fumarate seen along the c̲ axis: (b) pure; (c) grown in the presence of protein (5 μg/ml). The crystal is an {010} plate viewed edge-on. Stereoselective adsorption of the protein on the {010} faces drastically decreases the rate of crystal growth in the direction perpendicular to these faces and results in a change in the morphology of the plate.

Polymer conformation is influenced to some extent by polymer size, in that below a certain molecular weight the polymer loses its ordered conformation. Circular dichroism spectra show, for example, that poly-Asp of 6000MW in the presence of calcium adopts a conformation comprising approximately 40% β-sheet and 60% random coil, whereas the 20000MW polymer under the same conditions is predominantly in the β-sheet conformation (Figure 1). These differences influence the manner in which the polymer affects crystal growth, and may be partly responsible for the variations observed in the rate of growth of calcite crystals in the presence of poly-Asp of different molecular weights, as well as between poly-Asp and poly-Asp-poly-Ala copolymers (7).

**Controlled Nucleation Off Synthetic Substrates. The Importance of Rigidity and Conformation on Nucleation.** It has been repeatedly demonstrated that acidic proteins from mineralized tissues adsorbed onto solid substrates induce more rapid nucleation of mineral than the substrate alone (15,16). A general explanation for this effect may be that adsorption reduces the motion of the macromolecules inducing a concomitant increase in the local cation concentration, due to the polyanionic character of the protein. At the same time, the inhibiting effect manifested by the same proteins in solution is eliminated because the proteins are bound to a substrate and cannot be adsorbed onto the growing crystals or nuclei. What are the structural and chemical attributes of these macromolecules, if any, which permit them not only to induce nucleation, but as has been observed (18,19) to control the orientation of the induced crystals relative to their crystallographic axes? To address this question a series of experiments was performed *in vitro* using calcite crystal formation on a variety of synthetic substrates (19). Polystyrene films sulfonated to different extents were used as non-ordered substrates with increasing levels of charge and variable rigidity. The extent of sulfonation was assessed by attenuated total reflection - Fourier transform infrared (ATR-FTIR) spectrometry, and is directly proportional to the time of exposure of the films to sulfuric acid. Films treated for up to 8h are increasingly efficient nucleators of calcite, as judged from the number of crystals nucleated on the films per unit area. The percentage of these crystals nucleating from the 001 calcium plane also increases with sulfonation, reaching a maximum of 50% on films treated for 8 h. For reaction times longer than 8h, the parameters of the system change. The amount of sulfonation increases, but the reaction penetrates into the film etching it in depth and destroying the rigidity of the surface. This results in a fluid-like highly sulfonated surface that is a less efficient nucleator of calcite, and the percentage of (001) oriented crystals formed decreases to 5-8%. The relative degrees of rigidity and fluidity were inferred from quantitative measurements of the amount and depth of penetration of sulfonate in the films, and by comparison of the inhibitory behaviour of the highly sulfonated film surface with polystyrene sulfonate in solution. Nucleation is thus dependent not only on the charge of the nucleating surface, but also on its rigidity.

We note that the 001 layers of calcite crystals off which they nucleate are composed entirely of calcium or carbonate layers. It has been observed that sodium chloride crystals (37) and vaterite crystals (38) nucleated off compressed monolayers of stearate, also grow from layers composed entirely of sodium or calcium ions respectively. Fixed anionic charge on a rigid surface can thus induce oriented nucleation of mineral crystals from planes that are composed of positive charges. It is difficult to determine to what extent this type of effect is dependent on the precise locations of the charged groups, as the nucleation planes have an homogeneous potential.

Poorly nucleating polystyrene films sulfonated for 24 h were then used as substrates for adsorption of different carboxylate-rich polymers: polyaspartate, polyglutamate and polyacrylate, to test for their ability to enhance calcite nucleation. Polyacrylate does not adopt an ordered conformation. The conformation of the polypeptides adsorbed on the substrate in the presence of calcium were evaluated by ATR-FTIR, by monitoring the position of amide I band. Poly-Asp, with $\lambda$ max=1641 cm$^{-1}$, is approximately 50% $\beta$-sheet and poly-Glu, with $\lambda$ max=1650 cm$^{-1}$ has a lower content of $\beta$-sheet (Figure 3). Under the same adsorption conditions, polyacrylate did not induce oriented nucleation of calcite to any greater extent than the reference sulfonated film itself (8%). Adsorbed poly-Glu induced twice this amount, and poly-Asp three times the amount of (001) oriented calcite crystals. The mollusk shell acidic glycoproteins adsorbed on the same substrate also induced about 25% oriented calcite crystals. This emphasizes not only the importance of the polypeptide conformation in nucleation, but demonstrates that oriented nucleation of calcite may occur as a result of two cooperating factors together. Polyaspartate adsorbed in the $\beta$-sheet conformation on non-sulfonated films is not a nucleator of calcite, yet it is a nucleator when adsorbed on a sulfonated substrate. This must be the result of the substrate and the polymer cooperating to concentrate calcium ions (the purported role for the disordered sulfonates) and order them to some extent (the purported role for the polypeptide). This does not necessarily imply that the mechanism of nucleation is epitaxial. The 001 plane of calcite, off which nucleation occurs, is composed, as noted, of a continuous layer of calcium cations (Figure 4), and we do not know to what extent they need to be ordered in the plane for nucleation to be induced. The use of synthetic peptides of different sequences may provide some insight into this fundamental question.

The possibility of cooperative effects between different moieties attached to the same polymeric backbone had been proposed by Lee and Veis (39) for dentin phosphophoryns. A study of a synthetic peptide of 30 residues with a repeating Glu-Ser sequence, showed that partial phosphorylation of the serine residues induced an increase in the calcium binding constant, with a concomitant conformational transition of the polypeptide from random coil to $\beta$-sheet.

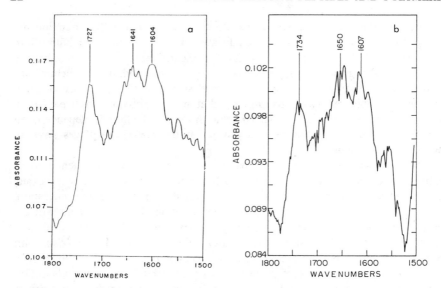

Figure 3. ATR-FTIR spectra of polypeptides adsorbed on polystyrene films treated with sulfuric acid: (a) Polyaspartate (MW 20,000). (b) Polyglutamate (MW 20,000). Absorptions around 1730 cm$^{-1}$ are due to carboxyl groups; around 1605 cm$^{-1}$ to carboxylate and around 1640 to 1650 cm$^{-1}$ to amide I. Amide I absorption for $\beta$-sheet is around 1630 cm$^{-1}$ and for $\alpha$-helix or random coil, around 1655 cm$^{-1}$ (42).

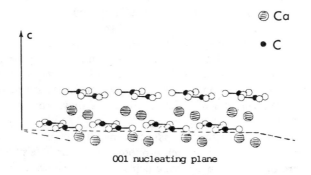

Figure 4. Packing arrangement of calcite showing the alternating layers of calcium and carbonate in the 001 plane. For clarity, the 001 plane has been tilted from the view perpendicular to the plane of the paper.

**Specific Versus Non-Specific Interactions.** We have repeatedly referred to specific interactions with polymers as opposed to non-specific ones. It is appropriate here to define and justify the differences between these two types of interactions.

Polyelectrolytes are known to act as very efficient inhibitors of the nucleation and growth of ionic crystals. In general, the electrostatic interaction arising between charged moieties on the polymer and on the crystal are sufficient to induce adhesion of the polymer to the crystal or nucleus surface, and thereby cause inhibition. This type of interaction, which we define as non-specific, may result in total blocking or in delayed growth of the crystals. Typically the crystals grown under these conditions lose their well defined morphology, and are characterized by rounded edges and stepped faces. The steps and kinks are delimited by the same stable faces which develop in the pure crystal. In contrast to this behaviour, specific inhibitors are selectively adsorbed on crystal planes which are not necessarily the most stable planes of the crystal and may not even be expressed by the natural crystal morphology. This effect is accompanied by well defined changes in morphology that may be unequivocally correlated to the structure of the faces on which adsorption takes place.

We have referred above to experiments involving the growth of calcium fumarate crystals in the presence of poly-Asp and poly-Glu, where specific and non-specific interactions were differentiated based on their effect on crystal morphology. Similar observations have been made using acidic glycoproteins from mollusk shells. When crystals of calcium fumarate, maleate and malonate are grown in their presence, at concentrations of 5 $\mu$g/ml, new faces are developed, whereas for calcium tartrate no change in morphology was observed. At 10 $\mu$g/ml protein concentration, the crystallization of calcium tartrate is delayed and the crystals begin to lose their well developed edges. At 50 $\mu$g/ml the crystals are heavily deformed and rounded, and have large pits on their flat faces. These observations were interpreted in terms of recognition of the protein for specific stereochemical motifs on the crystal surfaces of calcium maleate, malonate and fumarate, but not tartrate (18,21). The fact that the non-specific effect on calcium tartrate crystals is manifested at concentrations at least one order of magnitude higher than the concentration required for specific interactions to occur on crystals, is by no means unique to this system. It has been observed in other totally unrelated systems (40), and indicates a higher affinity of the crystals for additives which recognize and match certain structural motifs on the crystal surface.

Analogous to the acidic glycoproteins from Mytilus, polyaspartate, which has a specific effect on the growth of calcium fumarate crystals (see section (ii)), behaves towards the growth of calcite as a strong non-specific inhibitor. We interpret these results as indicating that stronger polyelectrolytes lose selectivity towards specific surfaces of high charge density. The interactions are dominated rather by electrostatic factors, independent from stereochemistry.

## Discussion

The synthetic analog experiments reviewed here demonstrate the following: (1) Polymers are much more effective at influencing crystal nucleation and growth than monomers (2). The $\beta$-sheet conformation is favored for obtaining specific interactions between carboxylate-rich polypeptides and a variety of crystals (3). The polypeptide size is important, in particular in so far as it affects conformation (4). Rigid substrates are much better nucleators than fluid-like substrates (5). Different ligands associated with the polymer and/or substrate can cooperate to induce crystal nucleation or modulation.

Perhaps the most important attribute of the stereochemical approach is its ability to differentiate between specific and non-specific macromolecule-crystal interactions. In this way it has been shown that a variety of acidic glycoproteins from mineralized tissues are able to specifically influence aspects of crystal formation. The observations using synthetic analogs reviewed here, provide key insights into some underlying basic mechanisms. Specificity involves a certain degree of matching between the interacting crystal and protein surfaces. This interaction is regulated by the stereochemistry of the functional group distribution and is, therefore, dependent upon the protein conformation and on the structure of the affected crystal face. In fact all the crystal faces affected by acidic glycoproteins studied to date possess a common structural motif (41). Specific recognition is also a function of total electrostatic interaction. A heavily charged protein or polypeptide may still be selective towards crystal surfaces of low charge density, or vice versa. Selective interactions can, however, be overwhelmed by densely charged crystal surfaces interacting with heavily charged macromolecules.

Specific interactions also require one or both of the two surfaces to be rigid. Thus, for nucleation to be induced, when the crystal nuclei are not stable, the protein counterpart must be rigid, or rendered rigid by adsorption on a matrix substrate. When the formed crystals themselves provide a rigid substrate, selective and cooperative adsorption of protein may result in the inhibition of growth in specific directions. This in turn can lead to control of crystal morphology and/or crystal properties.

The perspective described herein provides some insight into the manner in which the numerous macromolecules associated with the mineral phase in hard tissues may function. It obviously highlights the importance of the widespread carboxylate-rich acidic glycoproteins in mineralized tissues. Much more information on the macromolecules themselves is needed if we are to understand, for example, why some macromolecules from mollusk shells are nucleators of calcite, whereas others are strong inhibitors (unpublished results). Information on the primary, secondary and if at all possible tertiary structures of both the protein and oligosaccharide moieties are needed to explain such phenomena.

Purified fragments could be tested using the available assays and again, synthetic analogs could be produced to focus in on the basic mechanisms involved.

Understanding the modes of action of isolated macromolecules *in vitro* is only one step towards achieving the ultimate objective: determining their functions *in vivo*. *In vivo* they operate in a specific environment, either out of solution or off structured substrates; a fundamental difference in terms of their modes of action. The precise locations of adsorbed molecules and the identities of the other macromolecules with which they may be associated also need to be known, as different macromolecules may well cooperate to produce the necessary *in vivo* function. Very little direct information on the *in vivo* environment of crystal formation is available. To provide a perspective on just how sophisticated the *in vivo* solution may be, it is worthwhile in conclusion to reflect on aspects of the molecular architecture of the mollusk shell nacreous layer (42). The flat aragonite nacreous tablets with their extensive unstable 001 planes are quite different in shape when compared to the acicular inorganic crystals of aragonite. In nacre crystals not only are the c-axes oriented perpendicular to the plane of nucleation, but the a and b axes in the plane of nucleation are well aligned at the molecular level with some of the macromolecules of the underlying substrate. It is not even understood why aragonite and not one of the other polymorphs of $CaCO_3$ is always formed in this shell layer. We will need the combined information from molecular scale ultrastructural studies, exhaustive biochemical characterizations and well designed *in vitro* experiments, to ultimately solve these very challenging problems.

## Literature Cited

1. Lowenstam, H. A.; Weiner, S. On Biomineralization; Oxford University Press: New York, 1989; Chapters 2,3.
2. Crenshaw, M. A. In Skeletal Growth of Aquatic Organisms; Rhoads, D.C.; Lutz, R.A., Eds.; Plenum Press: New York, 1980; p 115.
3. Termine, J. D. In Cell and Molecular Biology of Vertebrate Hard Tissues. Ciba Foundation Symposium 136; John Wiley and Sons: Chicester, U.K., 1988; p 178.
4. Boskey, A. L. Bone and Mineral 1989, 6, 111-23.
5. Veis, A. In Cell and Molecular Biology of Vertebrate Hard Tissues. Ciba Foundation Symposium 136; John Wiley and Sons: Chicester, U.K., 1988; p 161.
6. Veis, A.; Perry, A. Biochemistry 1967, 6, 2409-16.
7. Sikes, C. S.; Wheeler, A. P. In: Chemical Aspects of Regulation of Mineralization; Sikes, C. S.; Wheeler, A. P., Eds.; Univ. South Alabama Publication Services: Mobile, Alabama, 1988; p 15.
8. Nancollas, G. H.; Mohan, M. S. Arch. Oral. Biol. 1970, 15, 731.
9. Cuervo, L. A.; Pita, J. C.; Howell, D. S. Calcif. Tissue Res. 1973, 13, 1-10.

10. Termine, J. D.; Eanes, E. D.; Conn, K. M. Calcif. Tissue Res. 1980, 31, 247-51.

11. Wheeler, A. P.; George, J. W.; Evans, C. A. Science 1981, 212, 1397-8.

12. Menanteau, J.; Neuman, W. F.; Neuman, M. W. Metab. Bone Dis. & Rel. Res. 1982, 4, 157-62.

13. Romberg, R. W.; Werness, P. G.; Riggs, B. L.; Mann, K. G. Biochemistry 1986, 25, 1176-80.

14. Moreno, E. C.; Kresak, M.; Kane, J. J.; Hay, D. I. Langmuir 1987, 3, 511.

15. Lussi, A.; Crenshaw, M. A.; Linde, A. Archs. Oral Biol. 1988, 33, 685-691.

16. Linde, A.; Lussi, A.; Crenshaw, M. A. Calcif. Tissue Int. 1989, 44, 286-95.

17. Boskey, A. L.; Maresca, M.; Appel, J. In The Third International Conference on Chemistry and Biology of Mineralized Tissues: Glimcher, M. J.; Ed.; 1989; p 175 (abstract).

18. Addadi, L.; Weiner, S. Proc. Natl. Acad. Sci. U.S.A. 1985, 82, 4110-4.

19. Addadi, L.; Moradian, J.; Shay, E.; Maroudas, N. G.; Weiner, S. Proc. Natl. Acad. Sci. U.S.A. 1987, 84, 2732-6.

20. Berman, A.; Addadi, L.; Weiner, S. Nature 1988, 331, 546-8.

21. Addadi, L.; Weiner, S. Mol. Cryst. Liq. Cryst. 1986, 134, 305-22.

22. Addadi, L.; Weiner, S. in Biomineralization Chemical and Biochemical Perspectives; Mann, S.; Webb, J.; Williams R.J.P., Eds.; VCH Verlagsgesellschaft: Weinhem, 1989; p 133.

23. Weiner, S.; Hood, L.; Science 1975, 190, 987-9.

24. Krippner, R. D.; Nawrot, C. F. J. Dent. Res. 1977, 56, 873.

25. Wheeler, A. P.; Rusenko, K. W.; Sikes, C. S. In Chemical Aspects of Regulation of Mineralization; Sikes, C. S.; Wheeler, A. P., Eds.; University South Alabama Publication Services: Mobile, Alabama, 1988; p 9.

26. Greenfield, E. M.; Wilson, D. C.; Crenshaw, M. A. Amer. Zool. 1984, 24, 925-32.

27. Wells, A. F. Phyl. Mag. 1946, 37, 180; ibid 217; ibid 605.

28. Buckley, H. E. In Crystal Growth, Wiley: New York, 1951.

29. Sarig, S.; Kahana, F.; Leshem, R. Desalination 1975, 17, 215.

30. Birchall, J. D.; Davey, R. J. J. Cryst. Growth 1981, 54, 323.

31. Addadi, L.; Berkovitch-Yellin, Z.; Weissbuch, I.; van Mil, J.; Shimon, L. J. W.; Lahav, M.; Leiserowitz, L. Angew. Chem. Int. Ed. Engl. 1985, 24, 46-85.

32. Sikes, C. S.; Wheeler, A. P. In Biomineralization and Biological Metal Accumulation. Westbroek, P.; de Jong, E. W.; Eds.; Reidel: Dordrecht, 1983; p 285.

33. Staab, E.; Addadi, L.; Leiserowitz, L.; Lahav, M. Advanced Materials 1990, 2, 40-43.

34. Zbaida, D.; Weissbuch, I.; Shavit-Gati, E.; Addadi, L.; Leiserowitz, L.; Lahav, M. Reactive Polymers 1987, 6, 241.

35. Keith, H.D.; Giannoni, G.; Padden, F. J. Biopolymers 1969, 7, 775.

36. Keith, H.D. Biopolymers, 1971, 10, 1099.

37. Landau, E.M.; Popovitz-Biro, R.; Levanon, M.; Leiserowitz, L.; Lahav, M.; Sagiv, J. Mol. Cryst. Liq. Cryst. 1986, 134, 323.

38. Mann, S.; Heywood, B. R.; Rajam, S.; Birchall, J. D. Proc. R. Soc. Lond. 1989, A423, 457-71.

39. Lee, S. L.; Veis, A. Int. J. Peptide Protein Res. 1980, 16, 231-40.

40. Weissbuch, I.; Zbaida, D.; Addadi, L.; Leiserowitz, L.; Lahav, M. J. Amer. Chem. Soc. 1989, 109, 1869.

41. Addadi, L.; Berman, A.; Moradian-Oldak, J.; Weiner, S. Conn. Tissue Res. 1989, 21, 457-465.

42. Weiner, S.; Traub, W. Phil. Trans. R. Soc. Lond. 1984, B304, 425-34.

43. Greenfield, N.; Fasman, G. D. Biochemistry 1969, 8, 4108.

44. Parker, F. S. Application of Infrared Spectroscopy in Biochemistry, Biology and Medicine; Plenum Press, New York, 1971.

RECEIVED August 27, 1990

# Chapter 3

# Crystal Engineering of Inorganic Materials at Organized Organic Surfaces

Stephen Mann, Brigid R. Heywood, Sundara Rajam, and Justin B. A. Walker

School of Chemistry, University of Bath, Bath BA2 7AY, United Kingdom

The nucleation and growth of the mineral, CaCO₃, under compressed Langmuir monolayers has been studied by optical and electron microscopy and electron diffraction. The structure and crystallographic orientation of the mineral formed is dependent on a complex interplay of surface and solution parameters. Negatively charged stearate films induce oriented calcite or vaterite nucleation depending on the Ca concentration in supersaturated solution. The crystals are aligned with the [1$\bar{1}$0] and [001] axes perpendicular to the monolayer surface, respectively. The common features of these interactions are Ca accumulation at the stearate headgroups, partial geometric matching of lattice distances and stereochemical correspondence between carboxylate and carbonate groups at the monolayer/ crystal interface. By contrast, crystallization under positively charged octadecylamine monolayers resulted in almost complete vaterite nucleation independent of Ca concentration. Two distinct crystallographic alignments were observed, viz. [001] and [110]. Although Ca binding is absent, there is the possibility of stereochemical matching between the binding motif of bicarbonate and the crystal surfaces nucleated under the film. These results indicate the potential of organized organic substrates to control inorganic crystal nucleation and suggest that this may be a viable system for modeling biomineralization processes and as a new approach to controlled materials synthesis.

The investigation of surface-reactive peptides and polymers involves a wide range of fields and interests. One rather esoteric relationship is that between inorganic solid state chemistry and organic macromolecular chemistry. The impetus for this apparently unrelated connection arises from studies of biological mineralization in which the nucleation and growth of inorganic materials are controlled by

0097–6156/91/0444–0028$06.00/0

the structural and chemical nature of organic macromolecules. Thus the mimicking of these processes through the use of synthetic peptides and polymers has potential applications in the controlled fabrication of advanced materials.

Several approaches are currently being investigated. These include the modification of crystal growth properties by water soluble macromolecules (1-6), activation of nucleation by macromolecules adsorbed onto rigid substrates (6-8) and onto functionalized polymer surfaces (8,9) and crystallization within membrane bound phospholipid vesicles (10-12). A further approach, developed initially in the study of oriented crystal growth of amino acids and NaCl (13-15), is the use of compressed two-dimensional Langmuir films as organized templates for nucleation. We have utilised this method in the study of the crystallization of calcium carbonate from supersaturated aqueous solutions (16-18. Our aims have been to investigate the influence of headgroup charge and stereochemistry and packing density, as well as solution conditions, on the structure and orientation of crystals nucleated under the organic surface. This paper summarizes our recent results.

## Methods

The general experimental approach is outlined in Figure 1. Monolayer films of stearic acid ($CH_3(CH_2)_{16}COOH$, $pK_a = 5.6$) and octadecylamine ($CH_3(CH_2)_{17}NH_2$; $pK = 10.6$) were spread from chloroform solutions onto surface-cleaned supersaturated Ca bicarbonate subphases. The films were compressed to surface pressures of 45 mN m$^{-1}$ (fully compressed) and 20 mN m$^{-1}$ (partial compression) by means of a constant perimeter barrier device. The supersaturated solutions were prepared by purging stirred aqueous suspensions of $CaCO_3$ with $CO_{2(g)}$ at a rate of approximately 0.18 m$^3$ h$^{-1}$ for 1 hour. The suspension was then filtered and the filtrate purged with $CO_2$ gas for 0.5 h to dissolve any remaining crystals. The resulting supersaturated solution had a pH of 5.8-6.0. Experiments were done at total Ca concentrations of 8.5-9.0 mM (estimated by EDTA titration) and successive dilutions of these solutions. At total [Ca] = 9 mM, the supersaturation, defined as $S = [H_2CO_3^*] / Kp(CO_2)$ where $[H_2CO_3^*]$ represents the overall concentration of aqueous carbon dioxide and carbonic acid, K the solubility of $CO_2$ in gmol l$^{-1}$ atm$^{-1}$ at an ionic strength of 2.34 x 10$^{-2}$ M, and p($CO_2$) the partial pressure of $CO_2$, was 3.7 x 10$^3$ atm.

Crystals were examined *in situ* by optical microscopy and on coverslips dipped through the films. These were also mounted on scanning electron microscopy (SEM) stainless steel stubs. The mature crystals were studied after a period of 21 hours. Crystals at early stages of growth were mounted on carbon-coated formvar-covered copper transmission electron microscopy (TEM) grids by dipping procedures. Imaging and electron diffraction studies were undertaken. Bulk samples for X-ray diffraction (XRD) were obtained by collecting the crystals on glass slides dipped through the air/water interface. Quantitative XRD was undertaken using calibration curves established for known calcite/vaterite mixtures recorded

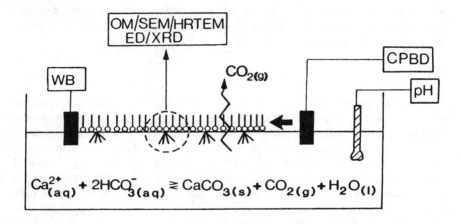

Figure 1. Experimental procedure for growth of $CaCO_3$ crystals under compressed Langmuir monolayers. OM = optical microscopy, SEM = scanning electron microscopy, HRTEM = high resolution transmission electron microscopy, ED = electron diffraction, XRD = X-ray diffraction, WB = Wilhelmy balance, CPBD = constant perimeter barrier device.

on a Philips PW 1710 X-ray diffractometer. Intensities measurements were made at the (104) and (118) reflections of calcite and vaterite respectively.

It was found that experiments involving fully compressed monolayers could be readily reproduced in rigorously cleaned glass crystallization dishes. In these experiments compressed films were formed by adding known amounts of surfactant to generate a solid phase film at the vacuum cleaned air/water interface. The advantage of this approach was that a range of experiments could be performed under identical conditions of system preparation. In particular, the effect of successive dilutions of the stock supersaturated solution could be readily assessed.

## Results

**Control Experiments.** In all experiments, crystallization was governed by the slow loss of $CO_2$ gas from unstirred supersaturated solutions according to the reaction:

$$Ca^{2+}_{(aq)} + 2HCO_3^-{}_{(aq)} = CaCO_{3(s)} + CO_{2(g)} + H_2O_{(l)}$$

In the absence of monolayer films crystals grew randomly at the air/water interface. The crystals were calcite, intergrown, non-oriented and had rhombohedral and truncated rhombohedral morphologies (Figure 2a). The particle size distribution was heterogenous (mean = 30 $\mu$m, $\sigma$ = 12.5 $\mu$m). Vaterite was present at the air/water interface in varying amounts (maximum 25 wt%). Crystals formed at the bottom of the containers were invariably discrete non-oriented rhombohedral calcite. The above observations were not significantly changed by lowering the total Ca concentration.

**Stearic Acid Monolayers.** *(a) [Ca] = 9 mM.* Crystallization under fully and partially compressed stearic acid monolayers at total [Ca] = 9 mM resulted in a white sheet of oriented calcite crystals (70 wt% from XRD). The crystals were of two related morphological types (Figure 2b). Type I crystals were discrete capped rhombohedral plates of pseudo $C_{2v}$ symmetry with four rhombohedral {104} basal edges (Figure 3a) and a roughened upper (104) surface. The mean particle size was variable in different experiments, typically, 60 $\mu$m with a relatively narrow size distribution ($\sigma$ = 10.5 $\mu$m). Type II crystals were triangular in projection with only two {104} basal edges (Figure 3b). Again, the surface apposed to the monolayer was elevated but no upper plate-like (104) face was expressed. The relative ratio of type I to type II crystals was variable. Crystallization under liquid phase stearic acid monolayers at [Ca] = 9 mM gave similar results to those observed for fully compressed films except that the nucleation density was reduced by approximately 30%.

Detailed SEM studies suggested that the type I crystals arose from a realignment and subsequent secondary growth of type II crystals at the monolayer surface. The evidence for this was as follows. Firstly, although the type I crystals had smooth rhombohedral {104} side faces they were wedge-shaped (Figure 3a).

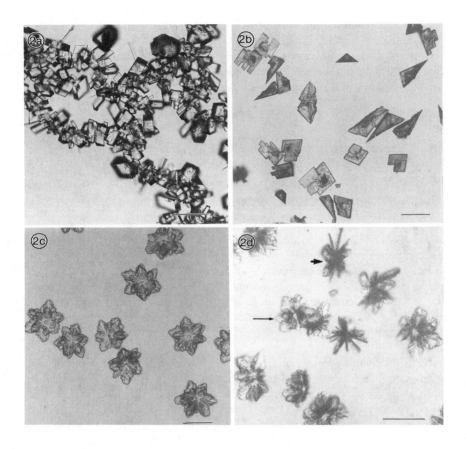

Figure 2. Optical micrographs of (a) control crystals, bar = 50 $\mu$m, (b) oriented
calcite under stearate monolayers [Ca] = 9 mM; arrows highlight different
morphological types, bar = 100 $\mu$m, (c) oriented vaterite under stearate
monolayers [Ca] = 4.5 mM, bar = 50 $\mu$m, (d) oriented vaterite under
octadecylamine monolayers; arrows highlight different morphological
types, bar = 100 $\mu$m.

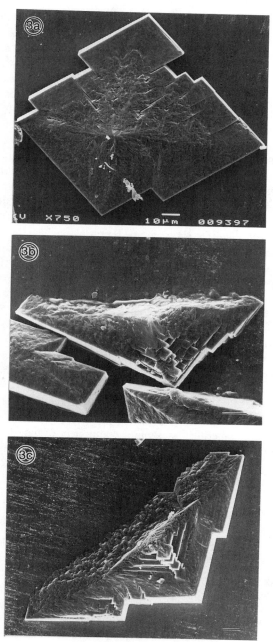

Figure 3. SEM micrographs of calcite crystals grown under stearate monolayers. (a) type I, (b) type II and (c) intermediate type crystal. Bars in all micrographs = 10 $\mu$m.

Furthermore, there were consistent differences in crystal texture at different sides of the long diagonal of the upper surface; the thicker half was well defined whilst the thinner side contained cracks and stepped edges (Figure 3a). As these effects were not symmetry related, they must arise from anisotropic growth effects due to the (changing) spatial relationships of the crystals and the membrane surface.

Secondly, the extensive elevation of type II crystals was identical to that of the corresponding smaller elevated feature on the upper surface of type I crystals. The elevation comprised three inclined faces, two of which were related by reflection symmetry. The ridges formed by intersection of these three faces ran parallel to the short and long diagonals of the basal rhombohedral plate of the type I crystals. Thus, if we assume that the apex of the elevated faces represents the initial point of attachment of the type I and II crystals, it is clear that both morphological forms nucleate along the same crystallographic direction.

Thirdly, some crystals were imaged which were intermediates between the type I and II end members. Figure 3c shows a type II crystal in which there is secondary growth perpendicular to the $a$ axis (long diagonal). The consequence of this further growth is to establish two additional {104} edges and the plate-like morphology of type I crystals. The filling-in of one side of the crystal in this way gives rise to the wedged {104} side faces and the localised structural irregularities seen in the mature type I crystals (Figure 3a).

The orientation of the central elevated features of type I and II crystals was determined by morphological examination of mature crystals (SEM) and imaging and electron diffraction studies of early crystals (TEM). Both these approaches (data not shown) gave results consistent with the $[1\bar{1}0]$ crystallographic axis aligned perpendicular to the monolayer surface.

*(b) [Ca] = 4.5 mM.* A marked change in the structure of the crystals was observed when the [Ca] concentration was reduced to 4.5 mM. Whereas the metastable polymorph, vaterite, was a minor component of the crystals nucleated under stearic acid at 9 mM, dilution of the solution resulted in oriented vaterite almost exclusively across the monolayer surface (Figure 2c). (At intermediate [Ca] levels, 5-8 mM, both oriented calcite and vaterite were observed). The crystals were discrete, of narrow particle size distribution (e.g., mean=50 $\mu$m, $\sigma$=7.6 $\mu$m) and exhibited hexagonal symmetry with the six-fold rotation axis perpendicular to the monolayer surface. This latter observation was consistent with the crystallographic $c$ axis being aligned perpendicular to the monolayer surface.

The mature crystals were floret-shaped with a central disk-like core and pseudohexagonal arrangement of plate-like radial outgrowths (Figures 4a and b). The direction of outgrowth was into solution such that the surface of the central disk in contact with the monolayer surface was smooth and partially elevated (Figure 4a). Vaterite crystals at early stages of growth were studied by electron diffraction (16,17) and shown to be single crystals oriented with the $c$ axis aligned perpendicular to the monolayer surface.     The size distribution of vaterite disks removed from the trough after 35 minutes was influenced by the degree of film compression. Interestingly, the size distribution for crystals

Figure 4. SEM micrographs of oriented vaterite. (a) view from above a stearate monolayer, [Ca] = 4.5 mM, (b) view from below a stearate monolayer, [Ca] = 4.5 mM, (c) type II crystal viewed from above an octadecylamine monolayer, (d) type II crystal, octadecylamine monolayer, side view. Bars in all micrographs = 10 $\mu$m.

grown under compressed films was increased and the mean size decreased compared with crystals grown under partially compressed monolayers (17). These results suggest that films of increased dynamical freedom may aid uniform nucleation.

**Octadecylamine Monolayers.** Substituting a positively charged amine headgroup for the negative carboxylate headgroup of stearic acid had a pronounced affect on the crystallization process. At [Ca] = 4.5-9 mM, oriented vaterite was the predominant polymorph deposited. Some calcite (<5 wt%) was formed but this was non-oriented and rhombohedral in habit. The formation of vaterite was independent of the extent of monolayer compression. The crystals were discrete, of uniform size and of two distinct morphological types (Figure 2d). Type I vaterite crystals (mean = 70 $\mu$m, $\sigma$= 8.5 $\mu$m), although usually more complex, were the same as those formed under compressed stearate films at [Ca] = 4.5 mM. In contrast, type II crystals (mean 87 $\mu$m, $\sigma$ = 14 $\mu$m) comprised similar crystal outgrowths to those of the vaterite type I crystals, but these were oriented along different directions with respect to the monolayer surface. The ratio of type I/II vaterite crystals was variable although both forms were present at significant levels (>30%) in all experiments. Observations of crystals growing *in situ* showed that both the type I and II crystals maintained their respective orientations throughout the growth process. These crystal types, therefore, develop independently, unlike the type I and II calcite crystals on stearate films.

Type I crystals consisted of a series of overlying radial outgrowths with hexagonal end faces comparable to the vaterite florets formed under stearate films (Figure 4a and b). Viewed from below the monolayer, the outgrowths were observed to originate from the peripheral edge of a central disk. The symmetry of these crystals is consistent with the crystallographic $c$ axis aligned perpendicular to the monolayer surface.

Figures 4c and d show SEM micrographs of type II vaterite crystals imaged from above the monolayer and from the side, respectively. The type II crystals had a pseudo $C_{2v}$ symmetry when viewed from above the monolayer (Figure 4c). The crystals also contained a central disk which was elongated and extended in four directions ($\pm$ 12° and 24° to the disk long axis) in the plane of the monolayer. The outgrowths were aligned almost directly into the solution subphase in the form of thin hexagonal plates (Figure 4d). These plates were oriented along the crystallographic $a$ axis of vaterite indicating that the type II crystals were oriented with the vaterite (110) face parallel to the monolayer surface. This morphological inference was confirmed by electron diffraction studies of type II crystals at early growth stages 18).

## Discussion

**Stearate films.** The above results clearly show that the crystallization of calcium carbonate from supersaturated solution is profoundly influenced by the presence of organized charged surfactant monolayers. Both oriented calcite and vaterite can be nucleated under stearate films whilst oriented vaterite is predominant under monolayers comprising amine headgroups.

Although two morphological types of calcite are observed under stearate films, the SEM results indicate that these are related through a realignment of the initial orientation (type II) followed by secondary growth to give the plate-like form (type I). The crystals are nucleated with the $[1\bar{1}0]$ face parallel to the monolayer surface such that rhombohedral growth gives rise to stable {104} faces aligned initially at 45° to the monolayer/solution interface. As the surface area of these faces increases, there is the possibility of realignment due to surface tension effects acting on the well-developed crystal faces with the result that the {104} faces become opposed to the monolayer surface. Why some crystals undergo this realignment whilst others do not is unclear.

Nucleation of calcite on the $[1\bar{1}0]$ face can be explained in terms of charge, stereochemical and geometric factors. Changes in the compression isotherms (17) were indicative of Ca binding prior to nucleation. Furthermore, the limiting area of these films was 22 Å$^2$ which is consistent with a hexagonal packed layer of molecules with an interheadgroup spacing of ca. 5Å. This distance is commensurate with a Ca-Ca distance in the plane of the $[1\bar{1}0]$ face of calcite. Thus ion-binding may aid nucleation of this face by restricting the arrangement of Ca atoms in two-dimensions. However, there are other calcite faces with 5Å Ca-Ca spacings implying that additional factors are responsible for the preferential stabilization of the $[1\bar{1}0]$ face.

An important feature of the $[1\bar{1}0]$ face is the presence of alternate rows of carbonates lying perpendicular to the crystal surface. All these anions are equivalent with a bidentate motif at the surface. Thus the stereochemistry of the carboxylate headgroups match those of the anions in the $[1\bar{1}0]$ crystal face (Figure 5). In this respect the Ca-stearate layers represent a sub-unit cell motif which is closely related to the structure of the $[1\bar{1}0]$ face.

The switch from calcite to vaterite nucleation on stearate films at low [Ca] suggests that the extent of Ca binding is important for calcite nucleation. Furthermore, there is the possibility of $HCO_3^-$ intercalation in the Stern layer at higher concentrations. This would require a change in the bonding stoichiometry of Ca at the headgroups from bridging (CaSt$_2$; St = stearate) to non-bridging (CaSt$^+$). with the excess positive charge balanced by $HCO_3^-$. The presence of carbonate in the first layer under the film would assist $[1\bar{1}0]$ nucleation as carbonates are almost coplanar with Ca atoms in this face.

At lower [Ca], both the reduced Ca binding and absence of intercalated $HCO_3^-$ favour vaterite nucleation on the (001) face. This face contains a unicharged layer of Ca atoms with a second layer of carbonates oriented in a bidentate motif perpendicular to the surface (Figure 6). However, the in plane Ca-Ca distances are 4.2Å which is incommensurate with the inter-headgroup spacing (5 Å). This mismatch can be accomodated particularly at low [Ca] when Ca binding is not extensive over the whole monolayer. It may also be offset by the close stereochemical

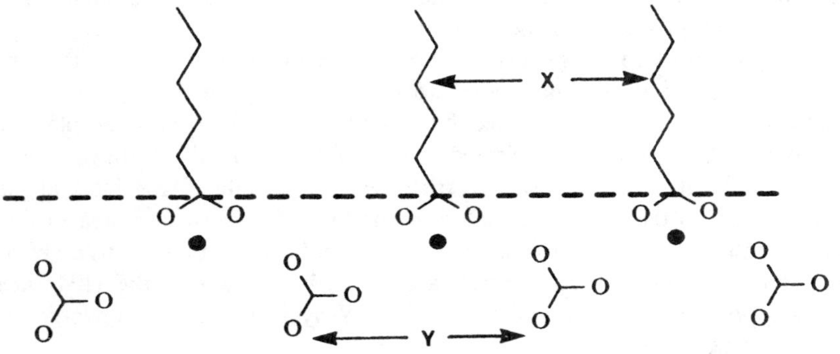

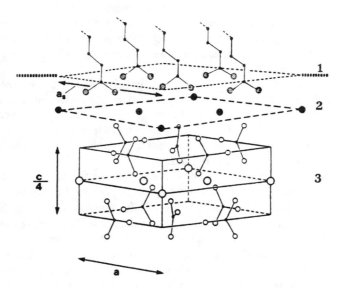

Figure 5. Possible structural and stereochemical relationships between the calcite (1Ī0) face and a compressed stearate monolayer. Inter-headgroup spacing, X=5Å, Ca-Ca and carbonate-carbonate distances, Y=4.96Å.

Figure 6. Possible organization at the interface between a vaterite (001) face and stearate headgroups. Note that the Stern layer Ca atoms (●) have a spacing equivalent to that of the stearate molecules ($a_s$ = 8.6 Å) whereas the corresponding Ca atoms in the unit cell (o) have $a$ = 7.1 Å. The distance between adjacent Ca or $CO_3^{2-}$ layers, 1/4 $c$ = 4.25 Å.

correspondence of the carboxylates and carbonates in the second layer of the incipient crystal surface.

The increased uniformity of the immature vaterite crystals on partially rather than fully compressed stearate films suggests that nucleation on the former can proceed by localized organization of the surfactant molecules (induced by localized Ca localized binding ?) and that this is a single autocatalytic event. Compressed films, on the other hand, possibly remain catalytic over a relatively long time course resulting in episodic nucleation and a range of particle sizes. This is an important observation because it suggests that a degree of dynamical freedom may be beneficial in controlling nucleation on organized organic surfaces. One can envisage a synergistic effect in which ion binding induces local conformational changes in the organic film which in turn induce further ion binding and consequent oriented nucleation. Such effects may be important in biomineralization and in the development of synthetic organic substrates for controlled crystallization.

**Octadecylamine Films.** The formation of oriented vaterite under positively charged amine monolayers indicates that Ca binding is not a prerequisite for the deposition of this metastable polymorph. This suggests that interactions involving $HCO_3^-$ and the amine headgroups may inhibit calcite and/or promote vaterite nucleation. Interestingly, vaterite formation in aqueous solution is favoured under conditions of high $HCO_3^-/Ca^{2+}$ ratio ([19]). Under such conditions, the normally positive charged surfaces of $CaCO_3$ are rendered negative by the surplus of adsorbed anions. Lippmann ([20]) has suggested that this effect could preferentially stabilise vaterite nuclei over calcite clusters due to the more open structure of vaterite and the increased tolerance of the carbonates in this lattice to disorientation. In this regard accumulation of carbonate under amine films would be consistent with the preferential stabilization of negatively charged nuclei and hence vaterite formation.

Why are two distinct vaterite orientations, viz. the $a$ and $c$ axes, observed on the amine films? Clearly, there is no direct stereochemical correspondence between headgroup molecules and lattice ions. However, at pH = 6, the mode of carbonate binding will be through bidendate $HCO_3^-$ interactions with the amine $-NH_3^+$ headgroups. Thus there is the possibility of indirect stereochemical control mediated by the structure of the underlying carbonate-containing boundary layer. Significantly, the common feature of both the (001) and (110) faces nucleated parallel to the organic film lies in the perpendicular orientation of their carbonates with respect to the crystal surfaces. The (001) face has all carbonates perpendicular whilst (110) has a subset in this orientation. Thus, the possibility of weak bidentate binding of $HCO_3^-$ provides both the general requirement of orthogonal carbonates and the stereochemical flexibility to accommodate the nucleation of both (110) and (001) faces.

## Conclusions

The results presented in this paper indicate that charged compressed Langmuir monolayers have the ability to control the oriented nucleation of CaCO₃ from supersaturated solution. That this is in essence a charge effect is shown by the absence of oriented nucleation under neutral alcohol films. Calcium binding is important for oriented calcite formation. In all cases to date, the orientation of the crystals is such that the carbonate groups are oriented perpendicular to the organic surface and this may be augmented by direct stereochemical matching with surfactant headgroups and through bidentate binding of anions. Geometric relationships may play a role but are secondary effects. Finally, we note that the use of organic films of precise molecular design could be a potential method of tailoring crystal synthesis in general. Thus surface reactive organic surfaces may have an important role in future advances in materials science.

## Acknowledgments

We thank Professor J.D. Birchall and Dr. R. J. Davey for interesting discussions and SERC and ICI plc for financial support.

## Literature Cited

1. Wheeler, A. P.; Rusenko, K. W.; Swift, D. M.; Sikes, C. S. Mar. Biol. 1988, 98, 71-80.
2. Sikes, C. S.; Wheeler, A. P. In Chemical Aspects of Regulation of Biomineralization, Sikes, C. S.; Wheeler, A. P., Eds.; Univ. of South Alabama Publication Services: Mobile, 1988; pp 15-20.
3. Wheeler, A. P.; Sikes, C. S.; In Biomineralization: Chemical and Biochemical Perspectives, Mann, S.; Webb, J.; Williams, R. J. P., Eds.; VCH Publishers: Weinheim, 1989; pp 95-132.
4. Berman, A.; Addadi, L.; Weiner, S. Nature 1988, 331, 546-8.
5. Hay, D. I.; Moreno, E. C.; Schlesinger, D. H. Inorg. Persp. Biol. Med. 1979, 2, 271-85.
6. Addadi, L.; Weiner, S. Proc. Natl. Acad. Sci. USA 1985, 82, 4110-4.
7. Campbell, A. A.; Ebrahimpour, L.; Perez, S. A.; Smesko, A.; Nancollas, G. H. Calcif. Tissue Int. 1989, 45, 122-8.
8. Addadi, L.; Moradian, J.; Shay, E.; Maroudas, N. G.; Weiner, S. Proc. Natl. Acad. Sci. USA 1987, 84, 2732-6.
9. Kallitsis, J.; Koumanakos, E.; Dalas, E.; Sakkopoulos, S.; Koutsoukas, P. G. J. Chem. Soc. Chem. Commun. 1989, 1146-7.
10. Mann, S.; Williams, R. J. P. J. Chem. Soc. Dalton Trans. 1983, 311-6.
11. Mann, S.; Hannington, J. P. J. Colloid. Interface Sci. 1988, 122, 326-35.
12. Bhandarkar, S.; Bose, A. J. Colloid. Interface Sci. 1990, 135, 531 .

13. Landau, E. M.; Levanon, M; Leiserowitz, L.; Lehav, 8 M.; Sagiv, J. Nature 1985, 318, 353-6.
14. Landau, E. M.; Popovitz-Bior, R.; Levanon, M.; Leiserowitz, L.; Lehav, M.; Sagiv, J. Molec. Cryst. Liq. Cryst. 1986, 134, 323-35.
15. Landau, E. M.; Grayer, Wolf, S.; Levanon, M.; Leiserowitz, L.; Lehav, M.; Sagiv, J. J. Am. Chem. Soc. 1989, 111, 1436-45.
16. Mann, S.; Heywood, B. R.; Rajam, S.; Birchall, J. D. Nature 1988, 334, 692-5.
17. Mann, S.; Heywood, B. R.; Rajam, S.; Birchall, J. D. Proc. R. Soc. Lond. A. 1989, 423, 457-71.
18. Mann, S.; Heywood, B.R.; Rajam, S.; Walker, J.B.A.; Davey, R.J.; Birchall, J.D. Adv. Materials 1990, 2. 257-261.
19. Turnball, A. G. Geochim. Cosmochim. Acta 1973, 37, 1593-1601.
20. Lippmann, F. Estudios geol. 1982, 38, 199-208.

RECEIVED August 27, 1990

# Chapter 4

# Role of Membranes in De Novo Calcification

B. D. Boyan[1], Z. Schwartz[1,2], and L. D. Swain[1]

[1]Department of Orthopaedics, University of Texas Health Sciences Center at San Antonio, San Antonio, TX 78284–7823
[2]Hebrew University Hadassah Faculty of Dental Medicine, Jerusalem, Israel

The association of lipids with hydroxyapatite deposition has been well characterized. This is particularly true of the acidic phospholipids, which require mineralized tissues to be decalcified before they can be extracted. In addition, phospholipid films can serve as nucleating sites for crystal formation. Mineral deposits in calcifying cartilage are initially observed in membrane-bound extracellular matrix vesicles. The earliest mineral crystals are, in fact, formed in association with the vesicle membrane. Specific membrane proteins, termed proteolipids, have been shown to structure phosphatidylserine, which has an extremely high affinity for $Ca^{2+}$, in a conformation conducive to hydroxyapatite deposition. Proteolipids have also been shown to be involved in ion transport, possibly facilitating the export of protons and import of Ca and $P_i$ required for crystal formation. It is increasingly clear that phospholipid metabolism in the growth plate is important to the regulation of matrix vesicle-dependent mineralization.

Initial hydroxyapatite formation in cartilage has been associated with matrix vesicles (1,2,3). Matrix vesicles are membrane bound extracellular organelles present in the territorial matrix of the cartilage cells and are derived from the plasma membrane of these cells (1). Transmission electron micrographs of the matrix vesicles have demonstrated that they are approximately 20-50 nm in diameter and contain an electron dense, amorphous interior (4,5). They appear to form by budding from the plasma membrane (6) in a process that has been termed non-inflammatory cell deletion (7). This implies that the process of matrix vesicle production is regulated by the cell and that it does not occur in response to inflammation but is a normal cellular event.

The mechanisms by which the cells regulate events in their matrix are complex. While it is now well accepted that mineral formation begins in or on matrix vesicles (8), it is clear that the bulk phase of hydroxyapatite deposition does not require the matrix vesicle or its constituents (9,10). To ensure that calcification of cartilage is under cellular control, the initial events are tightly regulated. This paper will describe the important role matrix vesicles play in this process.

## The Growth Plate

The growth plate of the epiphyseal cartilage (Figure 1) is a highly organized structure which permits the longitudinal expansion of the bone under tremendous load. At the top of the growth plate, the cartilage cells (chondrocytes) are arranged in a random fashion, dispersed in a proteoglycan aggregate-rich extracellular matrix. This area is called the reserve zone or resting zone cartilage because these cells are the reservoir for the chondrocytes which will populate the growth plate.

In response to a number of differentiation signals (11), the resting zone cells begin to proliferate and align into columns which are oriented parallel to the axis of the long bone. After a set number of divisions, the cells begin to hypertrophy. The effect of this series of events is to increase the length of the cartilage. The cells in the lower proliferative cell zone and in the hypertrophic cell zone produce matrix vesicles which have an amorphous, granular appearance. In the lower hypertrophic zone, hydroxyapatite crystals are seen along the inner leaflet of the matrix vesicle membrane. At the base of the growth plate, the intercellular matrix is calcified. The crystals extend out of the matrix vesicles and are deposited in the collagenous matrix.

We now have considerable information about the biochemical events which are associated with the differentiation of the growth plate. As the cells hypertrophy, the proteoglycan aggregate is degraded (12). In addition to the action of chondroitinases (13) and hyaluronidases (14), a number of proteases break down the proteoglycan core protein and link protein (15). Collagenase degrades the collagen as well (16). At the same time, the activity of alkaline phosphatase, an enzyme associated with calcification, increases. Alkaline phosphatase is enriched in matrix vesicles (17) and is often used as a marker enzyme to monitor their purification. There is a concomitant increase in phospholipase $A_2$ activity (18), suggesting that this enzyme is involved in the breakdown of the matrix vesicles. Once the cartilage begins to calcify, a new group of proteins is found in the matrix including the C-propeptide of cartilage-specific type II (19) and type X collagen (20) and osteocalcin (21,22), which is considered to be a bone matrix protein.

## Lipids in Cartilage Calcification

Early studies by Irving and Wuthier (23) indicated that acidic lipids were involved in mineralization of the growth plate. Histologic analyses of the tissue using Sudan Black B to stain the acidic lipids demonstrated that the stain was localized to the cells in the resting zone but was present in the extracellular matrix in the hypertrophic zone. It was necessary to repeatedly demineralize the tissue before all of the stainable material could be extracted with chloroform:methanol 2:1 suggesting that it was intimately associated with the mineral phase.

More recently, Boskey and Posner (24) demonstrated that calcium:phospholipid:phosphate complexes (CPLX), consisting of 1 mole of calcium per mole of total phosphate (phospholipid phosphate plus inorganic phosphate), increase in content in the hypertrophic zone of the growth plate. CPLX concentration is greatest just before bulk phase mineralization occurs in the calcifying zone. In a related series of studies, Boyan and Boskey (25,26) found that CPLX formation is related to the presence of specific proteolipids, a class of membrane proteins involved in ion transport (27). The concentration of these proteins also increases in the hypertrophic zone of the growth plate (28) and both CPLX (24) and proteolipids (29) have been isolated from matrix vesicles.

It is becoming increasingly clear that phospholipid metabolism in the growth plate is important to the regulation of matrix vesicle-dependent mineralization. Shapiro et al. (30) have observed that the oxidative metabolism of fatty acids in the growth plate varies with the zone of maturation. Matrix vesicle phosphatidylethanolamine content decreases whereas phosphatidylserine content is increased (31), probably through selective susceptibility to resident phospholipases. Phosphatidylethanolamine is a substrate for phospholipase $A_2$ (32) while CPLX containing phosphatidylserine is resistant to phospholipase $A_2$ (Boskey, personal communication).

Numerous investigators have shown that phospholipids (33) and proteolipids (34) can play a structural role in promoting hydroxyapatite formation. Liposomes formed from phosphatidylinositol-4,5-bisphosphate and egg white lysozyme (35) serve as excellent substrates for mineral deposition *in vitro*. When liposomes are constructed from phosphatidylcholine, dicetyl phosphate and cholesterol (36) and pre-loaded with phosphate, they do not support hydroxyapatite formation unless an ionophore is used to permit uptake of ions. The paper by Mann et al. in this volume (37) also shows that lipid films can serve as nucleating sites for crystal formation.

Perhaps the most conclusive experiment demonstrating the role of lipids in mineral formation is that performed by Raggio et al. (38). CPLX, proteolipids and non-proteolipid phospholipids were extracted from rabbit bone, placed in diffusion chambers and implanted in rabbit peritoneum. CPLX and proteolipids supported hydroxyapatite deposition whereas the non- proteolipid phospholipids bound calcium but did not calcify. In addition, CPLX formed on the proteolipid-associated phospholipids. Experiments performed in collaboration with Dr. David Howell at the University of Miami School of Medicine demonstrate that a phospholipid-containing material can be isolated from micropuncture fluid obtained from rat hypertrophic cartilage which supports hydroxyapatite formation *in vitro*.

Most calcification studies are performed under buffered conditions where the evolution of protons during hydroxyapatite crystal formation is not an issue.

However, *in vivo*, the lipid macromolecules described above are found in matrix vesicles which are delimited by a membrane. Thus, proteolipids may have two functions; to permit influx of calcium and phosphate ions and efflux of protons (39), and to serve as a structural entity in CPLX formation and hydroxyapatite deposition (40).

## Regulation of Matrix Vesicle Lipids

Since matrix vesicles are originally derived from chondrocyte plasma membranes, their production and maturation are under cellular regulation. As shown in Figure 2, there are a number of changes that take place in the matrix vesicles over time once they are released into the matrix. Cellular control of these events may occur at various points. There may be genomic regulation resulting in new gene transcription. Messenger RNA levels for matrix vesicle proteins may be differentially regulated and there may be post-translational regulation of these proteins as well. In addition, the cell may release factors into the matrix which influence activity of matrix vesicle enzymes in the matrix itself.

To determine whether one or more of the mechanisms function in cartilage, we developed a chondrocyte culture model for assessing the regulation of matrix vesicles at two different stages of chondrogenic maturation. Resting zone and growth zone chondrocytes are cultured separately in Dulbecco's modified Eagle's medium (DMEM) containing 10% fetal bovine serum (FBS) and the factor of interest. These chondrocytes retain their phenotypic markers in culture including production of matrix vesicles with distinctive lipid compositions and enzyme activities (41). The resting zone cells respond primarily to the vitamin $D_3$ metabolite, $24,25\text{-}(OH)_2D_3$. Non-collagenase digestible protein synthesis is stimulated (42). Matrix vesicle alkaline phosphatase is increased but phospholipase $A_2$ activity is decreased. In contrast, growth zone chondrocytes respond primarily to $1,25\text{-}(OH)_2D_3$. Collagen synthesis and matrix vesicle alkaline phosphatase and phospholipase $A_2$ activities are increased.

In culture, plasma membrane alkaline phosphatase and phospholipase $A_2$ activity are not changed by either metabolite in either cell type. This suggested that the vitamin D metabolites might be acting directly on the matrix vesicle membrane. To test this, we isolated matrix vesicles and plasma membranes and incubated them *in vitro* with $1,25\text{-}(OH)_2D_3$ and $24,25\text{-}(OH)_2D_3$ (43). Under these experimental conditions no gene transcription or protein synthesis are possible. $1,25\text{-}(OH)_2D_3$ stimulated alkaline phosphatase and phospholipase $A_2$ activity in matrix vesicles and plasma membranes isolated from the growth zone chondrocyte cultures but had no effect on these enzymes in the membrane fractions isolated from resting zone chondrocyte cultures. $24,25\text{-}(OH)_2D_3$ stimulated alkaline phosphatase activity, inhibited phospholipase $A_2$ activity in the matrix vesicles and plasma membranes isolated from the resting zone chondrocytes, but had no effect on these enzymes in membranes isolated from the growth zone chondrocyte cultures. Since the plasma membrane enzymes were responsive to

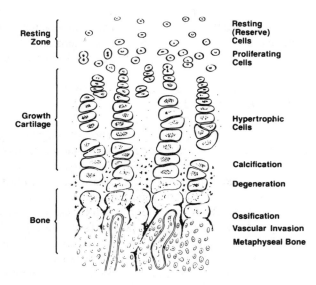

Figure 1. Schematic drawing of a cartilage growth plate.

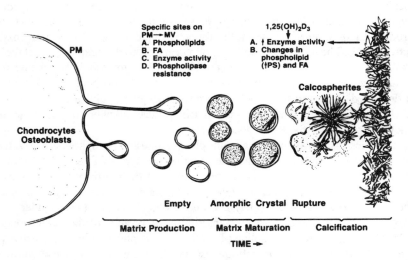

Figure 2. Schematic drawing depicting a calcifying chondrocyte or osteoblast during matrix vesicle production and maturation. Events are occuring in time rather than space.

hormone *in vitro*, it is possible that they are able to down regulate the effect of 1,25-$(OH)_2D_3$ and 24,25-$(OH)_2D_3$ in culture or that any effect of these vitamin D metabolites on the plasma membranes is completed prior to harvest.

The matrix vesicle data are summarized in Table I. They demonstrate that vitamin D metabolites can affect a response in matrix vesicles that is independent of the genome or protein synthetic machinery in the cell. This suggests that one mechanism by which the cell can regulate events in the matrix is by secreting the appropriate vitamin D metabolite. This hypothesis requires that chondrocytes be able to metabolize vitamin D. Hydroxylase activity has been reported in cartilage (44). Preliminary studies in our lab indicate that the cultured chondrocytes can metabolize vitamin D as well.

Table I. Effect of vitamin D metabolites on matrix vesicle alkaline phosphatase and phospholipase $A_2$ specific activity in culture and *in vitro*

| | Alkaline Phosphatase | | Phospholipase $A_2$ | |
| | Culture | In Vitro | Culture | In Vitro |
|---|---|---|---|---|
| Resting Zone Chondrocytes | | | | |
| Control | 127.8±11.1 | 54.4±12.8 | 16.7±2.3 | 18.1±4.3 |
| $10^{-7}$M 24, 25-$(OH)_2D_3$ | 194.4±33.3* | 108.8±27.2* | 2.7±0.7* | 3.5±0.9* |
| Growth Zone Cells | | | | |
| Control | 217.4±21.7 | 85.1±11.1 | 15.0± 3.3 | 17.1±1.8 |
| $10^{-8}$M 1,25-$(OH)_2D_3$ | 408.7±78.3* | 195.3±20.4* | 96.7±16.7* | 62.0±1.4* |

## Implications for the Role of Membranes in Calcification

One of the matrix vesicle enzymes that is influenced by 1,25-$(OH)_2D_3$ and 24,25-$(OH)_2D_3$ is phospholipase $A_2$. Phospholipase $A_2$ activity is the rate limiting step in production of prostaglandins (45), which are important local hormones involved in regulating cellular response to a number of factors (46). Phospholipase $A_2$ activity is also required for fatty acid turnover, resulting in changes in the degree of saturation of membrane phospholipids and the fluidity of the membrane itself. This can have consequences for ion transport (47) and membrane enzyme activity (48). Finally, the fluidity of the membrane can alter its structural characteristics (49), ultimately changing the ability of the membrane to support CPLX formation and hydroxyapatite deposition.

## Acknowledgments

The authors acknowledge the invaluable contributions of Brian Brooks, Ruben Gomez, and Virginia Ramirez to this research. Studies performed in our lab were supported by PHS Grants DE-05932, DE-08603, and DE-05937.

## Literature Cited

1. Anderson, H. C. J. Biol. Chem. 1969, 41, 59-72.
2. Ali, S. Y. Fed. Proc. 1976, 35, 135-42.
3. Ali, S. Y.; Wisby, A.; Gray, J. J. Metab. Bone Dis. Rel. Res. 1978, 1, 97-103.
4. Anderson, H. C. In The Biochemistry and Physiology of Bone; Bourne, G. H., Ed.; Academic Press: Orlando, FL, 1976; p 135-57.
5. Anderson, H. C. Metab. Bone Dis. Rel. Res. 1978, 2, 83-8.
6. Rabinovitch, A. L.; Anderson, H. C. Fed. Proc. 1976, 35, 112-6.
7. Kardos, T. B.; Hubbard, M. J. Prog. Clin. Biol. Res. 1982, 101, 45-60.
8. Ali, S. Y. Cartilage: Structure, Function, and Biochemistry; Hall, B. K., Ed.; Academic Press: Orlando, FL, 1983; p 343-78.
9. Glimcher, M. J.; Krane, S. M. In Treatise on Collagen; Ramachandran, G. N.; Gould, B. S., Eds.; Academic Press: New York, 1968; Vol. 2, Part B.
10. Hohling, H. J.; Althoff, J.; Barckhaus, R. H.; Krefting, E. R.; Lissner, G.; Quint, P. In International Cell Biology; Schweiger, H. G., Ed.; Springer-Verlag: Berlin, 1981.
11. Hall, B. K. Am. Sci. 1988, 76, 174-181.
12. Howell, D. S.; Pita, J. C. Clin. Orthop. 1976, 118,208-29.
13. Hirschman, A.; Dziewiatkowski, D. D. Science 1966, 154, 393-5.
14. Silbermann, M.; Frommer, J. Histochemie. 1973, 36, 185-92.
15. Hirschman, A.; Deutsch, D.; Hirschman, M.; Bab, I. A.; Sela, J.; Muhlrad, A. Calcif. Tissue Int. 1983, 35, 791-7.
16. Dodge, G. R.; Poole, A. R. J. Clin. Invest. 1989, 83, 647-61.
17. Schwartz, Z.; Knight, G.; Swain, L.; Boyan, B. J. Biol. Chem. 1988, 263, 6023-6.
18. Schwartz, Z.; Boyan, B. Endocrinology 1988, 122, 2191-8.
19. Van der Rest, M.; Rosenberg, L. C.; Olsen, B. R.; Poole, A. R. Biochem. J. 1986, 237, 923-5.
20. Schmid, T. M.; Linsenmayer, T. F. J. Cell Biol. 1985, 100, 598-605.
21. Hauschka, P. V.; Reddi, A.H. Biochem. Biophys. Res. Commun. 1980, 92, 1037-41.
22. Nishimoto, S. K.; Price, P. A. J. Biol. Chem. 1979, 254, 437-40.
23. Irving, J. T.; Wuthier, R. E. Clin. Orthop. 1968, 56, 237-60.
24. Boskey, A. L.; Posner, A. S. Calcif. Tissue Int. 1976, 19, 273-83.
25. Boyan-Salyers, B. D. and Boskey, A. L. Calcif. Tissue Int. 1980, 30, 167-74.
26. Boyan, B.; Boskey, A. Calcif. Tissue Int. 1984, 36, 214-8.
27. Swain, L. D.; Boyan, B. D. J. Dent. Res. 1988, 67, 526-30.
28. Boyan, B. D.; Ritter, N. M. Calcif. Tissue Int. 1984, 36, 332-7.

29. Boyan, B. D.; Martinez, S.; Guillory, G.; Taverna, R.; Norman, A. J. Dent. Res. 1984, 63, 323.
30. Shapiro, I.; Burke, A.; Schattschneider, A.; Golub, E. In Matrix Vesicles; Ascenzi, A.; Bonucci, E.; DeBernard, B., Eds.; Wichtig Editore: Milano, 1981; p 33-39.
31. Peress, N. S.; Anderson, H. C.; Sajdera, S. W. Calcif. Tissue Int. 1974, 14, 275-81.
32. Newkirk, J. D.; Waite, M. Biochem. Biophys. Acta 1981, 225, 224-33.
33. Wuthier, R. E. Clin. Orthop. 1982, 169, 219-42.
34. Ennever, J.; Boyan-Salyers, B.; Riggan, L. J. J. Dent. Res. 1977, 56, 967-70.
35. Hollinger, J. Biomat. Med. Dev. Artif. Organs 1982, 10, 71-83.
36. Eanes, E. D.; Hailer, A. S.; Costa, J. L. Calcif. Tissue Int. 1984, 36, 421-30.
37. Mann, S.; Heywood, B. R.; Rajam, S.; Walker, J. B. A. In Surface Reactive Peptides and Polymers: Discovery and Commercialization; Sikes, C. S.; Wheeler, A. P., Eds.; ACS Books: Washington, 1990.
38. Raggio, C. L.; Boyan, B. D.; Boskey, A. L. J. Bone Min. Res. 1986, 1, 409-15.
39. Swain, L. D.; Renthal, R. D.; Boyan, D. B. J. Dent. Res. 1989, 68, 1097-9.
40. Boyan, B. D.; Swain, L. D.; Boskey, A. L. In Dental Calculus: Formation, Epidemiology and Prevention; IRL Press: Amsterdam, 1989, pp 29-35.
41. Boyan, B. D.; Schwartz, Z.; Swain, L. D.; Carnes, D. L.; Zislis, T. Bone 1988, 9, 185-94.
42. Schwartz, Z.; Schlader, D. L.; Ramirez, V.; Kennedy, M. B.; Boyan, B. D. J. Bone Min. Res. 1989, 4, 199-207.
43. Schwartz, Z.; Schlader, D. L.; Swain, L. D.; Boyan, B. D. Endocrinology 1988, 123, 2878-84.
44. Garabedian, M.; DuBois, B.; Corvol, M. T.; Pezant, E.; Balsan, S. Endocrinology, 1978, 102, 1262-8.
45. Bell, R. L.; Kennerly, D. A.; Standford, N.; Majerus, P. W. Proc. Natl. Acad. Sci. U.S.A. 1979, 76, 3238-41.
46. Hakeda, Y.; Yoshino, T.; Natakani, Y.; Kurihara, N.; Maeda, N.; Kumegawa, M. J. Cell Physiol. 1986, 128, 155-61.
47. Nemere, I.; Yoshimoto, Y.; Norman, A. W. Endocrinology 1984, 115, 1476-83.
48. Rasmussen, H.; Matsumoto, T.; Fontaine, O.; Goodman, D. B. Fed. Proc. 1982, 41, 72-7.
49. Duzgunes, N.; Papahadjopoulas, D. In Membrane Fluidity in Biology; Aloia, R. C., Ed.; Academic Press: New York, 1983, Vol. 2, pp 187-216.

RECEIVED August 27, 1990

# Chapter 5

# Inhibition of Calcium Carbonate and Phosphate Crystallization by Peptides Enriched in Aspartic Acid and Phosphoserine

## C. Steven Sikes, M. L. Yeung, and A. P. Wheeler[1]

### Mineralization Center, University of South Alabama, Mobile, AL 36688

Polyanionic peptide analogs of matrix proteins from biominerals were synthesized by automated, solid-phase methods. The peptides were used to evaluate the chemical requirements for inhibition of calcium carbonate and phosphate crystallization as measured using pH-drift and constant-composition assays. Continuous runs of negatively charged residues were required for maximum inhibitory activity. $Asp_{15}$ was the optimum size for inhibition of $CaCO_3$ crystal growth, but $Asp_{(30\ to\ 40)}$ exhibited maximal inhibition of $CaCO_3$ crystal nucleation. An hydrophobic domain of $Ala_8$ added to $Asp_{15}$ led to increased inhibition of $CaCO_3$ crystal nucleation but had no effect on crystal growth. Adding the hydrophobic domain to $Asp_{40}$ did not enhance inhibition of either $CaCO_3$ nucleation or crystal growth. The results suggest that crystal nucleation can be suppressed through diffusion-limitation related to the presence of a barrier at the crystal/solution interface. $Asp_{40}$ molecules seemed to fill both the crystal-binding sites and the zone of diffusion around them, with additional anionic or hydrophobic residues simply contributing excess mass without enhancing performance. $CaCO_3$ crystal growth appeared not limited by diffusion but rather by some other process such as incorporation of already bound ions into the lattice. Phosphorylation of N-terminal serine residues by use of monochlorophosphoric acid produced $H-PSer_{(1\ to\ 3)}-Asp_{20}-OH$, the most powerful inhibitor of both calcium carbonate and calcium phosphate formation yet discovered. This supports the importance of phosphorylated residues to the function of protein inhibitors of mineralization

[1]Current address: Department of Biological Sciences, Clemson University, Clemson, SC 29634-1903

0097–6156/91/0444–0050$06.50/0

and suggests that the terminal residues may play a significant role in the mechanism of inhibition.

Proteins that inhibit calcium carbonate and phosphate formation *in vitro* occur as part of the organic matrix of biominerals and as components of saliva (1). These proteins are thought to function in part to stop or start crystal formation as needed in the case of matrix, depending on whether they are in solution or immobilized (2-4), or to stabilize enamel and saliva in the case of the salivary proteins (5).

The attempt to understand how these molecules interact with crystal surfaces is aided by consideration of their common structural features. For example, recent evidence suggests that several matrix inhibitors of crystallization may have polyaspartic acid domains (6-11). Moreover, virtually all soluble matrices are at least enriched in negatively-charged aspartic and glutamic acids (1,12). One hypothesis holds that aspartic acid residues may alternate with small, neutral amino acids in matrix (13). Consequently, it is of interest to determine whether polyaspartic acid domains are required for inhibition of crystallization by matrix and to what extent the length of the domain influences this activity.

Phosphoserine is also a prominent component of many matrix molecules (14-15), and in some cases there may be continuous runs of this residue as well (16-17). The importance of phosphoserine to activity of proteins as regulators of crystallization has clearly been seen in studies of the salivary proteins. In the case of the proline-rich phosphoproteins (PRP) from saliva (18-19), removal of phosphate groups by treatment with alkaline phosphatase greatly diminished the activity of the molecules. Similar observations had been made for phosvitin, the phosphoprotein from egg yolk (20), and more recently for proteins from oyster shell (1). Therefore, another point of interest is the extent to which phosphoserine is needed for regulation of crystallization by proteins that may already be polyanionic, especially those including runs of aspartic residues.

The possible importance of the placement of phosphoserine residues is also unknown. For example, in some proteins like the PRP's (21), phosphoserine may be dispersed, but in statherin (22), the two phosphoserine residues are at the N-terminus. Accordingly, the effect of having phosphoserine residues at specific locations on the ability of a protein to regulate crystallization is another topic for study.

Finally, the polyanionic protein inhibitors of crystallization invariably also contain hydrophobic domains that may be quite extensive. One possible function, among several suggested functions, of these domains has been formulated from the effects of statherin on calcium hydroxyapatite formation (23). The hydrophobic portion of statherin, consisting in the bulk of the 43-residue molecule, was required for maximum inhibition of crystallization. It seemed possible that statherin bound to crystal surfaces via the polyanionic N-terminus, blocking crystal growth there, while the rest of the molecule extended from the surface of the crystals, disrupting movement of lattice ions to the surface.

This concept has been tested in part by studying the effects of polyaspartate molecules with and without hydrophobic domains of polyalanine on $CaCO_3$ crystallization (24). The results suggested that the presence of a hydrophobic region could improve the inhibitory activity of polyaspartic-containing molecules. However, it was not known exactly what relative lengths of polyanionic and hydrophobic domains were required for effective interaction with forming crystals.

In view of these observations, the present study was undertaken 1) to determine if polyaspartic domains optimize inhibitory activity of proteins with respect to crystallization, 2) to determine the size of polyaspartic domains that interact most effectively with crystal nuclei and preformed crystal surfaces, 3) to demonstrate the influence of phosphoserine residues on inhibition of crystallization by proteins, including the possible effect of the location of phosphoserine residues, and 4) to further evaluate the possible influence of hydrophobic domains on crystallization inhibitory activity by peptides of different sizes. The general approach was to synthesize model peptides and to measure effects on crystal nucleation and crystal growth of calcium carbonate and calcium hydroxyapatite.

## Materials and Methods

**Preparation of Peptides.** An automated, solid-phase peptide synthesizer (Applied Biosystems, Model 430A) was used to prepare peptides. Aspartic acid was supplied as t-Boc-L-aspartic acid with beta carboxyl protection by O-benzyl linkage, serine as t-Boc-L-serine-O-benzyl, alanine as t-Boc-L-alanine, and glycine as t-Boc-L-glycine (all from Applied Biosystems). Standard reaction cycles were used.

In all cases, the coupling efficiency of each residue was checked by automated sampling of peptide resin for measurement of unreacted free amine by the ninhydrin method (25). Coupling efficiencies routinely were greater than 99% per cycle of synthesis.

**Cleavage.** Following synthesis, peptide-resin was repeatedly washed with methanol, then air-dried and weighed. Next, peptides were cleaved from the resin using a modification of the trifluoromethyl sulfonic acid (TFMSA) procedure at -10°C as a precaution to suppress aspartimide formation (26), in which the beta carboxyl group of aspartic acid undergoes a dehydration reaction with the secondary amine of an adjacent residue along the peptide backbone to form a 5-membered ring (27). For 100 mg samples, peptide-resin in a scintillation vial was treated for 10 minutes with 150 ml of anisole to swell the resin, making it more accessible for reaction. Then 1.0 ml of neat trifluoroacetic (TFA) was added with magnetic stirring and allowed to react for 10 minutes. Next, 100 $\mu$l of concentrated TFMSA (Aldrich Chemical Co.) were added with cooling using a salt-water ice bath at -10°C with cleavage for 30 minutes, followed by 30 minutes at room temperature. For cleavage of other amounts of peptide-resin, the amounts of the reagents were changed proportionally.

**Washing**. Following cleavage, 20 ml of methyl butyl ether (MBE, Aldrich) were added to the vial to ensure precipitation of the peptide, which already was relatively insoluble in the acidic reaction medium because of the acidic nature of the peptides. After stirring for 1-2 minutes, the entire slurry was passed through a 4.25 cm glass fiber filter (Fisher G4) using a filter funnel and vacuum at 15 psi. This began removal of TFA, TFMSA, anisole, and any soluble reaction products, leaving the cleaved peptide and resin on the filter. The peptide on the filter was washed several times with a total volume of 100 ml of MBE.

**Extraction**. The insoluble acid form of the peptide on the filter was converted into a soluble sodium salt by extraction into a clean, dry flask with 10 ml of $Na_2CO_3$ (0.02 M, pH 10.2), using 5 successive rinses of 2 ml, with at least 1 minute extraction on the filter prior to applying the vacuum each time. This produced a filtrate of about pH 5 that contained the solubilized peptide.

**Purification**. The filtrate was dialyzed twice with stirring against 2 liters of distilled water for 2 hours each time using dialysis tubing (Spectrapor, nominal MW cutoff of 1000 daltons). The dialysate was frozen and lyophilized, yielding white flakes or powders. The average yield of peptides was 40%.

Purity of peptides was checked by high performance liquid chromatography (Varian 5500 LC) using gel permeation columns designed for peptides (Toya Soda 2000 SW and 3000 PWXL). Single, sharp peaks at the appropriate MW's were obtained (24). Both anion-exchange and reverse-phase liquid chromatography of the peptides also produced well-defined, single peaks. The anion-exchange column (Synchrome, Inc., Synchropak AX300) was run with a continuous gradient of distilled water to 2.0 M NaCl in 1 hour at 25 °C. The reverse-phase column (Phenomenex, ultracarb 5 0DS 30) was run with a continuous gradient of 0.05 M sodium acetate (pH 6.5) to 100% acetonitrile in 20 minutes at 25 °C. Because there were no peaks in the hydrophobic portion of the gradient, few if any peptides having uncleaved benzyl protecting groups were present.

**Phosphorylation of Peptides**. Monochlorophosphoric acid ($H_2PO_3Cl$), produced by reacting one equivalent of phosphorus oxychloride ($POCl_3$, Aldrich), with 2 equivalents of water, was used as the phosphorylating agent (28-29). Because this reagent has only one reactive site, it is not as prone to inappropriate reactions as may occur with use of $POCl_3$ (30). Numerous attempts to phosphorylate polyanionic peptides by use of diphenylchlorophosphate (Aldrich) were unsuccessful due to irreversible binding of the peptides to the catalysts during the hydrogenation step to remove diphenyl protecting groups. In addition, there was evidence that some cleavage of the polyanionic peptides occurred with use of this reagent. Although diphenylchlorophosphate has been used successfully in synthesis of phosphopeptides having one or two serine residues (31-32), its use appears to lead to problems with larger, more polyanionic peptides.

An example protocol follows. Phosphorus oxychloride (1.872 ml, 20 mmole) was added over a period of one hour to water (0.72 ml, 40 mmole), followed by stirring for 30 minutes to allow completion of formation of monochloro-phosphoric acid. Next, approximately 100 mg of peptide (in the case of $Asp_{20}Ser_2$, for example, this represented about 0.096 mmoles of serine residues) were added, followed by stirring at 60 °C for 4 hours. The reaction was ended by dropwise addition of 0.36 ml (20 mmole) of water to degrade any unreacted monochlorophosphoric acid to orthophosphate. Next, 1N HCl (approximately 1.3 ml) was added to give a solution of 0.3 N HCl. This was heated in a boiling water bath for at least 20 minutes to destroy any polyphosphates that may have formed. Upon cooling, 10 ml of MBE was added to precipitate the phosphopeptide, followed by centrifugation and multiple washings with 5 ml of methanol. After lyophilization, the peptide was dissolved in distilled water, adjusted to about pH 7 with 1N NaOH. This solution was dialyzed twice against 2 liters of distilled water for 4 hours. The dialysate was lyophilized to give the final product.

The extent of phosphorylation of peptides was measured spectrophotometrically upon formation of the phosphomolybdate complex (33). All phosphate present was bound, requiring release by digestion in persulfate-$H_2SO_4$: there was no soluble reactive phosphate in the purified peptide samples.

**Verification of Peptide Structures**. a) Peptide Sequencing. An automated protein sequencer (Applied Biosystems, model 477A with an on-line amino acid analyzer, ABI model 120A) was used to confirm the primary structures of the peptides. Standard cycles based on Edman degradations were used. Amino acids were identified as their phenylthiohydantoin derivatives using PTH standards (ABI).

b) Alkalimetric Titrations. Solutions of peptides (5 ml at 1 mg/ml) were titrated with 0.1N NaOH over pH 2 to 12. The amount of base required to neutralize the carboxyl groups (pKa approx. 4.0) and phosphate groups (pKa=7.10) was used to calculate the amounts of these groups present (34).

c) β-elimination of phosphate. Phosphate O-linked to serine is removable by a base-catalyzed B-elimination reaction in which there is an increase in absorbance at 240 nm due to the production of dehydroalanine as a reaction product (35-36). Samples of peptides at 0.1 mg/ml in 0.5 N NaOH at 24 °C were monitored in a quartz cuvette continuously at 240 nm (Perkin Elmer UV/VIS, model lambda 4A).

**Crystallization Assays**. a) $CaCO_3$ Screening Assay. An assay fashioned after Hay et al. (5) was used to allow comparison of general activity of multiple inhibitors in one set of measurements. Solutions supersaturated with respect to $CaCO_3$ were prepared by separately pipetting 0.2 ml of 1.0 M $CaCl_2$ dihydrate and 0.4 ml of 0.4 M $NaHCO_3$ into 19.4 ml of artificial seawater (0.5 M NaCl, 0.011 M KCl). This resulted in initial concentrations of 10 mM $Ca^{2+}$ and 8.0 mM dissolved inorganic carbon (DIC). Actual concentrations of $Ca^{2+}$ were confirmed by atomic absorption and of DIC by Gran titration (37). The reaction vessels were 30 ml bottles with screwtops. These were partially immersed in a 1-gallon, plastic aquarium fitted for flow-through to a thermostated recirculating water bath (VWR model 1115) at 20 °C containing a submersible magnetic stirplate with 15 stirring locations (Cole Parmer). Crystallization was initiated by adjusting pH upward to 8.3 by titration of $\mu l$ amounts of 1 N NaOH. Final pH was read after 24 hours. Following addition of Ca but before addition of DIC, peptides were added to reaction vessels from stock solutions of 1.0 or 0.1 mg peptide/ ml. After experiments, reaction glassware was washed with 0.1 N HCl for at least 10 minutes followed by a treatment of at least 10 minutes with a solution of sodium hypochlorite (5.25 $\mu g$ NaOCl/ml, a 1/1000 dilution of commercial bleach). These treatments were necessary to dissolve crystals that formed on the surface of the glassware and to destroy any peptides that may adhere to the glass. Failure to perform either treatment led to abnormal results in control experiments. Following HCl and NaOCl treatments, reaction glassware was rinsed with several volumes of distilled water and dried for use in subsequent experiments. Results in control experiments demonstrated the importance of using the same reaction vessels for a set of measurements.

b) $CaCO_3$ Nucleation. The reaction conditions were the same as above. The reaction vessel was a 50 ml, 2-necked, round-bottom flask partially immersed in a thermostated water bath at 20 °C. The reaction vessel was closed to the atmosphere to minimize exchange of $CO_2$. The reaction was started by adjusting the pH upward to 8.3 by titration of $\mu l$ amounts of 1 N NaOH by digital pipette. The reaction was monitored by pH electrode and recorded by strip chart.

In these assays, there is first an induction period of stable pH of about 6 minutes during which crystal nuclei form, followed by a period of crystal growth when pH drifts downward due to removal of carbonate from solution. Inhibition of nucleation is indicated by the length of the induction period.

c) Seeded $CaCO_3$ crystal growth. The reaction conditions were modeled after Kazmierczak et al. (38) and consisted in 50 ml of artificial seawater as above with 2 mM each of $CaCl_2$ and DIC at pH 8.5. The reaction vessel was a 100 ml, water-jacketed, glass cylinder with a stoppered top fitted with a pH electrode and 2 burette delivery tips. The reaction was thermostated at 20 °C. The solution was stable until initiation of the reaction by addition of 2.5 mg of $CaCO_3$ seeds (Baker Analytical) which began to grow immediately. Seeds were aged prior to use by stirring as a slurry of 100 g $CaCO_3$ in 1 liter of distilled water for 3 weeks in a closed vessel. This aging of seeds was necessary to provide reproducible control curves. In some cases, 1 ml batches of aged seeds were sealed in ampules until opening for use. It also was acceptable to pipette seeds directly from the 1 liter slurry. However, frequent opening of the slurry to the atmosphere may lead to changes in the solution with resultant changes in the seeds, probably due to exchange of atmospheric $CO_2$. Peptides were added to the reaction vessel prior to the addition of the seeds. Reaction conditions during growth of seeds were held constant by autotitration of stocks of 0.1 M each of $CaCl_2$ and $Na_2CO_3$

(pH 11.1) from separate burettes controlled by a computer-assisted titrimeter (Fisher CAT). This replenished the lattice ions to the solution as they were removed due to crystallization and kept the pH at $8.50 \pm 0.02$.

d) Calcium phosphate crystallization. An assay modified after Termine and Conn (20) was used. A solution supersaturated with respect to calcium phosphate was prepared by separately pipetting 0.1 ml of 1.32 M $CaCl_2$ dihydrate and 0.1 ml of 0.90 M $NaH_2PO_4$ into 29.8 ml of distilled water. This yielded initial concentrations of 4.4 mM $Ca^{2+}$ and 3.0 mM dissolved inorganic phosphorus (DIP). The reaction vessel was a 50 ml, round-bottom, 2-necked flask partially immersed in a thermostated water bath at 20 °C. The reaction vessel was closed to the atmosphere. The reaction began upon mixing the reactants with an initial pH of 7.4. In experiments, peptides were added after the calcium and before the DIP.

Amorphous calcium phosphate (ACP) nucleates immediately upon addition of DIP and slowly grows as indicated by a modest decrease in pH during the first 30 minutes or so of the assay. Following this, ACP begins to transform to calcium hydroxyapatite (HAP), $Ca_{10}(PO_4)_6(OH)_2$, as indicated by a marked acceleration in the downward pH drift. The reaction ceases as reactants are depleted and the pH is lowered.

## Results

**Polyaspartate and Polyaspartate-alanine Molecules.** The enhancement of inhibition of $CaCO_3$ nucleation associated with an hydrophobic domain of alanine residues attached to a polyaspartate molecule is shown in Figure 1. An effect on crystal nucleation was clearly seen as an increase in the induction period prior to crystal nucleation as a function of length of hydrophobic domain. In the case of $Asp_{15}Ala_8$, addition of the hydrophobic domain was even more effective in these assays than increasing the number of Asp residues to 25.

In keeping with routine practice in polymer science, as pointed out in a prior study (24), results herein are reported on a weight basis rather than a molar basis because several smaller molecules of a homogeneous polymer may be functionally equivalent to one larger molecule of the same polymer. Therefore, the total mass of polymer rather than the total number of molecules can be the functionally important parameter.

A possible effect of the polyaspartate-alanine molecules on crystal growth was obscured in pH-drift measurements due to the changes that occur in pH and concentrations of lattice ions during crystal growth in the assays. Therefore, a constant composition, seeded-crystal approach to demonstrate effects on crystal growth was taken. An example experiment is shown in Figure 2 in which a control rate of crystallization was established, followed by sequential addition of increasing doses of a polyaspartate until crystallization ceased. This type of experiment was run for a set of molecules including polyaspartate$_{15}$ alanine$_{(2 \text{ to } 10)}$ and polyaspartate$_{(15 \text{ and } 25)}$ (Figure 3). In contrast to the findings for nucleation,

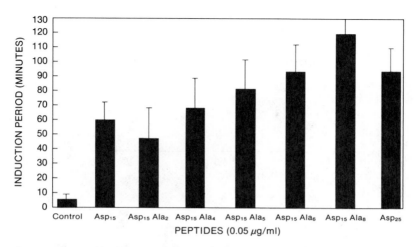

Figure 1. The effects of $Asp_{15}$, and $Asp_{25}$, and $Asp_{15}Ala_{(2to8)}$ on $CaCO_3$ nucleation. Means $\pm$ standard deviations, n=3 to 9.

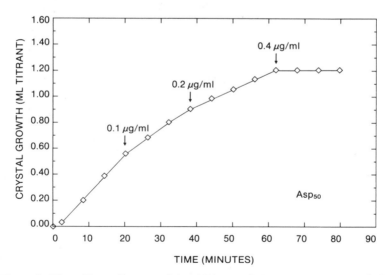

Figure 2. The effect of sequential additions of $Asp_{50}$ on growth of $CaCO_3$ seed crystals.

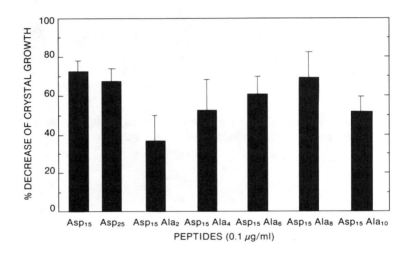

Figure 3. The effects of $Asp_{15}$, $Asp_{25}$, and $Asp_{15}Ala_{(2to10)}$ on growth of $CaCO_3$ seed crystals. Means $\pm$ standard deviations, n=3.

there was no enhancement of inhibition of crystal growth related to the presence of an alanine domain, indicating that inhibition of crystallization by these peptides was due only to an effect on nucleation.

The general inhibitory effect of a family of polyaspartate molecules on $CaCO_3$ formation as measured by use of the screening assay is shown in Figure 4. At a molecular size of about 30 to 40 residues, polyaspartate was most effective as an inhibitor as shown by the lowest dose (0.15 mg/ml for $Asp_{30}$ and $_{40}$) that stabilized the pH of the solutions for 24 hours. To confirm these results and to distinguish between effects on crystal nucleation and crystal growth, both nucleation assays (Figure 5) and crystal growth assays (Figure 6) were run using these peptides. Again, the most active polyaspartate versus crystal nucleation was in the range of 35 residues. Surprisingly, however, the most active versus crystal growth was $Asp_{15}$. Although $Asp_{10}$ had moderate activity (24), $Asp_5$ was ineffective as an inhibitor of $CaCO_3$ crystallization (data not shown).

Knowing that $Asp_{40}$ was a particularly effective size of polyaspartate for general inhibition of crystallization, it was of interest to determine the effect of adding a polyalanine domain to this longer polyanion, especially for comparison to the $Asp_{15}$ series. As seen in Figure 7, $Asp_{40}$ still performed better than $Asp_{40}Ala_6$ in the $CaCO_3$ screening assay, although the $Asp_{40}Ala_6$ seemed to perform better than $Asp_{45}$ molecules. A lack of effect on crystallization resulting from addition of the alanine domain to $Asp_{40}$ was confirmed by use of both the $CaCO_3$ nucleation assay (Figure 8) and the crystal growth assay (Figure 9) in that neither showed enhancement of inhibition imparted by the alanine domain.

**Phosphoserine-polyaspartate molecules.** The addition of one to three phosphoserine residues onto $Asp_{20}$ molecules resulted in a remarkable enhancement of both $CaCO_3$ (Figure 10) and calcium phosphate crystallization (Figure 11) relative to polyaspartate molecules of appropriate sizes. The effect on calcium phosphate formation indicated a surprisingly broad activity for these molecules in that although polyaspartate is a strong inhibitor of $CaCO_3$ formation, it is not particularly effective versus calcium phosphate formation. The effect of the PSer-containing $Asp_{20}$ molecules on calcium phosphate crystallization apparently was mainly at the level of conversion to apatite since amorphous calcium phosphate still formed early in the experiments and grew steadily but the sharp breaks in the curves indicative of apatite formation did not occur (Figure 11).

A more detailed analysis of the performance of the PSer-containing $Asp_{20}$ molecules is given in Table 1. Notice that both the $Asp_{20}PSer$, and $Asp_{20}PSer_2$ molecules were essentially completely phosphorylated based on the analysis of % phosphate by weight of the peptides. However, for reasons that are not clear, the $Asp_{20}PSer_3$ molecules on the average had only about 2 out of 3 serine residues phosphorylated.

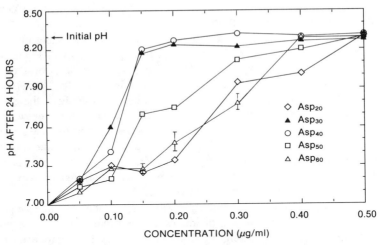

Figure 4. Comparison of the performance of polyaspartate molecules of different sizes in the $CaCO_3$ screening assay. Means $\pm$ standard deviations (representative values shown only for $Asp_{60}$), n=3.

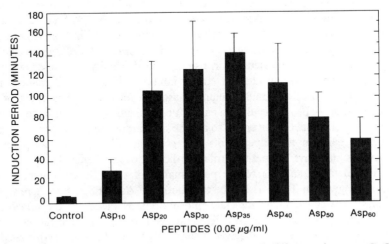

Figure 5. The effects of polyaspartate molecules of different sizes on $CaCO_3$ nucleation. Means $\pm$ standard deviations, n=4 to 10.

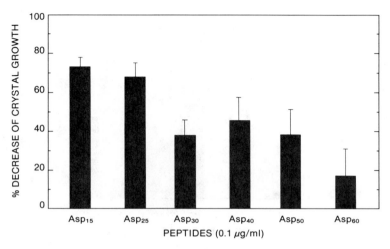

Figure 6. The effects of polyaspartate molecules of different sizes on growth of $CaCO_3$ seed crystals. Means ± standard deviations, n=3.

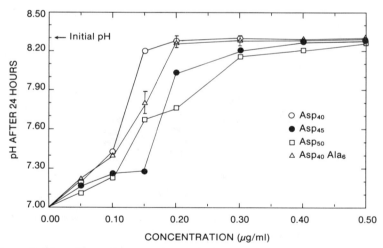

Figure 7. The effect of an alanine domain added to $Asp_{40}$ on performance in the $CaCO_3$ screening assay. Means ± standard deviations (representative values shown only for $Asp_{40}Ala_6$), n=3.

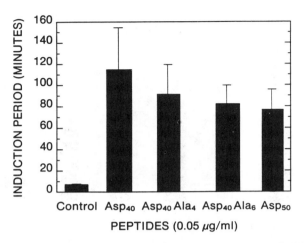

Figure 8. The effect of an alanine domain added to $Asp_{40}$ on $CaCO_3$ nucleation. Means ± standard deviations, n=6 to 10.

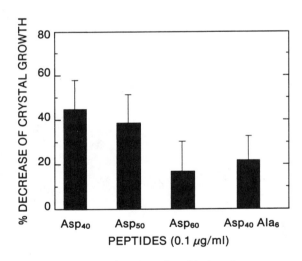

Figure 9. The effect of an alanine domain added to $Asp_{40}$ on growth of $CaCO_3$ seed crystals. Means ± standard deviations, n=3.

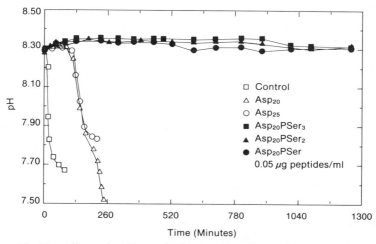

Figure 10. The effect of a N-terminal phosphoserine residues added to $Asp_{20}$ on $CaCO_3$ nucleation.

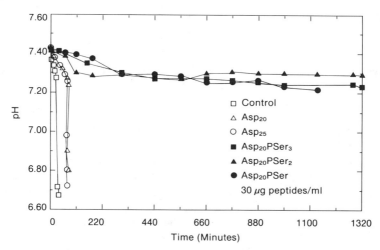

Figure 11. The effect of N-terminal phosphoserine residues added to $Asp_{20}$ on calcium phosphate formation.

Table 1. Effects of Phosphorylation of Terminal Residues of Polyanionic Peptides on Calcium Carbonate and Phosphate Crystallization. Means ± standard deviations, n = 70 for controls, n =3 to 5 for experiments with peptides.

| Peptide | % Ser as Pser | Calcium Carbonate Nucleation Assay | | Calcium Phosphate Crystallization Assay | |
|---|---|---|---|---|---|
| | | (ug/ml) | Induction period (minutes) | (ug/ml) | Period to Apatite formation (minutes) |
| Control | | | 5.86 ± 0.76 | | 20.7 ± 2.09 |
| $Asp_{20}$ | | 0.05 | 87.50 ± 19.00 | 30 | 59.7 ± 6.40 |
| $Asp_{20}Ser$ | | 0.05 | 65.00 ± 11.10 | 30 | 47.3 ± 3.05 |
| $Asp_{20}PSer$ | 97.6 | 0.05 | 20 hrs. | 30 | 20 hrs. |
| | | 0.02 | 84.00 ± 39.30 | 15 | 30.0 ± 0.10 |
| $Asp_{20}Ser_2$ | | 0.05 | 48.80 ± 16.20 | 30 | 54.5 ± 3.50 |
| $Asp_{20}PSer_2$ | 99.8 | 0.05 | 20 hrs. | 30 | 20 hrs. |
| | | 0.02 | 28.50 ± 11.80 | 15 | 37.8 ± 1.78 |
| $Asp_{20}Ser_3$ | | 0.05 | 31.30 ± 11.10 | 30 | 53.0 ± 0.75 |
| $Asp_{20}PSer_3$ | 59.0 | 0.05 | 20 hrs. | 30 | 20 hrs. |
| | | 0.02 | 44.00 ± 24.20 | 15 | 43.5 ± 4.12 |
| Phosphocitrate | | 0.01 | 163.30 ± 23.10 | 5.0 | 110.0 ± 1.00 |

For comparative purposes, the effect of phosphorylation on inhibition of crystallization by aspartic-enriched peptides having spacer amino acids is shown in Table 2. Polyserine (Sigma Chemical Company) is an unusual material in that although it might be expected to be polar and therefore soluble in polar solvents, it is essentially insoluble in a broad range of both polar and non-polar solvents, perhaps due to H-bonding to itself via adjacent OH groups. However, it is also insoluble in 8 M urea, often used to dissolve H-bonded proteins. It was possible to prepare stocks of 0.1 mg polyserine/ml for use in the inhibition studies by brief treatment at pH 12 followed by neutralization to pH 7. The polyserine did dissolve somewhat during phosphorylation with monochlorophosphoric acid, and a material with 23% of serine residues phosphorylated was produced. Similarly, the other peptides that had multiple serine residues yielded relatively low % phosphorylation. However, in all cases, the phosphorylated forms of the peptides exhibited enhanced activity to some extent as crystallization inhibitors. The presence of uncharged spacer residues between negatively-charged residues of phosphoserine or aspartic acid resulted in peptides with low inhibitory activity compared to polyaspartate.

Table 2.  Effects of Phosphorylation of Polyanionic Peptides That Contain Spacer Residues on Calcium Carbonate and Phosphate Crystallization. Means ± standard deviations, n = 70 for controls, n =3 to 5 for experiments with peptides.

| Peptide | % Ser as Pser | Calcium Carbonate Nucleation Assay | | Calcium Phosphate Crystallization Assay | |
|---|---|---|---|---|---|
| | | (ug/ml) | Induction period (minutes) | (ug/ml) | Period to Apatite formation (minutes) |
| Control | | | 5.86 ± 0.76 | | 20.7 ± 2.09 |
| PolySer(MW 5100) | | 0.05 | 4.00 ± 0.28 | 30 | same as control |
| PolyPSer | 23.0 | 0.05 | 113 ± 50.60 | 10 | 33.5 ± 8.74 |
| (AspSerGly)$_{10}$ | | 0.10 | 9.10 ± 1.20 | 10 | same as control |
| (AspPSerGly)$_{10}$ | 30.0 | 0.10 | 7.50 ± 2.10 | 10 | 28.0 ± 1.70 |
| | | | | 30 | 49.3 ± 1.15 |
| (AspSer)$_{10}$ | | 0.05 | 33.5 ± 2.12 | 10 | same as control |
| | | | | 30 | 50.0 ± 5.60 |
| (AspPSer)$_{10}$ | 29.0 | 0.05 | 205 ± 14.40 | 10 | 31.0 ± 2.80 |
| | | | | 30 | 170 ± 14.4 |

**Verification of Peptide Structure.** a) Aspartimide Formation. The formation of the 5-membered aspartimide ring occurred, to some extent, in some peptides as shown by infrared absorption at 1720 and 1780 cm$^{-1}$ (39) of KBr discs prepared as 0.1 mg peptide/g KBr (data not shown). This problem was confirmed by alkalimetric titration of peptides which revealed fewer titratable COO$^-$ groups than predicted in some cases. Consequently, such peptides were treated under mild alkaline conditions (about pH 10) at 60 °C for 1 hour to open the rings, restoring the free COO$^-$ groups. Although the imide rings are readily hydrolyzed under mild alkaline conditions, this produces some $\beta$ aspartic residues (27) in addition to the normal $\alpha$ residues already present. This was not thought to affect significantly the activity of the peptides as inhibitors of crystallization in that comparative studies of otherwise equivalent polyaspartates with and without $\beta$-residues produced through alkaline hydrolysis of imide rings showed essentially equivalent activity. Treatment was necessary for imide-containing peptides, which otherwise exhibited poor solubility and decreased activity.

  b) Sequence analysis. Automated Edman degradations of peptides revealed the appropriate sequences in all cases. Peptides that were treated for imide formation, however, exhibited lower yields of aspartic residues, presumably due to the known resistance to sequencing of $\beta$-residues (40-41).

The aspartate$_{20}$ serine$_{(1 to 3)}$ molecules gave proper sequences on Edman analysis. However, accompanying phosphorylation, there was measurable aspartimide formation. This was treated by mild alkaline hydrolysis that restored all of the COO$^-$ groups. Sequence analysis of the Asp$_{20}$PSer$_{(1 to 3)}$ molecules again gave proper sequences but with low yields in the polyaspartate region, presumably due to the presence of $\beta$-aspartic residues.

The presence of phosphate as O-linkages to serine residues was confirmed by increases in absorbance at 240 nm in the $\beta$-elimination reaction. Another possibility for the location of phosphate was as an N-linked group at the terminal amine (29,30). However, this arrangement would not yield the $\beta$-elimination reaction and if N-PO$_3$ groups were present, they would have been removed by the acid hydrolysis step for destruction of polyphosphates that may have formed during the reaction with monochlorophosphoric acid.

## Discussion

The observation that polyaspartate is most effective on a weight basis as an inhibitor of CaCO$_3$ formation at a molecular size of about 30 to 40 residues is consistent with earlier studies that suggested an activity maximum in this size range (24,42). The significance of this size was clearly related to an interaction with the crystal nuclei rather than crystal surfaces as such because polyaspartate inhibited crystal nucleation most effectively at a size of about 40 residues but inhibited crystal growth most effectively at about 15 residues.

The results raise questions about the size and structure of the crystal nucleus and the crystal growth sites. Although thermodynamic calculations suggest a possible size of a stable nucleus as small as 8 ion pairs (37), an interaction with a peptide of 40 residues would imply a much larger crystal nucleus if all residues actually interacted with the nuclear surface. On the other hand, the effects of the polyalanine domains suggested that not all of the 40 residues are in contact with the crystal nucleus. That is, the enhancement by the Asp$_{15}$Ala$_x$ molecules of inhibition of crystal nucleation may be interpreted that the peptide was bound to the crystal by the polyaspartate portion of the molecule, with the polyalanine region extending from the surface, disrupting movement of lattice ions to the crystal. As discussed by Sikes and Wheeler (24), a polyalanine domain of 8 or so residues would be sufficiently large at about 3 nm to interfere with the zone of attraction between lattice ions and surface charges, which is also estimated at about 3 nm (43). In addition, hydrophobic molecules are thought to impart around them a zone of exclusion of ions of about 0.5 nm (43), further adding to the hypothesized ability of an hydrophobic domain to enhance the inhibition of crystallization.

Crystallization theory predicts that although crystal nucleation may be diffusion-limited, crystal growth is not, being limited instead by the rate of incorporation of already-bound lattice ions into the growth sites (1,24,44,45). Therefore, it could be expected that one type of molecule might be a better inhibitor of crystal nucleation while another type is a better inhibitor of crystal growth. Along these lines, the $Asp_{15}(Ala)_x$ molecules exhibited an enhancement of inhibition of crystal nucleation but not of crystal growth (24).

In support of these concepts was the observation that the optimum size for inhibition by polyaspartate of crystal growth was about 15 residues, suggesting that additional residues may not actually bind to the crystals. In addition, there is the finding that attachment of a polyalanine region to a polyaspartate of 40 residues had no particular beneficial effect on inhibition of crystallization, and in fact simply added bulk to the molecule. An interpretation that is consistent with these results is that about 15 residues interacted directly with the surface of a crystal nucleus or a seed crystal, with the rest of the molecule extending from the surface. Hence, the zone of diffusion around the crystal nuclei would already be occupied by an $Asp_{40}$ molecule, and addition of more residues, hydrophobic or otherwise, would have no further benefit. In the case of inhibition of growth of seed crystals, the unbound portions of peptides would also be ineffective in increasing inhibitory activity because the process of crystal growth is not diffusion limited.

In general, crystal growth sites are thought to consist in kinks and dislocations on the crystal surface where the energy requirements for ion binding are more favorable due to more coordination sites than occur on flat portions of the surface (44,46). Although to date, it has not been possible to visualize these growth sites, evidently, their dimensions in the present study were such that a peptide of 15 or so residues could interact optimally with them. Curiously, polyaspartates of smaller sizes did not appear to block the growth sites very well, suggesting that essentially all of the 15 residues were involved in the interaction. The nature of the growth sites and their interactions with peptides may become subjects for direct study through application of recent advances in atomic force microscopy (47) that involve visualization of solid surfaces and peptides at the angstrom level.

The present results support the concept that one function of the hydrophobic regions of protein inhibitors of mineralization may be in control of access of lattice ions to crystal surfaces. However, they also demonstrate that hydrophobic zones do not need to be as extensive for this purpose as they in fact are, and further other non-hydrophobic portions of proteins could have similar effects. Therefore, it seems clear that the regions of protein inhibitors that do not directly interact with crystal surfaces would have other functions. Prominent among these as suggested by other studies are 1)provision of cell recognition sites (48-50), 2)enhancement of mobility of matrix proteins to the mineralization front (42, 51), 3)promotion of interactions with more hydrophobic, structural, matrix proteins (1), and 4)suppression of secondary nucleation associated with immobilized matrix complexes (4). It is also important to keep in mind the possibility that secondary structural differences in the peptides could affect their activities. For example, as clearly shown by Addadi and coworkers (52-55), polyanionic peptides with β-sheet structure may interact better with some crystal surfaces than peptides that have more $\alpha$-helix or random-coil structures. However it is not known that the secondary structures of the small polyaspartate molecules of the present study would be very different from each other.

The general importance of phosphoserine to the inhibition of $CaCO_3$ and calcium phosphate formation has been demonstrated in a number of studies (op. cit.). The marked enhancement of inhibitory activity imparted by a single or a few phosphoserine residues added to the N-terminus of polyaspartate molecules observed herein raises the possibility that terminal residues may have particular importance in binding of proteins to certain crystal surfaces such as those of $CaCO_3$ nuclei or amorphous calcium phosphate. Indeed several of the protein inhibitors of mineralization are known to have polyanionic terminal regions (1,6,7,11), and statherin in particular has an N-terminus consisting in H-Asp-PSer-PSer-Glu-Glu- (22). In some cases, there is evidence that polyelectrolytic polymers do bind to crystal at one end, with the rest of the molecule extending from the surface (56).

The presence of uncharged spacer residues, such as glycine or serine, between or interspersed among anionic residues of peptides led to considerably reduced inhibitory activity in all of the assays. Although phosphorylation of serine residues improved this activity in all cases, optimum activity of protein crystallization inhibitors still seemed to require continuous runs of anionic amino acids. As discussed elsewhere (1,7,8,24), although it is possible that ordered copolymers of (Asp-X)$_n$ and (Asp-X-Y)$_n$, where X and Y are small, neutral residues (12,13,57), exist to some extent in matrix proteins, such sequences do not account for the observed activity of the proteins as crystallization inhibitors.

Finally, the study of biomineralization inhibitors has produced a number of molecules that have uses ranging from biomedical to industrial applications (58,59). For example, Williams and Sallis (60) established that phosphocitrate was the most potent inhibitor on a molar basis of hydroxyapatite formation among a group of materials including 1-hydroxyethane 1,1-diphosphonate (EHDP), perhaps the most widely commercialized organic crystallization inhibitor. As seen in the present studies, the polyanionic peptides with phosphoserine N-termini (MW 2500) although not as potent as phosphocitrate (MW 272 g) on a weight basis, are the most potent inhibitors of calcium carbonate and phosphate crystallization on a molar basis yet discovered.

The design of this study required analysis of the effects of the peptides on a weight basis. The analysis of molar binding characteristics, molar affinities, surface coverage of crystals, and inhibition of $CaCO_3$ crystallization by selected polyanionic peptides is the subject of a following paper (61).

## Acknowledgments

This work was supported in part by grants from the Alabama Research Institute, the Mississippi-Alabama Sea Grant Consortium, the National Science Foundation, and the Office of Naval Research. We thank J. D. Sallis for providing a sample of phosphocitrate.

## Literature Cited

1. Wheeler, A. P.; Sikes. C. S. In Biomineralization: Chemical and Biochemical Perspectives; Mann, S.; Webb, J.; Williams, R. J. P., Eds.; VCH Publishers: Weinheim, W. Germany, 1989, p 95.
2. Crenshaw, M. A.; Linde, A.; Lussi, A. In Atomic and Molecular Processing of Electronic and Ceramic Materials; Aksay, I. A.; McVay, G. L.; Stoebe, T. G.; Wager, J. F., Eds.; Materials Research Society: Pittsburgh, 1988; p 99.
3. Linde, A.; Lussi, A.; Crenshaw, M. A. Calcif. Tissue Int. 1989, 44, 286-95.
4. Gunthorpe, M. E.; Sikes, C. S.; Wheeler, A. P. Biol. Bull. 1990, in press.
5. Hay, D. I.; Schluckebier, S. K.; Moreno, E. C. Calcif. Tissue Int. 1986, 39, 151-60.
6. Butler, W. T.; Bhown, M.; Dimuzio, M. T.; Cothran, W. C.; Linde, A. Arch. Biochem. Biophys. 1983, 225, 178-86.
7. Rusenko, K. W. Ph.D. Dissertation, Clemson University, Clemson, South Carolina, 1988.
8. Wheeler, A. P.; Rusenko, K. W.; Sikes, C. S. In Chemical Aspects of Regulation of Mineralization; Sikes, C. S.; Wheeler, A. P., Eds.; University of South Alabama Publication Services: Mobile, AL, 1988; p 9.
9. Wheeler, A. P.; Low, K. C.; Sikes, C. S. In Surface Reactive Peptides and Polymers: Discovery and Commercialization; Sikes, C. S.; Wheeler, A. P., Eds.; ACS Books: Washington, D. C., 1990.
10. Gorski, J. P.; Shimizu, K. J. Biol. Chem. 1988, 263, 15938-45.
11. Robbins, L. L.; Donachy, J. E. In Surface Reactive Peptides and Polymers: Discovery and Commercialization; Sikes, C. S.; Wheeler, A. P., Eds.; ACS Books: Washington, D. C., 1990.
12. Weiner, S. Crit. Rev. Biochem. 1986, 20, 365-408.
13. Weiner, S. Biochemistry 1983, 22, 4139-45.
14. Veis, A. In The Chemistry and Biology of Mineralized Tissues; Butler, W. T., Ed.; Ebsco Media; Birmingham, AL, 1985; p 170.
15. Veis, A.; Sabsay, B.; Wu, C. B. In Surface Reactive Peptides and Polymers: Discovery and Commercialization; Sikes, C. S.; Wheeler, A. P., Eds.; ACS Books: Washington, D.C., 1990.
16. Williams, J.; Sanger, F. Biochem. Biophys. Acta 1959, 33, 294- 6.
17. Lechner, J. H.; Veis, A.; Sabsay, B. In The Chemistry and Biological of Mineralized Connective Tissues; Veis, A., Ed.; Elsevier North Holand: Amsterdam, 1981, p 395.
18. Aoba, T.; Moreno, E. C.; Hay, D. I. Calcif. Tissue Int. 1984, 36, 651-8.

19. Hay, D. I.; Carlson, E. R.; Schluckebier, S. K.; Moreno, E. C.; Schlesinger, D. H. Calcif. Tissue Int. 1987, 40, 126-32.
20. Termine, J. D.; Conn, K. M. Calcif. Tissue Res. 1976, 22, 149-157.
21. Schlesinger, D. H.; Hay, D. I. Int. J. Peptide Protein Res. 1986, 27, 373-9.
22. Schlesinger, D. H.; Hay, D. I. J. Biol. Chem. 1977, 252, 1689-95.
23. Hay, D. I.; Moreno, E. C.; Schlesinger, D. H. Inorg. Persp. Biol. Med. 1979, 2, 271-85.
24. Sikes, C. S.; Wheeler, A. P. In Chemical Aspects of Regulation of Mineralization; Sikes, C. S.; Wheeler, A. P., Eds.; University of South Alabama Publication Services: Mobile, AL, 1988, p 15.
25. Sarin, V. K.; Kent, S. B. H.; Tam, J. P.; Merrifield, R. B. Anal. Biochem. 1981, 117, 147-57.
26. Bergot, J. D.; Noble, R.; Geiser, T. Applied Biosystems 11 Bulletin, 1986.
27. Tam, J. P.; Riemen, M. W.; Merrifield, R. B. Peptide Res. 1988, 1, 6-18.
28. Neuhaus, F. C.; Korkes, S. Bioch. Prep. 1958, 6, 75-9.
29. Frank, A. W. CRC Crit. Rev. Biochem. 1983, 16, 51-101.
30. Perich, J. W.; Johns, R. B. In Surface Reactive Peptides and Polymers: Discovery and Commercialization; Sikes, C. S.; Wheeler, A. P., Eds.; ACS Books: Washington, D. C., 1990.
31. Schlesinger, D. H.; Buku, A.; Wyssbrod, H. R.; Hay, D. I. Int. J. Peptide Protein Res. 1987, 30, 257-62.
32. Arendt, A.; Palczewski, K.; Moore, W. T.; Caprioli, R. M.; McDowell, J. H.; Hargrave, P. A. Int. J. Peptide Protein Res. 1989, 33, 478-76.
33. Eisenreich, S. J.; Bannerman, R. T.; Armstrong, D. E. Environ. Lett. 1975, 9, 43-53.
34. Saudek, V. Biopolymers 1981, 20, 1625-33.
35. Plantner, J. J.; Carlson, D. M. Meth. Enzymol. 1972, 28, 46-8.
36. Martensen, T. M. Meth. Enz. 1984, 107, 3-23.
37. Stumm, W.; Morgan, J. J. Aquatic Chemistry, John Wiley & Sons: New York, 1981.
38. Kazmierczak, T. F.; Tomson, M. B.; Nancollas, G. H. J. Phys. Chem. 1982, 86, 103-7.
39. Pivcova, H.; Saudek, V.; Drobnik, J.; Vlasak, J. Biopolymers 1981, 20, 1605-14.
40. Bornstein, P. Biochemistry 1970, 9, 2408-21.
41. Allen, G. Sequencing of Proteins and Peptides; Elsevier: Amsterdam, 1983.
42. Sikes, C. S.; Wheeler, A. P. U. S. Patent 4 534 881, 1985.
43. Norde, W. Croatica Chemica Acta 1983, 56, 705-20.
44. Nancollas, G. H. Adv. Colloid Interface Sci. 1979, 10, 215-52.
45. Reddy, M. M. In Chemical Aspects of Regulation of Mineralization; Sikes, C. S.; Wheeler, A. P., Eds.; University of South Alabama Publication Services: Mobile, AL, 1988; p 9.

46. Mann, S. Structure and Bonding 1983, 54, 125-74.
47. Drake, B.; Prater, C. B.; Weisenhorn, A. L.; Gould, S. A. C.; Albrecht, T. R.; Quate, C. F.; Cannell, D. S.; Hansma, H. G.; Hansma, P. K. Science 1989, 243, 1587-9.
48. Oldberg, A.; Granzen, A.; Heinegard, D. Proc. Natl. Acad. Sci. 1986, 83, 8819-23.
49. Gibbons, R. J.; Hay, D. I. Infection and Immunity 1988, 56, 439-445.
50. Butler, W. T.; Prince, C. W.; Mark, M. P.; Somerman, M. J. In Chemical Aspects of Regulation of Mineralization; Sikes, C. S.; Wheeler, A. P., Eds.; University of South Alabama Publication Services: Mobile, AL, 1988; p 29.
51. Wheeler, A. P.; Rusenko, K. W.; Swift, D. M.; Sikes, C. S. Mar. Biol. 1988, 98, 71-80.
52. Addadi, L.; Weiner, S. Proc. Natl. Acad. Sci. 1985, 82, 4110-4.
53. Addadi, L.; Weiner, S. Mol. Cryst. Liq. Cryst. 1986, 134, 305- 22.
54. Addadi, L.; Moradian, J.; Shay, E.; Maroudas, N. G.; Weiner, S. Proc. Natl. Acad. Sci. 1987, 84, 2732-6.
55. Addadi, L.; Moradian-Oldak, J.; Weiner, S. In Surface Reactive Peptides and Polymers: Discovery and Commercialization; Sikes, C. S.; Wheeler, A. P., Eds.; ACS Books: Washington, D. C., 1990.
56. Juriaanse, A. C.; Arends, J.; Ten Bosch, J. J. J. Colloid Interface Sci. 1980, 76, 212-9.
57. Weiner, S.; Hood, L. Science, 1975, 190, 987-9.
58. Sikes, C. S.; Wheeler, A. P. CHEMTECH 1988, 18, 620-6.
59. Sikes, C. S.; Wheeler, A. P. U. S. Patent 4 868 287, 1989.
60. Williams, G.; Sallis, J. D. Calcif. Tissue Int. 1982, 34, 169-77.
61. Wheeler, A. P.; Low, K. C.; Sikes, C. S. In Surface Reactive Peptides and Polymers: Discovery and Commercialization; Sikes, C. S.; Wheeler, A. P., Eds.; ACS Books: Washington, D. C., 1990.

RECEIVED August 27, 1990

# Chapter 6

# CaCO$_3$ Crystal-Binding Properties of Peptides and Their Influence on Crystal Growth

**A. P. Wheeler[1], K. C. Low[2], and C. Steven Sikes[2]**

**[1]Department of Biological Sciences, Clemson University, Clemson, SC 29634–1903**
**[2]Mineralization Center, University of South Alabama, Mobile, AL 36688**

The interactions of matrix phosphoproteins isolated from the carbonate shell of oyster and synthetic polyanionic peptides with CaCO$_3$ seed crystals were determined using both adsorption and crystal growth assays. In most cases a high affinity binding event with a rate constant having a $t_{0.5}$ less than 5 minutes was identified. An excellent correlation existed between the maximum inhibition of crystal growth effected by peptides and capacity of crystals at the rapid binding (high affinity) sites for the proteins and peptides. This correlation suggests that the rapid binding occurs at discrete crystal growth sites. Because the maximum inhibition varied with primary structure of the synthetic peptides, it would appear that the growth sites are heterogeneous. The matrix phosphoproteins and the synthetic peptides having primary structures similar to those found in the matrix proteins bind to most, if not all, of the growth sites and effect 100% inhibition of crystal growth. In addition, the large matrix proteins seem capable of binding to multiple growth sites at once.

Morphological and other evidence indicate that protein molecules which can be isolated from biomineral structures are intimately associated with the mineral component *in situ* (1-4). Consequently, these organic matrix molecules figure in most schemes describing the regulation of biomineral crystal growth. In this context, it repeatedly has been shown that the rate of nucleation of both calcium phosphate and carbonate crystals can be increased and the ensuing morphology of the crystals can be influenced by immobilized matrix or matrix-like proteins (4-8). The same proteins acting from solution can inhibit crystal growth (for review see 4) and have been shown in a few cases to alter crystal morphology

0097–6156/91/0444–0072$06.00/0
© 1991 American Chemical Society

as well (9). The effects of proteins on rate processes may well have biological significance because nucleation from metastable solutions such as those making up biological fluids is no doubt very slow, and ultimately crystal growth must be inhibited (terminated). The control of crystal orientation and microstructure by matrix proteins is also highly relevant in that biological mineral is often made up of crystals with highly predictable orientation and microstructures that differ greatly from those grown inorganically (3).

With the assumption that the proteins in fact control biomineral morphology and given that the morphologies are highly variable, it seems relevant to ask what structural properties of the proteins might be correlated to regulation of growth. As an example, in general it has long been understood that molecules which interact best with carbonate or phosphate minerals are typically polyanionic and matrix proteins are in fact usually highly negatively charged, containing carboxyl, sulphate and phosphate groups (4,10).

It should be mentioned that the properties of peptides which can optimize interaction of structures with crystals and control mineral growth is of interest to workers outside the field of biomineralization as well. As an example, crystal growth inhibitors (anti-scalants) or dispersants are marketed as water treatment chemicals in thousands of ton quantities annually. Although those currently used are often polymeric and structurally analogous to peptides, they are not biodegradable. It is our intent to develop biodegradable peptide substitutes for the current technology (10). In addition, any molecules that can potentially control crystal morphology or size or promote composite formation will be of interest at least as models to various groups in materials research.

Recently we have been studying the importance of some more detailed aspects of protein primary structure on mineral growth (11,12). To do this we have synthesized peptides with a variety of amino acid sequences, some of which have been identified in matrix proteins and other peptide mineralization regulators. One of the models we have studied in some detail is a class of matrix phosphoproteins isolated from the calcitic calcium carbonate shell of the oyster, *Crassostrea virginica*, which in many respects appears to be analogous to the phosphophoryns obtained from tooth dentin (13).

The determination of the structure of a specific class of the oyster matrix proteins is outlined elsewhere (14). In summary, they are highly anionic being made up of roughly 30 mole % aspartic acid (Asp). In addition they are also 30 mole % serine (Ser) much of which is phosphorylated (PSer), suggesting a name for these proteins might be molluscan shell phosphoproteins (MSPP). Uncharged glycine (Gly) also represented about 30 mole % with the residual of the protein being made up of a few positively charged and a mixture of hydrophobic residues. It appears that Asp in large measure is deployed in runs; that is, as regions of polyAsp. These findings are especially interesting in light of the fact that polyAsp peptides are superior inhibitors of carbonate crystallization when compared to peptides having other deployments of the amino acid (11). However, there are some studies which suggest that Asp in molluscan matrix proteins is deployed in part as $(Asp-X)_n$ in which X is either Ser or Gly (15). Such an arrangement would result in a near-match of the spacing between the carboxyl groups of Asp and the $Ca^{2+}$ in the carbonate crystal lattice (16).

It is nearly self-evident that for proteins in solution in quantities that are far sub-stoichiometric when compared to free lattice ions to affect crystal growth, they must absorb to the crystal surface. In fact it has been shown that the inhibitory activity of salivary and enamel proteins against calcium phosphate capacities can be clearly correlated to their apatite crystal-adsorption properties (17,18). Accordingly, it is the purpose of this contribution specifically to describe the adsorption of oyster MSPP and peptide analogs containing various deployments of the three primary amino acids of this protein, to $CaCO_3$ crystals and then correlate these adsorption processes to their effect on crystal growth. It is evident from this line of work that composition and primary structure of the peptides does alter selectivity of peptides for crystals and reveals different classes of peptide binding and crystal growth sites.

### Peptides and Crystals Studied

A major class of oyster MSPP which was isolated by reverse phase HPLC (RP-1) and various low molecular weight peptides synthesized by solid phase methods (11) were radiolabelled at the amino terminus (or $\varepsilon$-amino groups of the lysine f RP-1) using reductive alkylation (19). Radiolabelling proved necessary in the adsorption assays because the high affinity of the peptides for seed and the low specific surface area of the seeds required the use of very low concentrations of peptides to detect binding characteristics. Three peptides $Asp_{19}$, $(Ser-Asp)_{10}$ and $(Gly-Ser-Asp)_7$) were chosen to compare peptides of nearly the same molecular weight but differing in deployment of aspartic acid, representing polyAsp, $(Asp-X)_n$ and $(Asp-X-Y)_n$ respectively in which X and Y would in fact most likely be Ser or Gly in the native protein. $(Gly-Ser-Asp)_{10}$ was used as a control for the fact that as spacers are introduced between Asp's the number of Asp's per molecule is decreased. Finally phosphorylation of about 30 percent of the $(Gly-Ser-Asp)_{10}$ was accomplished as described elsewhere (12). Although the distribution of phosphate in the native matrix is uncertain, if any sequences with Ser between the Asp exist, it is highly likely they are phosphorylated.

Analytical $CaCO_3$ (calcite) seed crystals were used only following extensive preincubation of a stock suspension (100 mg ml$^{-1}$) in the medium used in all experiments: 500 mM NaCl and 10 mM KCl at pH 8.5. The individual rhombohedral crystals had edge lengths varying from 1-10 $\mu$m but were typically aggregated in solution. The specific surface area of the crystals as determined by gas adsorption ranged from about 0.2-0.7 m$^2$g$^{-1}$. These areas are much lower than those of hydroxyapatite crystals used in similar studies (e.g. 12.5 m$^2$g$^{-1}$; 17), thus, as mentioned above, it was necessary to radiolabel peptides in order to measure binding properties.

## Crystal-Binding Properties of Peptides

Most of the binding (adsorption) studies were performed by incubating peptides with 10 mg ml$^{-1}$ of seed for 5 min at 25°C. Quantities of peptide bound per mg of seed (Q) for any total peptide concentration (T) was determined indirectly as the difference (T-C) in which C is the unbound (free) concentration of peptide determined from the supernatant following rapid centrifugal sedimentation of the crystals.

The direct plots of Q vs C (adsorption isotherms) for all the peptides appear more or less as simple exponential curves (Fig. 1), suggesting there are a discrete number of saturable sites. For convenience, data such as these are often fitted to the Langmuir equation which in the form:

$$C/Q = 1/(NK) + C/N$$

will yield a straight line when C/Q is plotted as a function of C. N (the maximum number of adsorption sites and K (an affinity constant) can be obtained from the slopes and intercepts.

The regressions for the data from most of the isotherms fit the equation with high correlation coefficients (Fig. 2) suggesting that there are in fact a fixed number of non-interacting sites for peptides on the crystal surface. However, the correlations can not be used to indicate that all the sites are identical or that an adsorbate binds to just one site.

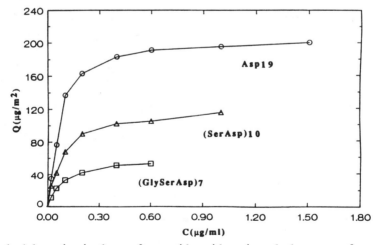

Figure 1. Adsorption isotherms for peptides with various deployments of aspartic acid. Seed crystals are incubated with C-14 labelled peptides for 5 min. The quantity bound to the crystal surface (Q, $\mu$g m$^{-2}$) is plotted as a function of the free peptide concentration (C, $\mu$g ml$^{-1}$) to obtain the isotherm. Note that the curves indicate saturation of sites for all the peptides. However, the quantity bound at saturation varies, with Asp$_{19}$ having the highest quantity bound.

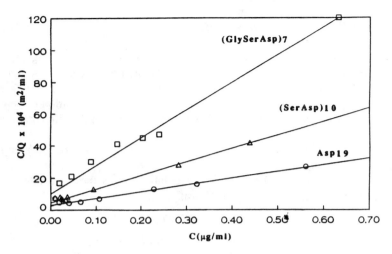

Figure 2. Linearized adsorption isotherms. In order to determine the capacity of crystals for a peptide (N) or the affinity of the peptide for the crystal (K), C/Q is plotted as a function of C (see text). The values of N are obtained as the reciprocals of the slope and K values from the reciprocals of the intercept divided by N. the regression coefficient for the curves shown vary from 0.96-0.99.

In examining the capacity and affinity constants obtained from the data, several trends emerge (Table I). First of all, $Asp_{19}$ has a higher capacity (N) on crystals than peptides such as $(Ser-Asp)_{10}$ and $(Gly-Ser-Asp)_7$ which have spacer amino acids between the Asp. Interestingly, there is no appreciable difference among these three peptides in terms of the affinity of the peptides for crystals. This

Table I. Capacity (N) and Affinity (K) Values for Peptide-Crystal Binding

| Peptide | $N\ (\mu g/m^2)^1$ | $N(\mu mol/m^2)^1$ | $K\ (ml/mg)$ | $K(1/\mu M)$ |
|---|---|---|---|---|
| $Asp_{19}$ | $227.5 \pm 45.9$ | 0.103 | $12.3 \pm 7.8$ | 27.0 |
| $(SerAsp)_{10}$ | $121.5 \pm 9.2$ | 0.060 | $11.6 \pm 2.7$ | 23.6 |
| $(GlySerAsp)_7$ | $49.5 \pm 10.9$ | 0.027 | $14.6 \pm 2.4$ | 26.7 |
| $(GlySerAsp)_{10}$ | $77.2 \pm 14.6$ | 0.030 | $11.2 \pm 3.9$ | 29.2 |
| $(GlyPSerAsp)_{10}^2$ | $237.9 \pm 55.3$ | 0.084 | $5.2 \pm 1.5$ | 14.5 |
| $RP\text{-}1^3$ | $1814.0 \pm 286.3$ | 0.034 | 0.034 | 345.6 |

[1]Specific surface area of $CaCO_3$ used for calculations was $0.7\ m^2g^{-1}$. Means $\pm$ standard deviations. Number of determinations ranged from 4-9.
[2]30% phosphorylated.
[3]Molecular weight estimated as 54,000 from gel permeation chromatography.

comparison might lead to speculation that polyAsp has a high affinity for most (or all) sites on the crystal whereas peptides with other configurations have an affinity similar to that of $Asp_{19}$ for a fraction of the $Asp_{19}$ sites but simply do not recognize the remainder of the sites. However, it is also possible theoretically that the various peptides are binding to completely different sites. In either case, it would appear that there is some diversity among peptide binding sites.

The difference in the binding behavior among the various Asp-containing peptides does not seem to be due to differences in the Asp content of the peptides. That is, the molar capacity of $(Gly-Ser-Asp)_{10}$ is not different from that of $(Gly-Ser-Asp)_7$.

Partial phosphorylation of $(Gly-Ser-Asp)_{10}$ increases the capacity of the peptide compared to the non-phosphorylated counterpart but, interestingly enough, appears to decrease the overall affinity. This observation may seem inconsistent with the known increase in activity of peptides as inhibitors when phosphorylated. However, in the present case phosphorylation did increase activity of the peptide against crystal growth (see below) by increasing its capacity to levels even greater than $(Ser-Asp)_{10}$, a peptide which has approximately the same charge density as the 30% phosphorylated $(Gly-Ser-Asp)_{10}$ (assuming phosphate groups have an average charge of 1.5).

The apparent lowering of affinity upon phosphorylation might be explained if one assumes that the recognition of addition $Asp_{19}$ sites by the partially phosphorylated peptides when compared to the non-phosphorylated peptide occurs, but that the affinity of phosphorylated $(Gly-Ser-Asp)_{10}$ for the additional sites is not as high as the affinity of the higher charge density $Asp_{19}$ for these same sites. Before any final conclusions regarding this hypothesis can be drawn, the adsorption properties of a more fully phosphorylated $(Gly-Ser-Asp)_{10}$ should be examined. An appropriate partial control for the importance of phosphate versus carboxylate groups in such a sequence then would be a peptide such as $(Gly-Asp-Asp)_{10}$.

The MSPP, RP-1, has a lower capacity on a mole basis than the small peptides. In contrast, the molar affinity of RP-1 is much higher than that for the peptides. The extreme divergence of these constants from those for the peptides is no doubt due to the high apparent molecular weight of the matrix protein. That is, because of its size, the protein may simultaneously bind to several sites, lowering its capacity but increasing its affinity for crystals.

## Sequential Binding Studies

To establish whether the various peptides in fact bind to the same sites, crystals were exposed in sequence to saturating concentrations of differing peptides. As an example, crystals pretreated with saturating levels of $Asp_{19}$ completely blocked binding by $(Gly-Ser-Asp)_7$. This finding suggests that this latter peptide has no unique sites compared to $Asp_{19}$, and that it in fact associates with only a fraction of the $Asp_{19}$ sites. In the converse experiment, $(Gly\text{-}Ser\text{-}Asp)_7$ blocks $Asp_{19}$ binding to an extent that is proportional to the relative N values of the two peptides. That is, N for the $Asp_{19}$ is reduced only by the N value for $(Gly\text{-}Ser\text{-}Asp)_7$, again suggesting that the latter peptide has no unique sites and does not recognize all the $Asp_{19}$ sites. The failure of $(Gly\text{-}Ser\text{-}Asp)_7$ to block additional $Asp_{19}$ sites was not due to exchange of part of the former peptide from the crystal in response to the presence of $Asp_{19}$. This was established in separate experiments in which saturating levels of cold $Asp_{19}$ failed to liberate radiolabelled $(Gly\text{-}Ser\text{-}Asp)_7$ from the crystals (Wheeler et al., unpubl.).

It is important to note that the binding of various peptides to hydroxyapatite has been observed to deviate from Langmuir models when the crystals are exposed to binary mixtures of the adsorbates (20). Therefore, the values for N obtained in the current study from Langmuir plots might not necessarily apply to studies involving more than one peptide. However, it should be reiterated that the peptides in this study were applied sequentially rather than simultaneously, and, given that peptides do not exchange readily from the surface of the crystals, the use of Langmuir N values to interpret the results might be justified.

## Time Course of Peptide Binding

Initial studies using the centrifugal method for determination of adsorption indicated that steady state in these crystal binding reactions was achieved by 5 minutes at 25 °C. Consequently, binding analyses were routinely performed using 5 min incubation of the peptides with the crystals. Such rapid rates of binding are not inconsistent with those reported in studies of peptide-apatite binding in which equilibria for the adsorption was approached in minutes (21). However, the binding rates for carbonate may be even greater than those reported for the apatite system in that, using independent rapid filtration experiments, we find that the equilibrium may be nearly achieved by one minute. This difference between apatite and carbonate in binding rate would be consistent with the apparent higher affinity of polyaspartate for carbonate when compared to apatite (4).    In so far as we discovered that peptides vary in their binding capacity depending on their primary structure, we wanted to be certain that the lower capacities of certain peptides was not due to a slower (lower affinity) process which might be revealed by extending the time of incubation. Therefore, suspensions of seed crystals were incubated with peptides for extended intervals and the quantity adsorbed was determined as usual. To control for changes in the seed during the extended incubation, separate experiments were run in which peptide was added at intervals to crystals that were continuously incubated in the medium but previously unexposed to peptide. In this case, adsorption was determined at each time interval after the normal 5 min of exposure of crystals to peptides.

From Figure 3 it is evident that in the absence of continuous incubation of the seed with peptide, no additional binding takes place. This suggests that in the absence of peptides the crystals do not change appreciably during incubation.

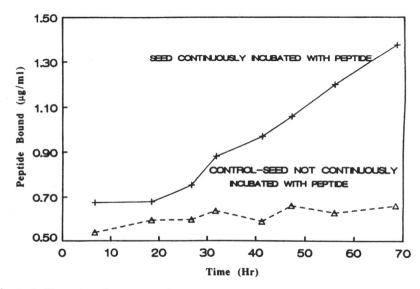

Figure 3. Extended time course of adsorption of (GlySerAsp)$_{10}$ to CaCO$_3$ seed crystals. Suspensions of seed were either continuously incubated with radiolabelled peptide or added at intervals to control seed which had been continuously incubated in adsorption media but previously unexposed to peptide. In the latter case, quantity bound was determined after a 5 min incubation of seed with peptide.

However, if the peptide is present throughout incubation the capacity of the crystals for the peptide will slowly increase. If this slow binding represents the presence of low affinity sites, it is not exclusively the result of binding of peptides to residual $Asp_{19}$ sites because $Asp_{19}$ binding also increases over the hours of incubation. Alternatively, it is possible that continuous presence of peptides induces a slow dispersion of the crystal aggregates which in turn would reveal more binding sites. That polyanions can induce dispersion is well-established ($\underline{22}$) and therefore this possibility should be further explored by examining crystals during the time course of incubation.

## Adsorption and Crystal Growth Inhibition

To correlate the adsorption data with the capacity of peptides to regulate crystal growth, the same seed crystals as used in the adsorption assays were grown in the presence of peptides in a constant composition assay. The ionic medium and temperature for crystal growth was the same as for adsorption except that the solution for crystal growth was supersaturated with respect to carbonate. The percent inhibition by peptides was determined by establishing a growth rate in each assay before and after addition of peptide (Fig. 4). The percent of the adsorption sites covered ($Q/N \times 100$) in each assay was determined either directly, using labeled peptide and the N values obtained from the independent adsorption assays for that peptide, or more indirectly using equilibrium euqations, the total peptide added to the assay, and both the N and K values obtained from independent adsorption assays ($\underline{23}$).

In general, the plots of percent decrease in crystal growth versus the percent binding sites covered were linear. This direct correlation between inhibition and adsorption agreed with the findings for the interaction of proteolytic fragments of salivary proteins with hydroxyapatite ($\underline{19}$). However, in this earlier study, the linearity only extended to Q equal to approximately 50% of N. Although a correlation between inhibition and coverage was clear for all the peptides in this study, the maximum percent inhibition varied from peptide to peptide. For example, although $Asp_{19}$ inhibited growth 100% when 100% of its sites were covered, the other peptides showed only a fractional inhibition when 100% of their sites were covered.

One possible interpretation for these observations is that peptides bind only to growth sites, with $Asp_{19}$ binding to all the growth sites, and the other peptides binding to only a fraction of the sites. If this conjecture were true, then the ratio of the capacity of any peptide X to that of $Asp_{19}$ ($N_x/NAsp_{19}$) should be equal to the same ratio for maximum growth inhibition (max. % decrease growth for peptide X/max. % decrease growth for $Asp_{19}$). Because the maximum decrease for $Asp_{19}$ was 100%, the above equality can be reduced to ($N_x/NAsp_{19}$) $\times 100 =$ max. % decrease for peptide X. In fact this relationship between maximum percent inhibition and percent of $Asp_{19}$ sites available to the adsorbate seemed to hold for the small, synthetic peptides (Table II). Therefore, excluding the unlikely

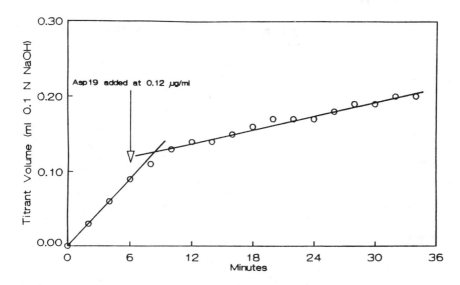

Figure 4. Seeded-crystal constant-composition mineralization assay. In this assay, 2.5 mg of the same crystal seeds as used in the adsorption assays are allowed to grow in 50 ml of a supersaturated solution of 500 mM NaCl and 10 mM KCl at $20.0 \pm 0.5\,°C$. The composition of the medium and the pH are held constant by simultaneous titration from 2 burettes, one containing 0.1 M $CaCl_2 \cdot 2H_2O$ and the other 0.1 M $Na_2CO_3$. After a control growth rate is established, the peptide is added and a new growth rate is determined. The figure shows a representative assay in which radiolabelled $Asp_{19}$ at a concentration of 0.12 $\mu g$ $ml^{-1}$ was added to the reaction vessel at the arrow. The percent decrease in crystal growth rate was 80.0% and the percent of the maximum binding sites occupied ($Q/N$ x 100%) was 86%. The Q value was determined directly from measurement of adsorption of radiolabelled peptides to seed crystals grown in the assay and N values were obtained from the independent adsorption assays.

coincidence that peptides can bind in identical multiples to sites or bind to other sites and still somehow maintain these ratios, it can be suggested that peptides in fact bind only to growth sites in a ratio of one peptide per site. Further, these growth sites appear to be heterogeneous in that many of the peptides do not recognize all the sites.

Table II. Percent Decrease in Crystal Growth and Percent $Asp_{19}$ Binding Sites Covered When Crystals are Saturated with Peptides.

| Peptide $100^2$ | Growth Decrease[1] | $(N_x/NAsp_{19})$ x |
|---|---|---|
| $Asp_{19}$ | 95% | 100% |
| $SerAsp_{10}$ | 52 | 60 |
| $(GlySerAsp)_7$ | 26 | 27 |
| $(GlySerAsp)_{10}$ | 36 | 30 |
| $(GlyPSerAsp)_{10}$ | 74 | 83 |
| RP-1 | 98 | 34 |

[1]Maximum percent decrease in crystal growth observed for peptides.
[2]The maximum percent of the total $Asp_{19}$ binding sites that can be occupied by peptide "X".

The relatively large molecular weight MSPP RP-1 is similar to $Asp_{19}$ in that it can produce 100% inhibition of crystal growth. However, the protein deviates from the small peptide in two ways. First, as mentioned above, it appeared capable of binding to more than one $Asp_{19}$ site thus effecting complete inhibition even though its capacity is only 34% that of $Asp_{19}$ (Table II). Secondly, the complete inhibition was achieved by the matrix protein at approximately one-half saturation of the crystals by the protein (Q = 0.5N). This latter observation suggests that either the proteins have binding sites unrelated to growth or, while binding to one set of sites, they can influence growth at adjacent sites.

## Conclusions

It appears that high affinity peptide binding sites on carbonate crystals are associated with crystal growth and may be exclusively growth sites. These sites appear to be heterogeneous in so far as peptides having primary structures other than those of simple polyanions recognize only a fraction of the sites. This differential binding may be biologically relevant in that some selection of which faces or regions of a crystal will grow must be made in order to produce crystal morphology different from crystal grown inorganically. That immobilized proteins or other polyanions can induce changes in normal crystal morphology has been demonstrated (8,24). Given the results described herein, it is not unreasonable to assume that proteins acting from solution might also have a predictable modifying influence.

Based on the observations for simple peptides, the MSPP has properties, such as runs of aspartic acid and phosphorylated serine, that would optimize protein-crystal interaction. However, the protein also has properties not observed for the smaller peptides. These include the apparent capability of the much larger protein to bind to several high affinity sites and, once bound, to interfere with growth at adjacent sites. To what extent these properties can influence crystal morphology is difficult to know, but bears further investigation. It is apparent that the interaction of matrix proteins with crystals can not be entirely modeled after those of small peptides.

## Acknowledgments

This work was supported in part by grants from the Alabama Research Institute, Mississippi-Alabama and South Carolina Sea Grant Consortia and from the National Science Fountation.

## Literature Cited

1. Watabe, N. J. Ultrastruct. Res. 1965, 12, 351-70.
2. Berman, A.; Addadi, L.; Weiner, S. Nature 1988, 331, 546-8.
3. Lowenstam, H. A.; Weiner, S. On Biomineralization: Oxford University Press; New York, 1989.
4. Wheeler, A. P.; Sikes, C. S. In Biomineralization: Chemical and Biochemical Perspectives; Mann, S.; Webb, J.; Williams, R. J. P., Eds.; VCH: Weinheim, 1989; p 95.
5. Termine, J. D.; Kleinman, H. K.; Whitson, W. S.; Conn, K. M.; McGarvey, M. L., Martin, G. R. Cell 1981, 26, 99-105.
6. Addadi, L.; Weiner, S. Proc. Natl. Acad. Sci. USA 1985, 82, 4110-4.
7. Linde, A.; Lussi, A.; Crenshaw, M. A. Calcif. Tissue Int. 1989, 44, 286-95.
8. Addadi, L.; Moradian-Oldak, J.; Weiner, S. In Surface Reactive Peptides and Polymers: Discovery and Commercialization. Sikes, C. S.; Wheeler, A. P., Eds.; ACS Books: Washington, 1990.
9. Wheeler, A. P.; Sikes, C. S. Amer. Zool. 1984, 24, 933-44.
10. Sikes, C. S.; Wheeler, A. P. CHEMTECH 1988, 18, 620-6.
11. Sikes, C. S.; Wheeler, A. P. In Chemical Aspects of Regulation of Mineralization; Sikes, C. S.; Wheeler, A. P., Eds.; Univ. South Alabama Publ. Ser.: Mobile, AL, 1988; p 15.
12. Sikes, C. S.; Yeung, M. L.; Wheeler, A. P. In Surface Reactive Peptides and Polymers: Discovery and Commercialization. Sikes, C. S.; Wheeler, A. P., Eds.; ACS Books: Washington, 1990.
13. Veis, A.; Sabsay, B.; Wu, C. B. In Surface Reactive Peptides and Polymers: Discovery and Commercialization. Sikes, C. S.; Wheeler, A. P., Eds.; ACS Books: Washington, 1990.

14. Rusenko, K. W.; Donachy, J. E.; Wheeler, A. P. In Surface Reactive Peptides and Polymers: Discovery and Commercialization. Sikes, C. S.; Wheeler, A. P., Eds.; ACS Books: Washington, 1990.
15. Weiner, S. Biochemistry 1983, 22, 4139-45.
16. Weiner, S.; Traub, W. Phil. Trans. R. Soc. Lond. 1984, 304B, 425-34.
17. Aoba, T.; Moreno, E. C.; Hay, D. I. Calcif. Tissue Int. 1984, 36, 651-8.
18. Aoba, T.; Moreno, E. C. In Surface Reactive Peptides and Polymers: Discovery and Commercialization. Sikes, C. S.; Wheeler, A. P., Eds.; ACS Books, 1990.
19. Rice, R. H.; Means, G. E. J. Biol. Chem. 1981, 246, 831-2.
20. Moreno, E. C.; Kresak, M.; Kane, J. J.; Hay, D. I. Langmuir 1987, 3, 511-9.
21. Moreno, E. C.; Kresak, M.; Hay, D. I. Calcif. Tissue Int. 1984, 36, 48-59.
22. Fivizzani, K. P. In Surface Reactive Peptides and Polymers: Discovery and Commercialization. Sikes, C. S.; Wheeler, A. P., Eds.; ACS Books: Washington, 1990.
23. Low, K. C. M.S. Thesis, Univ. South Alabama, Mobile, 1990.
24. Mann, S.; Heywood, B. R.; Rajam, S.; Walker, J. B. A. In Surface Reactive Peptides and Polymers: Discovery and Commercialization. Sikes, C. S.; Wheeler, A. P., Eds.; ACS Books: Washington, 1990.

RECEIVED August 27, 1990

# Chapter 7

# Structural Relationship of Amelogenin Proteins to Their Regulatory Function of Enamel Mineralization

**Takaaki Aoba and Edgard C. Moreno**

**Forsyth Dental Center, 140 Fenway, Boston, MA 02115**

Observations *in situ* and *in vitro* suggest that the regulatory mechanism of enamel mineralization in porcine amelogenesis involves the parent amelogenin (major matrix proteins secreted by the ameloblast) which is selectively adsorbed onto originally precipitating apatite crystals, thereby inhibiting apatite crystal growth. This phenomenon has been observed in supersaturated solutions having ionic composition similar to that of the liquid phase surrounding the enamel crystals *in vivo*. The adsorption affinity of amelogenin (and its inhibitory activity) is lost with cleavages of specific segments, suggesting that known enzymatic degradation of the secreted proteins plays a role in controlling the kinetics of enamel mineralization. Studies using fragments of the amelogenin (and a synthetic peptide) indicate that both the hydrophobic and hydrophilic sequences, at the N- and C-termini respectively, are essential for the proposed function of the amelogenin as a crystal growth regulator.

Most of the biomineralization processes are initiated by the secretion of specific matrix proteins, which seem to be involved in the regulation of the process itself. In general, matrix proteins in biomineralized tissues are classified into two groups: structural proteins (hydrophobic in nature) such as collagen in bone, and acidic proteins (hydrophilic in nature) such as osteonectin and other phosphoproteins (1). Recent studies suggested that the acidic proteins, if immobilized onto the hydrophobic proteins, may act as nucleators, while most of the acidic matrix proteins, if solubilized in the fluid surrounding the precipitating mineral, can act as inhibitors of crystal growth (2-4, see also Dr. Veis' paper in this volume). However, details of the regulatory mechanism, particularly the relationship between the structure of matrix proteins and their postulated functions, still remain to be elucidated.

0097–6156/91/0444–0085$06.50/0
© 1991 American Chemical Society

## Enamel Matrix Proteins: Enamelins and Amelogenins

Dental enamel is the hardest tissue in mammals and is composed of large carbonated calcium apatite crystals, elongated along their c-axis direction. It is believed that the secreted enamel matrix proteins play a regulatory role in the mineralization process, but the actual mechanism is not yet completely defined. In developing enamel, two major classes of matrix proteins have been characterized, namely, enamelins and amelogenins (5,6). The properties of each class of proteins are given in Table I.

Table I. Enamel Matrix Proteins: Enamelins and Amelogenins

| Enamelins | Amelogenins |
|---|---|
| minor matrix constituents | major matrix constituents |
| hydrophilic (acidic) | in general hydrophobic |
| 50-70 kD molecular masses | 20-30 kD molecular masses |
| Gly, Asp and Ser-rich | Pro, Leu, Gln, and His-rich |
| EDTA-4M Guanidine extractable | 4M Guanidine extractable |

The enamelins were originally characterized as acidic glycoproteins interacting strongly with the enamel crystals and, thereby, not being extractable from the tissue unless the enamel mineral is dissolved with, e.g., an EDTA solution (6-8). Several investigators proposed (9,10) that this class of proteins might play a major role in the nucleation of enamel crystallites at the initial stage of enamel mineralization. However, no confirmative evidence has been obtained for such a function. In the literature, it was reported (6) that the enamelins corresponded to about 10% of the total matrix proteins; however, it is becoming accepted that the quantity of enamelins is much smaller (about 1-2% of the total matrix proteins) in early developmental stages (11). It was also reported that other proteins, e.g., albumin (12), either are extracted with the enamelins or contaminate them. At present there is no information about the primary structure of any of the enamelins, although partial sequences of proteins classified as non-amelogenins (on the basis of their amino acid composition) are becoming available (13).

The amelogenins are major matrix constituents and the primary structure of secreted amelogenin proteins have already been determined in various mammalian species (14-17). As shown in Figure 1, the most striking feature of the amelogenin sequence is that the first 30-60 residues at the N-terminus and the last 10 residues at the C-terminus are homologous among mammalian species. Most of the amelogenin species are hydrophobic and the original amelogenin is degraded after secretion, decreasing its quantity dramatically with developmental advancement

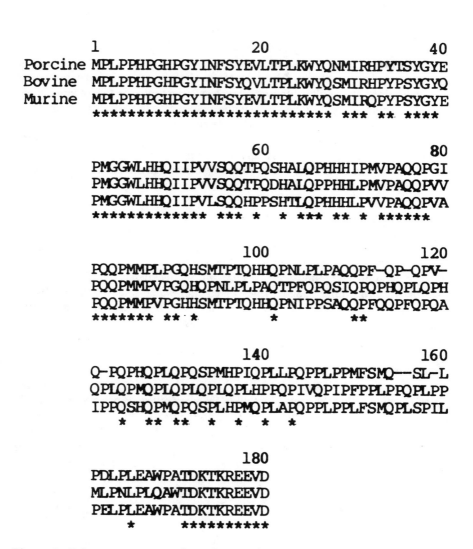

Figure 1. Primary structures of porcine, bovine, and murine amelogenins. The results are taken from references 17, 16, and 15, respectively. (*) indicates homology in these three sequences.

(or prior to the progress of the bulk of enamel mineralization) (18,19). In the literature, it has been considered that the amelogenins may contribute passively to the growth of enamel crystals by filling the intercrystallite space and providing a space for further growth by their massive removal (20,21).

Recent work in our laboratories has been concentrated on understanding the possible functions of the matrix proteins, particularly amelogenins, in early enamel mineralization. The results obtained give a new insight into the role of amelogenin proteins in controlling the kinetics of enamel crystal growth (22,23). The postulated function for the amelogenins is intimately related to their unique structure, including the hydrophobic and hydrophilic sequences at the N- and C-termini (24), which are conserved among mammalian species (Figure 1). We report here our recent work which was conducted using secretory enamel dissected from developing permanent incisors of 5-6 month old piglets.

## Enamel Mineralization in Early Developmental Stages

Figure 2A shows a typical microradiogram of a permanent incisor dissected from 5 to 6-month-old piglets. The tooth, at this age, was embedded in the mandible. For preparation of ground sections (100 $\mu$m in thickness), the dissected tooth was dehydrated with a graded series of ethanol/water solutions from 50% to 100% v/v and embedded in polyester resin. From the microradiogram, it was ascertained that most of the enamel tissue was in the secretory phase, except for the early maturing enamel displaying a higher mineral content (radiopacity) at the incisal end. The mineral content of the secretory enamel is generally lower than that of the underlying dentin, except for a narrow zone at the dentino-enamel junction (DEJ) showing high radiopacity. An important feature of enamel mineralization is that the mineralization in the secretory stage progresses gradually along two directions, from the cervical end to the incisal end and from the outer (close to the ameloblast) to the inner (close to DEJ) region.

The progress of enamel mineralization taking place in the secretory stage is more clearly displayed by high resolution electron microscopy. In Figures 2B through D are shown typical c-plane images of enamel apatite crystallites observed in the outer, middle-inner, and innermost secretory enamel, respectively. The observation of these enamel crystallites was conducted using JEOL 2000FX and 2010FX electron microscopes at 200 kV. In the outer zone, the crystallites were thin, ribbon-like in shape, having only a few unit cells in thickness in the cross section (Figure 2B). In the middle and innermost zones, the crystallites appeared plate-like, showing elongated-hexagonal cross sections (Figures 2C and 2D). These results, as well as those reported by others (25,26), indicate that the mineral accretion in the secretory phase, occurring at limited rates, is due mainly to the growth of crystallites on their prism planes or parallel to the a-axis. It is our working hypothesis that this growth process of enamel crystallites is regulated by the selective adsorption of matrix proteins, particularly amelogenins, onto the crystal surfaces.

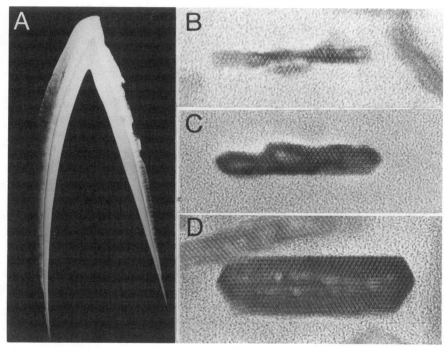

Figure 2. (A) Typical microradiogram of a permanent incisor from 5-6 month-old piglets. Note the low mineral content (radiolucency) of the secretory enamel. (B through D) high resolution electron micrograms of enamel crystallites obtained from the outer, middle, and inner zones of the secretory enamel, respectively.

## Enamel Matrix Proteins Soluble in Fluid Phase

It is currently accepted that the amelogenin(s) is degraded prior to its removal (21,27-30). Evidence has accumulated showing that the post-secretory degradation of the amelogenins occurs shortly after secretion (29-31) and proteolytic activity was found in the secretory enamel (31-37). Also, ultrastructural and histochemical work showed that the secretory ameloblast has the dual function of secreting and resorbing the proteins (38,39). It is reasonable to assume that peptides (or amino acids) cleaved from the parent protein are mobile in the fluid phase thus reaching the cell membrane or diffusing out of the tissue. Indeed, some of the degraded amelogenin products have been shown to be more soluble than the parent molecule; the latter is mostly present in the solid (condensed) state *in vivo* (40). This result was obtained by separating the fluid phase from the secretory enamel and analyzing the protein constituents. The procedures to separate the enamel fluid were published previously (41). Briefly, following the removal of the enamel organ tissue, the surface of the enamel was gently wiped with tissue paper and then covered with extra heavy mineral oil to minimize evaporation. The mineral oil had been presaturated with water vapor by bubbling $N_2$ gas that had passed through a 160 mM NaCl solution. The secretory enamel tissue, soft in consistency, was dissected with an excavator. The dissected enamel pieces were pooled, under the mineral oil, in an ultracentrifuge tube and then centrifuged at $2.4 \times 10^5$ g for 1 hr at 4°C. About 50 $\mu$L of the fluid was collected from 1 g of dissected enamel.

In Figure 3 are shown SDS-electrophoretograms of the proteins soluble in the enamel fluid (lane 1) and the proteins recovered from the centrifuged solid (lane 2). The main constituent of the soluble fraction, corresponding to about 5 %wt of the total proteins, corresponded to a 13 kD moiety. Analyses of amino acid composition and partial sequencings of its N- and C-termini (42) showed that this 13 kD protein corresponds to the 46-148 residue fragment of the parent amelogenin. The majority of the matrix proteins (about 95%) were sparingly soluble in the fluid and recovered by dissolving the tissue in 0.5 M acetic acid. Among these sparingly soluble proteins, the moieties with molecular masses of 5, 20, 23, and 25 kD on SDS-PAGE had an amino acid composition characteristic of amelogenins. Furthermore, sequencing work on these moieties (17,43) demonstrated that: a) the 25 kD amelogenin has the 173 residues shown in Figure 1; b) the 23 kD amelogenin lacks a 12- residue segment at the C-terminus of the parent amelogenin; c) the 20 kD amelogenin corresponds to the 1 through 148 residue segment, lacking the 25 residues at the C-terminus; and d) the 5 kD fragment corresponds to the 1 through 45 residue segment. The presence of a protease cleaving the amelogenins between Trp[45] and Leu[46] was verified in porcine and bovine secretory enamel (28,36,37).

Table II gives the results of amino acid analysis of the aforementioned amelogenins. The 25 and 20 kD porcine amelogenins are Pro, Leu, Gln and

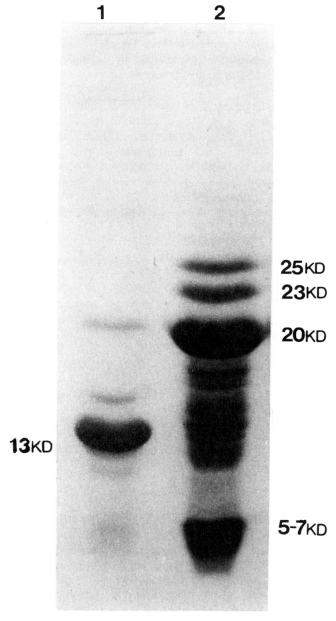

Figure 3. SDS-polyacrylamide gel electrophoretogram of matrix proteins of secretory porcine enamel. Lane 1, the proteins soluble in the enamel fluid. Lane 2, the proteins sparingly soluble in the fluid and isolated by dissolving the centrifuged enamel solid in 0.5 M acetic acid.

His rich, displaying characteristic amino acid composition common to those of
the amelogenins in other mammalian species. The sparingly soluble 5 kD fragment
is characterized by the high content of Pro, Tyr and Gly. The composition of
the soluble 13 kD amelogenin is characterized by the high content of Gln and
His residues, although the Pro content is also high.

Table II. Amino Acid Composition of Porcine Amelogenins (residues/1000 amino
acid residues)

| Amino acid | Amelogenins | | | | |
|---|---|---|---|---|---|
| | 25 kD | 23 kD | 20 kD | 13 kD | 5 kD |
| Asx | 37 | 25 | 21 | 15 | 48 |
| Thr | 40 | 32 | 35 | 20 | 47 |
| Ser | 42 | 51 | 45 | 51 | 41 |
| Glx | 177 | 184 | 187 | 223 | 74 |
| Pro | 248 | 247 | 262 | 304 | 186 |
| Gly | 43 | 44 | 50 | 26 | 113 |
| Ala | 31 | 25 | 25 | 34 | 0 |
| Val | 35 | 32 | 31 | 25 | 23 |
| Met | 42 | 63 | 67 | 64 | 71 |
| Ile | 37 | 44 | 36 | 19 | 46 |
| Leu | 99 | 95 | 78 | 98 | 71 |
| Tyr | 36 | 38 | 38 | 0 | 140 |
| Phe | 19 | 19 | 18 | 24 | 23 |
| His | 82 | 89 | 89 | 96 | 70 |
| Lys | 19 | 6 | 12 | 1 | 24 |
| Arg | 13 | 6 | 6 | 0 | 23 |
| Cys | 0 | 0 | 0 | 0 | 0 |

## Localization of Amelogenins in Secretory Enamel

Recent advances in protein separation techniques have disclosed the heterogeneous
nature of the enamel proteins (over 20-100 protein components) (6,16,44,45). Even
on the SDS-PAGE in Figure 3, numerous bands are distinguished below 25
kD. This multiplicity of enamel proteins is currently explained by (i) alternative
RNA splicings (11,16,46-48), (ii) posttranslation modification (glycosylation and
phosphorylation) (14,49), and (iii) postsecretory degradation (21,27-30,40). Most
of the matrix constituents having molecular masses of 25 kD and below belong
to the amelogenin class, although their exact origin remains to be determined.
The most extensive work in this area has been reported by Fincham et al. (11,48).
In the last several years, our effort was devoted to the understanding of the
possible functional roles of various amelogenin moieties and the modulation of

their functional roles with the postsecretory degradation or cleavages of specific segments.

In previous work using rat incisors (19,31), differences in the enamel matrix proteins have been studied from the cervical to the incisal end. As shown in Figure 2, the growth of enamel crystallites increases from the outer to the inner zones in porcine secretory enamel. This gradient in mineral content has also been verified by separating the secretory enamel with a fine razor blade into 5 or 6 layers from the flat labial surface to the DEJ of porcine incisors. The outermost zone (adjacent to the ameloblast) was 100-150 $\mu$m in thickness. Ca and P analyses of the dissected enamel samples indicated that the Ca and P contents of the outer two zones were about 10% and 7% by wt, respectively, and that the mineral content increased gradually toward the inner zones which had 20% by wt of Ca and 12% by wt of P.

Figure 4 shows the SDS-electrophoretograms of the enamel proteins isolated from the outer (I) through the innermost (VI) zones dissected from an incisor. Significant differences in the protein composition were obtained among the various zones. In the outermost zone, the 25 kD protein was the most prominent; the presence of high molecular mass proteins (60-90 kD) are also characteristic, although their quantities were limited. These high molecular mass proteins are classified as non-amelogenins, since their amino acid compositions are quite different (Glx, Ala, Leu and Asx-rich and low Pro) (50) from that of the amelogenin. In the second zone, the 20 kD protein as well as the 21-33 kD molecules increase in relative terms as the 25 kD amelogenin decreased. Advancing from the middle to the inner zones, the 25 kD protein almost disappears, but the 20 kD protein remained relatively constant. The 13 kD amelogenin, a major constituent of the enamel fluid (see Figure 3), was not apparent in zones I and II, but its quantity increased substantially in zones III through V. Proteins or peptides less than 5-7 kD also increased in the inner zones. In addition, the 60-90 kD proteins decreased substantially in the inner zones.

The foregoing results, as expected, indicate that the degradation of amelogenins occurs in the secretory stage and, importantly, that the degradation of the 25 kD amelogenin, as well as 60-90 kD proteins, seems to coincide with the increase of mineral content or the growth of enamel crystallites on the prism planes. The significance of the 25 kD amelogenin in enamel mineralization was further investigated in the following *in vitro* adsorption and crystal growth experimentation.

## Adsorption of Amelogenins onto Apatite Crystals

Until recently, it was believed that the enamelins were strongly adsorbed onto the enamel crystals, while the amelogenins in general were filling the intercrystalline space (20,21,51,52). However, recent work (53,54) using rabbit polyclonal antibodies against amelogenins and enamelins showing no cross reaction with each other, did demonstrate the distribution of both proteins on the crystals as well as in

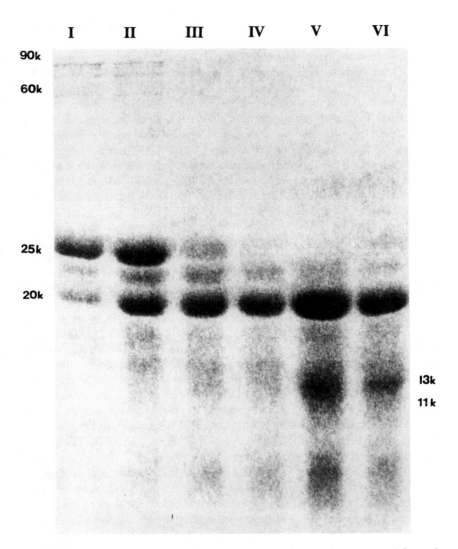

Figure 4. SDS-electrophoretogram of enamel matrix proteins extracted from the various layers (I through VI) of secretory porcine enamel. Note the difference in localization of the 25 kD, 20 kD, and 13 kD amelogenins.

the intercrystalline space adjacent to the ameloblast. We also investigated the possible interaction of the amelogenins, particularly the parent 25 kD amelogenin, with apatite crystals *in vitro* experimentation. For this purpose, enamel matrix proteins obtained from the outer zone (about 100 $\mu$m in depth from the surface) of porcine secretory enamel, were used as adsorbates. As mentioned above, this zone contains most of the originally secreted 25 kD amelogenin. Hydroxyapatite and fluorapatite (HA and FA, respectively) used as adsorbents were prepared in our laboratories (55,56) and their specific surface areas were 10.6 m$^2$/g for HA and 7.43 m$^2$/g for FA.

Figure 5 shows typical results of the selective adsorption of enamel proteins onto HA (lanes 1 through 3) and FA (lanes 4 through 6). The crystal-proteins equilibration was conducted using either 50 mM acetate buffer (pH 6.0 and at 0 °C) or 50 mM Tris buffer (pH 7.4 and at room temperature) containing 4 M guanidine to yield 0.1% wt/v adsorbate solutions. These solutions were carefully selected to completely solubilize the protein molecules, since most of the amelogenins are sparingly soluble at neutral pH and in the presence of high electrolyte concentrations (40). As shown in lane 3 of this SDS-PAGE, the 25 kD amelogenin, as well as the 60-90 kD non-amelogenins, were concentrated on the HA crystals after equilibration. In contrast, the 20 kD amelogenin and the other amelogenin moieties (23, 13 and 5-7 kD) were not adsorbed extensively onto the crystals and most of them remained in the solution. An interesting finding was that the adsorption of the 25 kD and 60-90 kD proteins was larger onto the FA and the adsorption of the other amelogenins (21-32 kD, 14-18 kD and 5-7 kD) also became prominent (lane 6). A previous study using fluorhydroxyapatites with various degrees of fluoride substitution (45) also showed that this enhancement of protein adsorption occurs with increasing degrees of fluoride substitution in the lattice, presumably due to a lower surface energy and therefore a weaker interaction of the water molecules with the surface of fluoridated apatite crystals, thus enhancing the competitive adsorption of the proteins.

The foregoing results of the adsorption studies were obtained using the whole matrix proteins, including the amelogenins and the enamelins (or nonamelogenins). In such adsorption studies, the question may arise as to whether the adsorption of the 25 kD amelogenin onto the apatite crystals was due to a direct interaction with the crystal surface, or was it mediated by coexisting enamelins or high molecular mass proteins. To elucidate this matter, adsorption experiments were also conducted using purified amelogenins (25, 20, and 13 kD). Isolation and purification of these amelogenins were performed using recyclings of gel-filtration and then ion-exchange chromatography (22,24,40). In Figure 6 is plotted the adsorption of the purified 25 kD and 20 kD amelogenins onto HA crystals as a function of the crystal surface area available. It is apparent that the adsorption of the 25 kD amelogenin increases with an increase of the total surface area of HA adsorbent, while the adsorption of the 20 kD amelogenin is suppressed at a lower adsorption value even with an increase of the available surface area

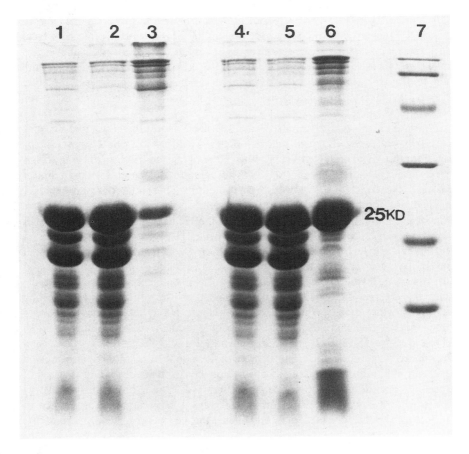

Figure 5. SDS-electrophoretogram showing the selective adsorption of enamel proteins onto hydroxyapatite (lanes 1 through 3) and fluorapatite (lanes 4 through 6). Lanes 1 and 4: the original enamel proteins used as adsorbates, which were isolated from the outer (young) secretory enamel of porcine incisors; Lanes 2 and 5: proteins remaining in solution after equilibration with the adsorbent crystals; Lanes 3 and 6: proteins adsorbed onto the crystals. Note the selective adsorption of the 25 kD amelogenin, as well as 60-90 kD moieties, onto both HA and FA. Lane 6: low-molecular weight standards (Bio-Rad).

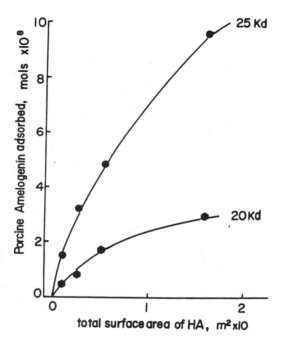

Figure 6. Plots of the amounts of the 25 kD and 20 kD porcine amelogenins adsorbed onto hydroxyapatite as a function of the total available surface area of the adsorbent.

from 0.5 to 2 m$^2$/g. In parallel experimentation, it was also confirmed that the 13 kD amelogenin did not show any significant adsorption onto the crystals and that the purified 5 kD fragment at the N-terminus showed only weak adsorption under solubilized conditions.

### Inhibition of Apatite Crystal Growth by Amelogenins

In previous studies (58), it was reported that bovine enamelins (nonamelogenins) can inhibit the growth of apatite crystals in *in vitro* supersaturated solutions, whereas the amelogenins isolated from bovine secretory enamel showed a marginal, if any, inhibition under the comparable conditions. It is important to point out that the amelogenin used in that report may not have been the parent protein but the degraded products, because the primary structure of bovine amelogenin reported by Takagi et al. (14) lacked the 10 residues at the C-terminus. The foregoing results obtained in our adsorption studies indicate that there is a significant difference in the adsorption affinity between the 25 kD amelogenin and the degraded products. Thus, we reevaluated the inhibitory activity of the amelogenins on apatite crystal growth.

The amelogenins tested were the purified 25, 20, and 13 kD proteins. Their effect on the growth kinetics of hydroxyapatite was investigated in a supersaturated solution having a composition similar to that of the enamel fluid separated from porcine secretory enamel (41). The initial pH of the experimental solution was 7.3 and the initial concentrations of Ca and P were 1.0 mM and 3.0 mM, respectively. The solution contained 160 mM NaCl so that an ionic strength similar to that of the enamel fluid would be maintained. The experimental solution (100 mL) was prepared fresh prior to every crystal growth run, from stock solutions of CaCl$_2$, K$_2$HPO$_4$, and NaCl. The two phosphate solutions were mixed in appropriate proportion to give the specified pH. The seed crystals used were from the same batch of HA used in the adsorption experiments. The freeze-dried protein sample was weighed and dissolved in the required volume of 50 mM acetate buffer (pH 6.0) at 0°C, to give a final concentration of 0.2 %wt/v. The seed crystals (30 mg of HA) were pretreated in 2 mL of each protein solution for 1 hr at 0°C. The suspension of HA seed crystals was added to the supersaturated solution (time zero), whereupon the precipitation commenced. In a control run, the seed crystals were dispersed in the NaCl solution, without any protein prior to their introduction into the experimental solution. The precipitation time course was monitored by recording continuously the pH value of the experimental solution. Also, aliquot solutions were withdrawn from the well-agitated experimental solution at specified intervals to determine the calcium and phosphorus concentrations.

Figure 7 shows the results of crystal growth experiments on HA seeds pretreated with the 25 kD (run A), 20 kD (run B), 13 kD (run C), and control (without protein-pretreatment). The precipitation course was plotted in terms of the pH and P concentration of the supersaturated solution as a function of the experimental

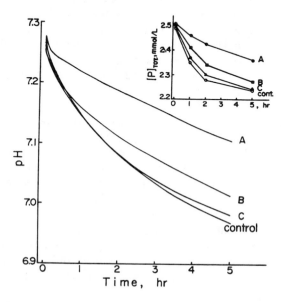

Figure 7. Kinetics of hydroxyapatite crystal growth in terms of solution pH vs. experimental time. In the control run, the HA crystals (30 mg/100 mL) were introduced into the supersaturated solution (time zero) without pretreatment with enamel proteins, while in runs A through C, the same amount of HA seeds was pretreated with the 25 kD, 20 kD, and 13 kD amelogenins, respectively. Insert: Plots of total phosphate concentration vs. time after addition of HA seeds.

time after addition of the apatite seeds. As illustrated by a slow decrease of the pH value of the solution (and a decrease of the P concentration corresponding to the pH changes), the 25 kD amelogenin showed the strongest inhibition of HA precipitation. Clearly, the 20 kD amelogenin caused less inhibition than the 25 kD protein and, as demonstrated by the superimposed plots of run C and the control, the 13 kD amelogenin did not display any significant inhibition of apatite crystal growth. At the end of the experimentation, it was determined, analytically, that the amounts of protein recovered from the crystal seeds were: 25 kD, 0.4 mg; 20 kD, 0.1 mg; and 13 kD, not detectable. Parallel experimentation with the 5 kD amelogenin showed no significant adsorption onto the crystal seeds, due to its sparingly soluble property, and thereby no inhibitory activity on apatite crystal growth. These results support the view that the parent amelogenin can be a potent inhibitor of enamel crystal growth, while the degraded products lose this functional role.

### Use of a Synthetic Peptide to Study the Protein Crystal Interaction

The differences in the adsorption affinity and crystal growth inhibitory activity of the parent 25 kD amelogenin and the rest of the amelogenins (23, 20, 13, and 5 kD) suggested a physiological importance for the hydrophilic sequence at the C-terminus. As illustrated in Fig. 1, the last 10 residues at the C-terminus, -TDKTKREEVD, are common to porcine, murine and bovine amelogenins. It was deemed probable that, in the amelogenin-enamel crystal interaction, this highly charged hydrophilic sequence might play an instrumental role, either by directly interacting with adsorption sites on the crystal surfaces or by affecting the molecular conformation. To study such possibilities, we synthesized the decapeptide and used it as adsorbate (24).

The high solubility of the decapeptide allowed us to obtain its adsorption isotherm onto HA at 37°C and at an ionic strength (160 mM NaCl) similar to that of the enamel fluid (41). The adsorption isotherm was described by a Langmuir model. As shown in Table III, the adsorption affinity, K, and the maximum number of adsorption sites, N, were 6.2 mL/$\mu$mol and 0.53 $\mu$mol/m$^2$, respectively. For comparative purposes, the adsorption parameters of single amino acids (aspartic acid and phosphoserine) and a salivary peptide (statherin) were determined under comparable conditions. A significant finding was that the K value for the decapeptide is very small being only one order of magnitude higher than that of Asp, of the same order of magnitude as that of phosphoserine, and having an affinity much lower than that of statherin. Furthermore, as expected by its low adsorption affinity, the decapeptide showed a weak inhibitory activity on seeded crystal growth.

These findings obtained using the decapeptide gave negative evidence for the association of the inhibition of crystal growth by the 25 kD amelogenin with any specific sequences in the parent molecule. From the current understanding of protein adsorption onto apatitic surfaces (59-61), it is quite possible that the adsorption of amelogenin onto enamel crystals may be an entropically driven process. Changes in the secondary structure upon adsorption, as well as transfer of water from the molecule and the adsorbent surface to the surrounding fluid, may provide the increase in entropy required for the adsorption to take place. If this is the case, the properties conducive to adsorption are related to the whole molecular structure. In this sense, both electrical charges and hydrophilicity of the C-terminal segment may play important roles in the stabilization of the secondary structure of amelogenin and its adsorption onto the crystal surfaces.

Table III.  Adsorption Parameters onto Hydroxyapatite

| Adsorbate | K | N |
|---|---|---|
| | mL/$\mu$mol | $\mu$mol/m$^2$ |
| C-peptide | 6.2 | 0.53 |
| Aspartic acid | 0.22 | 1.0 |
| Phosphoserine | 4.7 | 0.95 |
| Statherin | 21,090 | 0.80 |

## Current Understanding of the Mechanism of Early Enamel Mineralization

Table IV gives a summary of the characterization of various amelogenins found in porcine secretory enamel. The parent 25 kD amelogenin, secreted by the ameloblast, is sparingly soluble in the enamel fluid but, in the solubilized status, can function as a potent inhibitor of crystal growth through selective adsorption onto apatite crystals. The moiety is unstable *in situ*, as indicated by the substantial decrease of its quantity in the outer enamel. The 20 kD amelogenin, derived through cleavage of the C-terminal segment of the 25 kD protein (17), is also sparingly soluble; in contrast to the parent amelogenin, this degraded product shows less adsorption affinity onto the crystals and, consequently, less inhibition of crystal growth. Further enzymatic degradation of the 20 kD amelogenin yields both sparingly soluble (5 kD) and soluble (13 kD) moieties (28,37). These products seem not to participate directly in the regulation of enamel mineralization *in situ*. At present the C-terminal peptides cleaved from the 25 kD amelogenin have not yet been isolated from the secretory enamel.

Previous analysis of the chemical composition of the fluid (41) indicated that the fluid phase surrounding the secretory enamel mineral has a composition different from that found in porcine serum, particularly with respect to the concentrations of calcium, phosphate and carbonate which are major ions in the apatite lattice. These findings support the view that the enamel mineralization in the early developmental stage occurs in a compartmented microenvironment segregated from the circulating blood. In Figure 8 is illustrated schematically our working hypothesis of the regulatory mechanism in early enamel mineralization and the

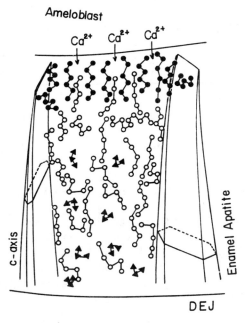

Figure 8.  Schematic illustration of enamel mineralization mechanism during the secretory stage of porcine amelogenesis. The ameloblast secretes the 25 kD amelogenin and supplies lattice ions, particularly regulating the Ca transport through the membrane. The secreted amelogenin interacts with surfaces of ribbon-like enamel crystallites blocking the growth in thickness. With enzymatic cleavages of original molecules, adsorption affinity onto the crystals and the solubility in the fluid of the degraded amelogenins change substantially. See text for details.

Table IV. Properties of Porcine Amelogenins

| Amelogenins | Solubility in enamel fluid | Adsorption onto HA | Inhibition of crystal growth |
|---|---|---|---|
| 25 kD | sparingly soluble | +++ | +++ |
| 23 kD | sparingly soluble | + | N.D. |
| 20 kD | sparingly soluble | + | + |
| 13 kD | soluble | - | - |
| 5 kD | sparingly soluble | +/- | +/- |
| C-peptide | soluble | +/- | +/- |

N.D.: not yet determined.

possible function of amelogenins in this process. The supply of lattice ions, particularly calcium (62), through the ameloblast, i.e., Ca active transport (63,64), increases momentarily the driving force for precipitation, enough to induce the formation of thin ribbon-like crystallites (Figure 1) in the vicinity of the ameloblast. However, the concentration of the potent inhibitors (e.g., the 25 kD amelogenin) in this zone, prevents further nucleation, or crystal multiplication (65), and limits the growth in the thickness of the enamel crystallites. Solid amelogenin (25 kD and 20 kD, sparingly soluble) maintains saturation in the enamel fluid; their concentration in the fluid is sufficient to block the crystal growth sites on the initially formed crystals. Amelogenins in solution are cleaved enzymatically giving rise to soluble fragments (13 kD) with very low adsorption affinity onto apatitic surfaces. This destructive process induces further dissolution of solid amelogenins to keep enamel fluid saturated and, therefore, the growth sites on enamel mineral blocked. When solid amelogenins are exhausted, further enzymatic cleavage brings about desorption of amelogenin from the surface of the mineral, and crystal growth starts (67).

## Acknowledgments

The authors are grateful to Drs. M. Fukae, T. Tanabe, and M. Shimizu for their stimulating discussion and collaboration in conducting the present work. They are also grateful to Dr. Y. Miake for his help in the electron microscopic study. This work was supported by the grants DE07623 and DE03187 from the National Institutes of Dental Research.

## Literature Cited

1. Weiner, S. CRC Crit. Rev. Biochem. 1986, 20, 365.
2. Addadi, L.; Moriadian, J.; Shay, E.; Maroudas, N. G.; Weiner, S. Proc. Natl. Acad. Sci. U.S.A. 1987, 84, 2732.
3. Lussi, A.; Crenshaw, M. A.; Linde, A. Arch Oral Biol. 1988, 33, 685.
4. Linde, A.; Lussi, A.; Crenshaw, M. A. Calcif. Tissue Int. 1989, 44, 286.
5. Eastoe, J. E. Arch Oral Biol. 1963, 8, 633.
6. Termine, J. D.; Belcourt, A. B.; Chritner, P. J.; Conn, K.; Nylen, M. U. J. Biol. Chem. 1980, 255, 9760.
7. Fincham, A. G.; Belcourt, A. B.; Lyaruu, D. M.; Termine, J. D. Calcif. Tissue Int. 1982, 34, 182.
8. Ogata, Y.; Shimokawa, H.; Sasaki, S. Calcif. Tissue Int. 1988, 43, 389.
9. Samuel, N.; Bessem, C.; Bringas, P. Jr.; Slavkin, H. C. J. Craniofacial Genetics Develop. Biol. 1987, 7, 371.
10. Slavkin, H. C.; MacDougall, M.; Zeichner-David, M.; Oliver, P.; Nakamura, M.; Snead, M. L. Am. J. Med. Genetics 1988, 4, 7.
11. Fincham, A. G.; Hu, Y.; Pavlova, Z.; Slavkin, H. C.; Snead, M. L. Calcif. Tissue Int. 1989, 45, 243.
12. Limeback, H.; Sakarya, H.; Chu, W.; Mackinnon, M. J. Bone Mineral Res. 1989, 4, 235.
13. Tanabe, T.;Aoba, T.; Moreno, E. C.; Fukae, M.; Shimizu, M. Calcif. Tissue Int. 1990, 46, 205.
14. Takagi, T.; Suzuki, M.; Baba, T.; Minegishi, K.; Sasaki, S. Biochem. Biophys. Res. Comm. 1984, 121, 592.
15. Snead, M. L.; Lau, E.C.; Zeichner-David, M.; Fincham, A. G.; Woo, S. L. C.; Slavkin, H. C. Biochem. Biophys. Res. Comm. 1985, 129, 812.
16. Shimokawa, H.; Sobel, M. E.; Sasaki, S.; Termine, J. D.; Young, M. F. J. Biol. Chem. 1987, 262, 4042.
17. Yamakoshi, Y.; Tanabe, T.; Fukae, M.; Shimizu, M. Proc. Enamel Symposium V 1989, Abst. #39.
18. Fukae, M.; Shimizu, M. Arch Oral Biol. 1974, 19, 381.
19. Robinson, C.; Briggs, H. D.; Atkinson, P. J.; Weatherell, J. A. J. Dent. Res. 1979, 58(B), 871.
20. Warshawsky, H. In The Chemistry and Biology of Mineralized Tissues; Butler, W. T., Ed.; EBSCO Media Inc.: Birmingham, AL, 1985, p 33.
21. Robinson, C.; Kirkham, J. In The Chemistry and Biology of Mineralized Tissues; Butler, W. T., Ed.; EBSCO Media Inc.: Birmingham, AL, 1985, p 249.
22. Aoba, T.; Fukae, M.; Tanabe, T.; Shimizu, M.; Moreno, E.C. Calcif. Tissue Int. 1987, 41, 281.
23. Aoba, T.; Tanabe, T.; Moreno, E. C. Adv. Dent. Res. 1987, 1, 252.
24. Aoba, T.; Moreno, E. C.; Kresak, M.; Tanabe, T. J. Dent. Res. 1989, 68, 1331.

25. Nylen, M. U.; Eanes, E. D.; Omnell, K. A. J. Cell Biol. 1963, 18, 109.
26. Kerebel, B.; Daculsi, D.; Kerebel, L. M. J. Dent. Res. 1979, 58B, 844.
27. Seyer, J. M.; Glimcher, M. J. Calcif. Tissue Res. 1977, 24, 253.
28. Shimizu, M.; Tanabe, T.; Fukae, M. J. Dent. Res. 1979, 58(B), 782.
29. Sasaki, S.; Shimokawa, H.; Tanaka, K. J. Dent. Res. 1982, 61 (Special Issue), 1479.
30. Strawich, E.; Glimcher, M. J. Biochem. J. 1985, 230, 423.
31. Smith, C. E.; Pompura, J. R.; Borenstein, S.; Fazel, A.; Nanci, A. Anat Rec. 1989, 224, 292.
32. Suga, S. Arch Oral Biol. 1970, 15, 555.
33. Crenshaw, M. A.; Bawden, J. W. In Tooth Enamel IV; Fearnhead, R. W.; Suga, S. Eds.; Elsevier: Amsterdam, 1984, p 109.
34. Overall, C. M.; Limback, H. Biochem. J. 1988, 256, 965.
35. Carter, J.; Smillie, A. C.; Shepherd, M. G. Arch Oral Biol. 1989, 34, 195.
36. Sasaki, S.; Takagi, T. Proc. 67th Int. Assoc. Dent. Res. Mtg. 1989, Abs. #663.
37. Tanabe, T. Tsurumi Univ. Dent. J. 1984, 10, 443.
38. Blumen, G.; Merzel, J. J. Biol. Buccale 1982, 10, 73.
39. Ozawa, H.; Yamada, M.; Uchida, T.; Yamamoto, T.; Takano, Y. In Mechanisms of Tooth Enamel Formation, Suga, S. Ed.; Quintessence: Tokyo, 1983, p 17.
40. Shimizu, M.; Fukae, M. In Mechanisms of Tooth Enamel Formation; Suga, S.; Ed.; Quintessence: Tokyo, 1983; p 125.
41. Aoba, T.; Moreno, E. C. Calcif. Tissue Int. 1987, 41, 86.
42. Aoba, T.; Tanabe, T.; Moreno, E. C. J. Dent. Res. 1987, 66, 1721.
43. Fukae, M.; Tanabe, T.; Ijiri, H.; Shimizu, M. Tsurumi Univ. Dent. J. 1980, 6, 87.
44. Robinson, C.; Kirkham, J.; Briggs, H. D.; Atkinson, P. J. J. Dent. Res. 1982, 61, 1490.
45. Tanabe, T.; Aoba, T.; Moreno, E. C.; Fukae, M. J. Dent. Res. 1988, 67, 536.
46. Young, M. F.; Shimokawa, H.; Sobel, M. E.; Termine, J. D. Adv. Dent. Res. 1987, 1, 289.
47. Yeh, J. H.; Takagi, T.; Sasaki, S. Adv. Dent. Res. 1987, 1, 276.
48. Fincham, A.; Belcourt, A. B.; Termine, J. D.; Butler, W. T.; Cothran, W. C. Biosci. Rep. 1981, 1, 771.
49. Glimcher, M. J. J. Dent. Res. 1979, 58B, 790.
50. Fukae, M.; Tanabe, T. Calcif. Tissue Int. 1987, 40, 294.
51. Bai, P.; Warshawsky, H. Anat. Rec. 1985, 212, 1.
52. Hayashi, Y.; Bianco, P.; Shimokawa, H.; Termine, J. D.; Bonucci, E. Basic Appl. Histochem. 1986, 30, 291.
53. Inage, T.; Toda, Y.; Shimokawa, H.; Teranishi, Y. J. Dent. Res. 1989, 68(Special Issue), 926.

54. Yanagisawa, T.; Sawada, T.; Miake, Y.; Shimokawa, H.; Takuma, S. J. Dent. Res. 1989, 68(Special Issue), 926.
55. Aoba, T.; Moreno, E. C. J. Dent. Res. 1984, 63, 874.
56. Moreno, E. C.; Kresak, M.; Zahradnik, R. T. Caries Res. 1977, 11, 142.
57. Belcourt, A. B. In Tooth Enamel IV, Fearnhead, R. W.; Suga, S. Eds.; Elsevier: Amsterdam, 1984, p 540.
58. Doi, Y.; Eanes, E. D.; Shimokawa, H.; Termine, J. D. J. Dent. Res. 1984, 63, 98.
59. Moreno, E. C.; Kresak, M.; Hay, D. I. J. Biol. Chem. 1982, 257, 2981.
60. Moreno, E. C.; Kresak, M.; Hay, D. I. Calcif. Tissue Int. 1984, 36, 48.
61. Aoba, T.; Moreno, E. C.; Hay, D. I. Calcif. Tissue Int. 1984, 36, 651.
62. Moreno, E. C.; Aoba, T. Adv. Dent. Res. 1987, 1, 245.
63. Bawden, J. W.; Wennberg, A. J. Dent. Res. 1977, 56, 313.
64. Crenshaw, M. A.; Takano, Y. J. Dent. Res. 1982, 61, 1574.
65. Glimcher, M. J. Anat. Rec. 1989, 224, 139.
66. Aoba, T.; Moreno, E. C. In Tooth Enamel V, Fearnhead, R.W. Ed.; Florence: Yokohama, 1989, p 163.

RECEIVED August 27, 1990

Chapter 8

# Purification and Characterization of a Shell Matrix Phosphoprotein from the American Oyster

Kirt W. Rusenko[1], Julie E. Donachy[2], A. P. Wheeler[3]

[1]Department of Dermatology, University of North Carolina, Chapel Hill, NC 27599
[2]Department of Biological Sciences, University of South Alabama, Mobile, AL 36688
[3]Department of Biological Sciences, Clemson University, Clemson, SC 29634–1903

The major protein component contained in the shell matrix of the American Oyster, *Crassostrea virginica*, was purified using size-exclusion and reverse phase HPLC chromatography. This protein was shown to be highly acidic by amino acid analysis which found 90% of the amino acids were comprised of aspartic acid, serine, and glycine in approximately equimolar amounts. The majority of the serine residues in this protein were present as *O*-phosphoserine. The remaining 10% of the amino acid residues were composed of predominantly non-polar and hydrophobic amino acids many of which appear to be located at the carboxy terminal end of the protein, preceded by polyaspartate, as revealed by carboxypeptidase digestion. The results of mild acid hydrolysis indicated the presence of domains of polyaspartic acid and polyphosphoserine. Functionally, this acidic protein was found to effectively inhibit the precipitation of calcium carbonate in vitro when compared to other less acidic protein fractions extracted from the shell, suggesting a role for this protein in the regulation of biomineralization in the oyster shell. The results of these analyses demonstrate remarkable similarities between this molluscan shell matrix protein and the mammalian phosphophoryns found in tooth dentin.

Although from one of the more commonly known bivalve species, the soluble matrix proteins from the shell of the American Oyster, *Crassostrea virginica*,

0097–6156/91/0444–0107$06.00/0
© 1991 American Chemical Society

have not often been characterized. Published analyses of these matrix proteins have generally been limited to amino acid analyses of unpurified or partially purified fractions extracted from demineralized shell (1,2). Further characterization of this matrix protein has been limited to mild acid hydrolysis analyses designed to demonstrate repeating amino acid sequences about the aspartic acid residues (2-5).

Purification of acidic shell matrix proteins from other bivalve species resulted in the observation that size-exclusion and ion exchange chromatography were not reliable methods for the separation of these unusually acidic proteins (6,7). The use of these techniques in other laboratories on other species of molluscs resulted in a confusing array of protein fractions with little demonstrable inter-species similarity other than the acidic nature of the fractionated proteins. From similar studies Weiner (8) concluded that the variety of protein fractions could be described as belonging to two distinct classes of proteins based on their amino acid compositions. Coincident with that report was the discussion by Wheeler and Sikes (9) suggesting that many variations observed in the different protein fractions were the result of different methods of preparation. To further complicate fractionation, these matrix proteins contain a large number of charged post-translational modifications (10-12) which enhance variable interactions with a variety of chromatographic media.

The purpose of this study was to reproducibly fractionate the soluble matrix proteins from the American Oyster in order to characterize the component responsible for the regulation of biomineralization utilizing common protein chemistry techniques. This purification resulted in the isolation of a new class of matrix proteins which, unlike the two classes described by Weiner (8), contains nearly 30% $O$-phosphoserine in addition to 30% aspartic acid. This phosphorylated class of proteins represents approximately 95% of the total soluble matrix proteins by weight. The presence of $O$-phosphoserine has not previously been demonstrated in the matrix proteins of molluscan shell, although the presence of phosphate was first reported in the oyster matrix proteins by Sikes and Wheeler (13). This new class of matrix phosphoproteins was shown to be a more effective inhibitor of calcium carbonate precipitation than less purified fractions of oyster soluble matrix previously separated, indicating that this class of proteins may play a role in the regulation of biomineralization of the oyster shell (9,14,15). That the phosphate appears to enhance the inhibition of calcium carbonate crystallization is consistent with results obtained for the mammalian dentin phosphoproteins (16-18). Therefore, besides demonstrated similarities with invertebrate matrix proteins, the oyster phosphoprotein class also shares some chemical as well as functional similarities with mammalian dentin phosphoproteins (19,20).

## Purification of the Soluble Matrix Proteins

Freshly shucked shells, commercially obtained, were cleaned, powdered, and demineralized in a solution of 10% EDTA, pH 8.0. Following centrifugation to remove the insoluble matrix proteins (IM), the soluble matrix proteins (SM) were dialyzed against distilled water and concentrated to a volume of 30 mls. This concentrated sample of SM was then subjected to size-exclusion chromatography on a column packed with Sephacryl S-300 equilibrated with 50 mM Tris-HCl, pH 7.4. The void volume peak (>600 kDa, SE I) was collected for later analysis and the second, broad peak (<600 to > 15 kDa, SE II) was collected as a bulk fraction, dialyzed against distilled water and lyophilized in order to obtain a dry weight on the material.

Determination of phosphate, carbohydrate, OD260/280 ratios, and amino acid analyses revealed that the SE I and SE II peaks are indistinguishable except by size. Similar analyses also demonstrate little significant differences in both the chemical and amino acid compositions of the light IM proteins (21) and the SE I and SE II proteins, indicating a high degree of polymerization of these components to remarkably high molecular weights. In support of this idea, when the SE I material is subjected to size-exclusion chromatography on Sephacryl S-500, the apparent molecular weights range from 600 kDa to greater than 20 million Daltons. Because of these high apparent molecular weights, only the SE II fraction was subjected to further purification.

For the final stages of purification, the lyophilized SE II proteins were rehydrated to a concentration of 4.0 mg/ml in distilled water and subjected to reverse phase high performance liquid chromatography (RP-HPLC) the results of which are seen in Figure 1. Three peaks were detected and labeled SE II RP-1, SE II RP-2, and SE II RP-3 based on their order of elution. When the SE II RP-2 fraction was subjected again to RP-HPLC under conditions identical to the original separation (see Figure 1), the peak was found to be separable into both the SE II RP-1 and SE II RP-3 fractions. Additional RP-HPLC of the separated SE II RP-1 and SE II RP-3 fractions was unable to resolve any additional components from either peak, indicating that each fraction represents a distinct class of proteins.

Chemical analysis of SE II RP-1 and SE II RP-3 (Table I) demonstrate significant differences between these components in all cases. Carbohydrate determination was performed by the methods of Dubois et al. (22), phosphate by the methods of Marsh (23), as modified by Wheeler and Harrison (24) and protein by the method of Lowry et al. (25) as modified by Miller (26).

The most notable difference between these fractions are the concentration of phosphate and carbohydrate contained in each. Further, most if not all, of the phosphate on SE II RP-1 is apparently O-linked as seen by its susceptibility to the $\beta$-elimination reaction carried out in 0.5 N NaOH. The phosphate found in the SE II RP-3 fraction on the other hand, is not apparently O-linked as determined under the same reaction conditions. Interestingly, the OD260/280 ratio is similar to that reported by Veis et al. (27) for the mammalian dentin phosphoprotein, phosphophoryn.

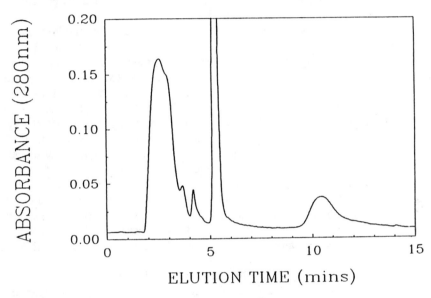

Figure 1. Preparative reverse phase HPLC separation of SE II proteins. One milliliter of SE II proteins (4.0mg/ml) was injected on a Protesil 300 octyl reverse phase column at a flow rate of 1.0 ml/min. The "A" buffer consisted of 0.1% TFA titrated to pH 4.9 with 0.5M ammonium carbonate and the "B" buffer consisted of the same formulation in 70% acetonitrile. The column was developed with a gradient of 15% "B" for 6 minutes, 15 to 100% "B" from 6 to 12 minutes, and 100% "B" from 12 to 15 minutes.

Table I. Chemical Analyses of Oyster SE II RP-1 and SE II RP-3 Fractions

| Chemical Analysis[1] | RP-1[2] | RP-3[2] |
|---|---|---|
| Lyophilized Concentration | 1.0 | 1.0 |
| Lowry Concentration | 0.309 ± 0.1(5) | 0.741 ± 0.2(5) |
| $OD_{235-280}$ [3] | 0.467 ± 0.0(6) | 0.658 ± 0.2(6) |
| $OD_{260/280}$ | 1.28 ± 0.1(6) | 0.998 ± 0.1(6) |
| Carbohydrate[4] | 14.44 ± 0.2(4) | 46.42 ± 10.7(5) |
| Phosphate[4] | 153.1 ± 22.0(6) | 49.05 ± 21.0(6) |
| Phosphate by $OD_{240}$ [4,5] | 141.3 ± 18.1(5) | 15.98 ± 8.3(4) |

[1] All chemical analyses based on 1.0 mg/ml by lyophilized weight.

[2] Mean ± SEM(n).

[3] Spectrophotometric assay for protein concentration of a 1mg/ml by weight sample (28).

[4] μg/mg of lyophilized weight.

[5] Determination of dehydroalanine at $OD_{240}$ (29).

Further differences between the SE II RP-1 and SE II RP-3 fractions are demonstrated by the amino acid composition analysis (Table II) in which aspartic acid, serine, and glycine make up 90% of the residues in the SE II RP-1 fraction and only 57% of the SE II RP-3 fraction. The charged residue to hydrophobic residue (C/HP, asp + glu + lys + arg/pro + ala + val + ile + leu + phe) ratio clearly demonstrates that the SE II RP-3 fraction contains significantly more hydrophobic residues than the SE II RP-1 fraction as expected from the relative elution positions from the reverse phase column. Interestingly, when analyzed by a mathematical method devised to compare the relatedness between proteins based on their amino acid composition (30), the amino acid composition of the SE II RP-3 fraction was found to be nearly identical to the composition of the extracted matrix from the clam, *Mercenaria mercenaria*, as published by Crenshaw (10) and Weiner and Hood (2) indicating that these proteins may be related.

The composition of the SE II RP-3 fraction is also similar to the "high aspartic acid" class of proteins described by Weiner (8) although it is important to emphasize that this fraction is only a minor component of the oyster matrix compared to the SE II RP-1. Indeed, the SE II RP-1 fraction appears to represent a new class of matrix proteins characterized by high concentrations of aspartic acid, serine, and glycine. Because of the relative abundance of the SE II RP-1 material, it was chosen as the subject for the biochemical analyses described in this publication.

Table II. Amino Acid Composition of Oyster SE II RP-1 and SE II RP-3
Fractions

| Amino acid | SE II RP-1[1,2] | SE II RP-3[1,3] |
|---|---|---|
| asp | 31.5 ± 1.5 | 21.6 ± 1.4 |
| thr | 0.5 ± 0.1 | 0.5 ± 0.04 |
| ser | 28.3 ± 1.7 | 13.5 ± 1.3 |
| glu | 4.2 ± 0.3 | 8.5 ± 0.8 |
| pro | 0.9 ± 0.3 | 5.6 ± 0.4 |
| gly | 30.0 ± 1.5 | 22.2 ± 2.6 |
| ala | 0.9 ± 0.2 | 4.9 ± 0.8 |
| cys | n.d.[4] | n.d. |
| val | 0.2 ± 0.06 | 2.3 ± 0.4 |
| met | n.d. | n.d. |
| ile | 0.1 ± 0.03 | 2.3 ± 0.4 |
| leu | 0.1 ± 0.04 | 3.8 ± 0.7 |
| dopa | n.d. | n.d. |
| tyr | 1.7 ± 0.4 | 2.7 ± 0.3 |
| phe | 0.1 ± 0.02 | 1.5 ± 0.3 |
| lys | 0.6 ± 0.1 | 1.8 ± 0.4 |
| arg | 0.3 ± 0.3 | 2.2 ± 0.2 |
| C/HP Ratio | 16.3 | 1.7 |

[1] Residues per 100.
[2] Mean ± SEM, n=5.
[3] Mean ± SEM, n=4.
[4] Not detected.

## Analysis of SE II RP-1 by Mild Acid Hydrolysis

Because of the high concentration of aspartic acid contained in the SE II RP-1 protein, mild acid hydrolysis, which preferentially cleaves aspartic acid from a peptide chain, was chosen as a technique to reveal information about the placement of the aspartic acid residues within the primary structure of the protein. As mentioned previously, this technique has been used to elucidate repeat sequences in both molluscan shell (2-5,31) and mammalian teeth (32-34); however, each experiment made use of relatively harsh conditions, long digestion periods, and collection of data at a single time point (i.e., 0.25M acetic acid, 48 hr). A review of the biochemical literature regarding the mild acid hydrolysis technique reveals that the release of aspartic acid is only preferential, not specific, and that small side chain amino acids such as serine, glycine, and alanine are released nonspecifically at all times during the hydrolysis (35). This release of small side chain amino acids was accentuated after 8 hours of hydrolysis leading investigators to conclude that the technique was not suitable for the selective degradation of whole proteins (36).

With these cautions in mind, our rationale was to utilize the milder conditions (2% formic acid) described by Inglis (37) with data collection at several time points of less than 12 hours. The results of these experiments are shown in Figure 2 where the release of aspartic acid is apparently complete after 3 hours of hydrolysis; however, the release of serine and glycine appears to continue at a linear rate (r=0.97) of 17 pmoles/hr. This linear rate of release for serine and glycine implies that a different mechanism of release is involved than that seen for the release of aspartic acid. This implies that the release of serine and glycine is unrelated to the release of aspartic acid possibly due to a nonspecific reaction rather than release from between two aspartic acid residues. Further, although the amount of total aspartic acid released (42%) agrees well with the total release seen by others in both molluscan (2) and mammalian tooth (34) systems, the release of both serine and glycine in this study is much lower than reported by others (3% of each residue vs. 14%). These data raise the question of whether the release of serine and glycine is the result of release from between two aspartic acid residues as previously suggested (2) or is the result of simple nonspecific hydrolysis of these small side chain residues. Indeed, even if the release of serine and glycine reported in this study is the result of release from between aspartic acid residues, the actual amounts of repeating sequences contained in the primary structure of the protein are much less than previously reported. Regardless of the release of serine and glycine, it is apparent that the large release of aspartic acid is consistent with the presence of domains of polyaspartic acid.

In order to more fully understand the nature of this reaction, synthetic peptides were constructed and subjected to mild acid hydrolysis the results of which are seen in Table III. Although the release of aspartic acid from the (-asp-ser-gly-)$_7$ peptide is the same as that seen in the authentic SE II RP-1 protein, the release of serine and glycine is clearly higher from the peptide. During the course of hydrolysis of this peptide, an unknown peak appeared on the chromatogram which comigrated with a ser-gly dipeptide standard indicating that aspartic acid is released first, as expected, followed by the nonspecific hydrolysis of the ser-gly dipeptide. At no point in the hydrolysis of the SE II RP-1 protein was this ser-gly dipeptide noted indicating that this sequence is not present in the native protein. Interestingly, the hydrolysis of the gly-ser dipeptide results in the release of free glycine and serine in amounts comparable to that seen in the (-asp-ser-gly-)$_7$ peptide. The release of glycine and serine comparable with that seen in the SE II RP-1 protein is seen in the (-gly-ser-)$_{10}$ peptide demonstrating that this polymer apparently undergoes a lower rate of nonspecific hydrolysis and could potentially represent a part of the primary structure of the SE II RP-1 protein.

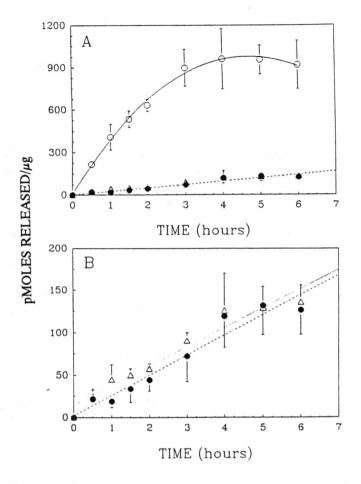

Figure 2. Mild acid hydrolysis of SE II RP-1.; (A) SE II RP-1 (20 μg) was subjected to mild acid hydrolysis in 100 μl of 2% formic acid in vacuo at 108 °C. At the specified time intervals, the samples were removed from heat, dabsylated, and 0.8 μg of protein was injected for amino acid analysis. The initial rate of aspartic acid (open circle) release is 400 pmol/hr and appears to level after three hours of hydrolysis (fit by second order polynomial regression, r=0.998). The rates of glycine (open triangle) and serine (closed circle) release are much lower (17.4 and 18.2 pmol/hr, respectively) and are linear over the course of the reaction. (B) Expanded plot of (A) excluding the aspartic acid data. First order regression lines fit with r=0.97 for both glycine and serine. The molar ratio of glycine:serine released is 1:0.78 ± 0.28.

Table III.  Nonspecific Release of Amino Acids from Peptides of Known Sequence by Mild Acid Hydrolysis After Four Hours

| Peptide | Amino Acids Released[1] |
|---------|------------------------|
| (-Asp-Ser-Gly-)$_7$ | Asp, 42%; Ser, 9%; Gly, 13% |
| (-Gly-Ser-)$_{10}$ | Gly, 2%; Ser, 2% |
| Gly-Ser | Gly, 13%; Ser, 13% |
| (-Ser-Asp-)$_{10}$ | Asp, 27%; Ser, 27% |

[1] Percent of total residues.

Because of the questionable nature of sequence data derived solely from mild acid hydrolysis, the technique was utilized as a method to study portions of the SE II RP-1 protein remaining following hydrolysis. The analysis of the SE II RP-1 hydrolysate by RP-HPLC is demonstrated in Figure 3 where the loss of the prevoid volume native protein during the course of hydrolysis is followed by the appearance of more hydrophobic peaks at approximately 6, 9.5, and 11 minutes. Because hydrophobic amino acids were not released individually in the reaction, these more hydrophobic peaks may represent areas in the SE II RP-1 protein that are enriched in these hydrophobic amino acids.

Analysis of the amino acids remaining in the SE II RP-1 protein after 4 hours of hydrolysis and following dialysis against distilled water to remove free amino acids is seen in Table IV. From these results it is clear that the loss of aspartic acid is concomitant with the increase in the levels of the hydrophobic and basic amino acids indicating that a significant amount of the primary structure of the SE II RP-1 protein remains intact following the removal of aspartic acid by the hydrolysis. Of interest is the observation that the amount of phosphate associated with the protein increases two-fold (15% to 30% w/w) following this treatment implying that the phosphoserine residues are present in sizable peptides consisting of polyphophoserine or polyphosphoserine-glycine copolymers. Amino acid sequence analysis using automated Edman degradation techniques (Applied Biosystems model 477A/120A, normal cycles) of the major reverse phase peak eluted at approximately four minutes after 6 hours of mild acid hydrolysis in fact shows an NH$_2$-terminal sequence consisting of

$$NH_2-(P)Ser-Gly-(P)Ser-Gly-\{(P)Ser\}_7-Ile\ ...-COOH$$

(P)Ser is used to indicate that the residue may be either serine or phosphoserine. Sequence analysis of the pre-void volume native protein after 6 hours of acid hydrolysis showed the presence of multiple NH$_2$-termini indicating a heterogeneous mixture of peptide fragments.

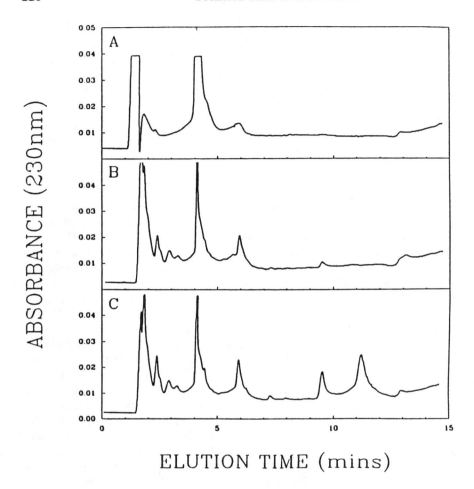

Figure 3. HPLC reverse phase analysis of SE II RP-1 proteins subjected to mild acid hydrolysis for various time periods.; (A) Untreated SE II RP-1 (50 $\mu$g) injected on PLRP-A 300 reverse phase column in 0.47% ammonium formate, pH 6.3, at a flow rate of 1.0 ml/min. Sample was eluted with a linear gradient of 5 to 100% B in 15 min with a "B" buffer of 0.33% ammonium formate in 70% acetonitrile at 50°C. (B) SE II RP-1 subjected to mild acid hydrolysis for three hours. (C) SE II RP-1 subjected to mild acid hydrolysis for 6 hours. The peaks eluting before two minutes and near 4 minutes both are reduced in quantity during the course of the reaction whereas peaks eluting later than 6 minutes begin to appear at 3 hours of hydrolysis and increase by 6 hours of hydrolysis.

Table IV. Amino Acid Composition of SE II RP-1 Fraction Subjected to Mild
Acid Hydrolysis for Four Hours Followed by Dialysis

| Amino acid | Residues per 100 |
|------------|------------------|
| asp  | 15.4 |
| thr  | tr.[1] |
| ser  | 26.3 |
| glu  | 12.2 |
| pro  | nd.[2] |
| gly  | 38.7 |
| ala  | 2.6 |
| val  | 0.8 |
| ile  | 0.5 |
| leu  | 0.9 |
| dopa | nd. |
| tyr  | 1.0 |
| phe  | 0.6 |
| lys  | 1.2 |
| his  | 0.5 |
| C/HP | 4.6 |

[1] Trace amount.

[2] Not detected.

The conclusions derived from these analyses are that few, if any, repeating sequences of the -(-asp-Y-)$_n$- type are present in the primary structure of the SE II RP-1 protein and that domains of polyaspartic acid and generally "hydrophobic" amino acids may be present. $NH_2$-terminal sequence analysis of a reverse phase peak isolated following mild acid hydrolysis demonstrates the presence of domains of polyphosphoserine and/or polyphosphoserineglycine present in the native structure of the SE II RP-1 protein. The persistence of glycine in the dialyzed sample of hydrolyzed protein implies that domains of polyglycine may also be present in the native protein, the presence of which would impart a large degree of rotational freedom to the polypeptide chain. Domains within proteins are known to sometimes be linked by a flexible hinge of glycine residues which is apparently important to the function of the molecule (38). Another distinct possibility is that the majority of the glycine seen in the amino acid compositions is actually the dehydroalanine by-product of the $\beta$-elimination reaction of O-phosphoserine, not authentic glycine. It is known that shell matrix proteins are susceptible to the effects of the $\beta$-elimination reaction *in situ* and apparently exposure to sea water over time results in higher concentrations of glycine in the amino acid compositions in conjunction with decreased concentrations of phosphate (39). This trend is seen in the analyses of fossil shell which are phosphorylated (5) indicating that at least a portion of the glycine detected in the amino acid composition is the result of a β-eliminated phosphoserine. Further analysis of the peptides resulting from mild acid hydrolysis should verify the presence of these domains within the protein.

## Enzymatic Analysis of the SE II RP-1 Protein

When subjected to digestion by chymotrypsin or *Staphylococcus aureus* V8 protease the SE II RP-1 protein appeared to be refractory to cleavage indicating that aromatic amino acids were not present deep within the primary sequence and the aspartic acid residues may exist as polyaspartic acid since the *Staphlococcus aureus* V8 protease is known not to cleave this sequence. The presence of lysine or arginine residues deep within the primary sequence of the SE II RP-1 protein is also doubtful as digestion with trypsin resulted in no detectable change in the apparent molecular weight of the protein as determined by size-exclusion chromatography on Sephadex G-50. On occasion, a small peak was seen in the inclusion volume of the column indicating that some small peptide may have been released as a result of the tryptic digestion.

The results of carboxypeptidase digestion of the SE II RP-1 protein were more definitive as seen in Figures 4 and 5 and in Table V. Digestion by carboxypeptidase B (Figure 4) plainly demonstrated that tyrosine was released at the highest rate and at a concentration nearly two-fold over the other released residues. Glycine is released next, followed by lysine indicating the sequence of the carboxy terminus of SE II RP-1 is $H_2N$-(?)-lys-gly-tyr-tyr-COOH. The placement of lysine and glycine is unambiguous as carboxypeptidase B cleaves glycine inefficiently but favors the cleavage of lysine.

Digestion with carboxypeptidase P (Figure 5) revealed a similar pattern of amino acid release for tyrosine, glycine and lysine in support of the carboxypeptidase B data. Interestingly, after 2 hours of hydrolysis, small (<1 pmole) amounts of alanine, valine, leucine, and isoleucine were released coincident with a rapid release of aspartic acid residues. To determine whether the hydrophobic amino acids are released before or during the release of aspartic acid, carboxypeptidase Y, which does not cleave polyaspartic acid (40), was used next. Table V displays the results of this digestion which apparently had proceeded to completion before the first hour of the reaction. Again, as with carboxypeptidase B, a two-fold excess of tyrosine is seen supporting the sequence of two tyrosines at the carboxy terminus of the SE II RP-1 protein. The appearance of nearly equimolar amounts of hydrophobic amino acids in addition to the released glycine, lysine, and tyrosine suggests that these hydrophobic residues are located in a domain at the carboxy terminus of the SE II RP-1 protein and that this domain is preceded by a more internal sequence of polyaspartic acid. Since only leucine is present in the carboxy-termini of the various carboxypeptidases, the release of amino acids seen is not the result of autocatalysis.

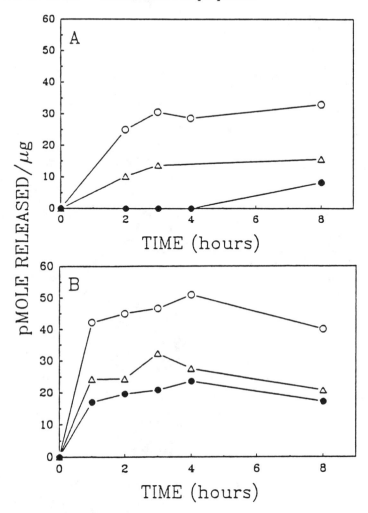

Figure 4. The release of carboxyl-terminal amino acids from SE II RP-1 by Carboxypeptidase B.; (A) SE II RP-1 (2.0 mg/ml, w/w) incubated with 0.45 U/ml carboxypeptidase B-DFP in 0.1 M Tris-HCl, pH 8.0, at 37°C. At the specified time intervals, 7 µl of sample was removed for amino acid analysis (0.55 µg SE II RP-1/injection). Tyrosine (open circle) is released rapidly to a two-fold molar excess over glycine (open triangle). Lysine (closed circle) is not released at all until late in the hydrolysis. (B) Identical conditions as in (A) except 3 U/ml of enzyme was present in the reaction mixture. Tyrosine remains at a two-fold excess over glycine whereas lysine now is approximately 1:1 with glycine.

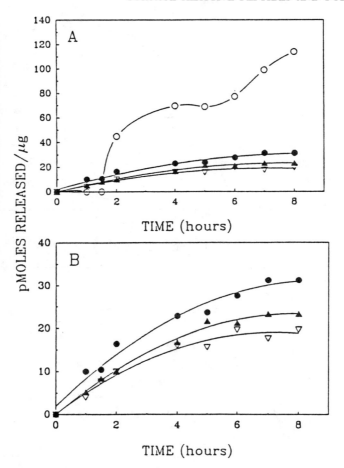

Figure 5. The release of carboxy-terminal amino acids from SE II RP-1 by
    carboxypeptidase P.; (A) SE II RP-1 (2.0 mg/ml, w/w) was incubated
    with 4 U/ml of carboxypeptidase P in 0.05 M sodium acetate buffer,
    pH 3.7, at 37°C. At each time point, 7 μl of this solution was removed
    for dabsylation (0.55 μg protein/injection). The release of aspartic acid
    (open circle) is clearly evident beginning at 2 hours of hydrolysis. Tyrosine
    (closed circle), glycine (closed triangle), and lysine (open triangle) are
    released at similar concentrations however, each of these amino acids
    is released before aspartic acid. (B) Expanded plot of (A) excluding
    the aspartic acid data. Each set of data points are fit with a second
    order regression curve. Up to 4 hours of hydrolysis glycine and lysine
    are released in equivalent concentrations whereas tyrosine is released
    approximately 1.5-fold in excess of glycine or lysine.

Table V. Amino Acid Residues Released From Oyster SE II RP-1 Fraction By Carboxypeptidase Y at 0, 1, and 24 Hours of Incubation

| Amino acid | Incubation Time[1,2] | | |
|---|---|---|---|
| | 0 | 1 | 24 |
| gly | 0.0 | 26.4 | 20.7 |
| ala | 0.0 | 29.9 | 21.8 |
| val | 0.0 | 23.1 | 22.1 |
| ile | 0.0 | 20.4 | 17.2 |
| leu | 0.0 | 29.2 | 25.3 |
| phe | 0.0 | 22.1 | 19.9 |
| lys | 0.0 | 20.2 | 17.1 |
| tyr | 0.0 | 44.9 | 39.8 |

[1] Hours at 37 °C, soluble enzyme.
[2] All amino acids reported as pmoles/$\mu$g.

The results of these enzymatic analyses have shown that the primary structure of the carboxyterminus of the SE II RP-1 protein is $H_2N$-(?)-lys-gly-tyr-tyr-COOH. This structure can be extended with less certainty to $H_2N$-(?)-(-asp-)$_n$-(val, ile, leu, phe, ala)-lys-gly-tyr-tyr-COOH where the hydrophobic amino acids have not been placed in any individual order. This evidence also explains why cleavages were not detected by chymotrypsin and trypsin as the susceptible residues are located close to the carboxy- terminus of the SE II RP-1 protein; hence, only small peptides would be expected to be released which were generally undetectable under our experimental conditions.

## Summary

The results of this study, intended to elucidate features of the primary structure of the SE II RP-1 protein, unexpectedly resulted in a preliminary analysis of the domain structure of this protein. Clearly, a domain of hydrophobic, or perhaps more specifically, nonacidic amino acids is present at the carboxy terminus of the SE II RP-1 protein. Independent studies by Sikes and Wheeler (41) have shown that synthetic polyaspartatepolyalanine peptides are more effective inhibitors of calcium crystallization than peptides consisting of pure polyaspartic acid, -(-X-asp-Y)$_n$-, or -(-asp-Y-)$_n$-. Of interest is the fact that synthetic peptides of the structure, asp15-ala8, were the most effective inhibitors of crystallization which is surprisingly consistent with the results of this study where the "hydrophobic" domain of the SE II RP-1 protein may consist of up to nine amino acid residues preceded by a polyaspartic acid. Preliminary studies by Donachy and Sikes (unpublished results) indicate that the SE II RP-1 protein contains runs of at least asp15 or, with less, certainty, up to asp20. Since the synthetic peptides and SE II RP-1 proteins are comparable in their relative effectiveness of crystal inhibition, the implication is that these acidic and "hydrophobic" domains are a feature that is important in the regulation of mineralization. As synthetic peptides containing repeating sequences such as, -(-X-asp-Y-)$_n$ or -(-asp-Y-)$_n$- are not as effective crystallization inhibitors as the SE II RP-1 protein, it is reasonable to presume that such sequences are not as widespread in the primary structure as previously thought. This is consistent with the results of our mild acid hydrolyzes which show that although the yield of aspartic acid is comparable with that obtained under harsher conditions, the yields of both serine and glycine are much lower than earlier reports.

The presence of O-phosphoserine in the SE II RP-1 protein of the oyster matrix is unique among bivalves at this point although the existence of phosphoprotein in the matrix of one other invertebrate, the sea urchin, has been reported (42). Evidence from the mild acid hydrolyses indicate that the O-phosphoserines are present in domains of polyphosphoserine as indicated by amino acid sequence analysis of the resulting reverse phase HPLC peaks. Further, the phosphate is apparently important in the regulation of biomineralization in the oyster since matrix proteins dephosphorylated with alkaline phosphatase are not as effective inhibitors of crystallization nor do they maintain inhibition as long as phosphorylated matrix proteins (15,21).

Perhaps the most remarkable feature of the SE II RP-1 protein is its general similarity to the mammalian dentin phosphoprotein, phosphophoryn as described by Veis (34). In addition to the presence of phosphoserine, located in domains of polyphosphoserine, both proteins appear to have a "hydrophobic" region at one end of the molecule which contains tyrosine and possibly lysine residues. Furthermore, both proteins appear to contain regions of polyaspartic acid as determined by mild acid hydrolysis (33,34). Interestingly, calcite mineral is an evolutionarily recent crystal structure among bivalves (43) implying that the similarity between these two proteins may be significant phylogenetically. Recent studies have shown that polyclonal antibodies to mammalian phosphoryns are reactive to the sea urchin matrix phosphoprotein, indicating that conserved sequences may be present in the proteins from these different animals (42,44). That the epitopes for this antibody appear to be in the trypsin-sensitive terminal domain of phosphophoryn raises the speculation that this antibody would cross-react with the SE II RP-1 protein also, which would confirm the apparent similarities between these two proteins.

## Literature Cited

1. Simkiss, K. Comp. Biochem. Physiol. 1965, 16, 427-35.
2. Weiner, S.; Hood, L. Science 1975, 190, 987-9.
3. Weiner, S.; Lowenstam, H. A.; Hood, L. Proc. Natl. Acad. Sci. 1976, 73, 2541-5.
4. Weiner, S. In The Chemistry and Biology of Mineralized Connective Tissues; Veis, A., Ed.; Elsevier: North Holland, 1981; p 517- 21.

5. Akiyama, M. In The Mechanisms of Biomineralization in Animals and Plants. Omori, M.; Watabe, N. Eds.; Tokai University Press: Tokyo, 1980; p 257-62.
6. Weiner, S. J. Chromatog. 1982, 245, 148-54.
7. Worms, D.; Weiner, S. J. Exp. Zool. 1986, 237, 11-20.
8. Weiner, S. Amer. Zool. 1984, 24, 945-51.
9. Wheeler, A. P.; Sikes, C. S. Amer. Zool. 1984, 24, 933-44.
10. Crenshaw, M. A. Biomineralization 1972, 6, 6-11.
11. Wheeler, A. P.; George, J. W.; Evans, C. A. Science 1981, 212, 1397-8.
12. Weiner, S. CRC Crit. Rev. Biochem. 1986, 20, 365-408.
13. Sikes, C. S.; Wheeler, A. P. In Biomineralization and Biological Metal Accumulation; Westbroek, P.; de Jong, E. W., Eds.; D. Reidel Publishing Co.: Dordrecht, Holland, 1983.
14. Wheeler, A. P.; Rusenko, K. W.; Swift, D. M.; Sikes, C. S. Mar. Biol. 1988, 98, 71-89.
15. Borbas, J. E. Master of Science Thesis, Clemson University, Clemson, SC, 1988.
16. Termine, J. D.; Conn, K. M. Calcif. Tiss. Res. 1976, 22, 149-57.
17. Termine, J. D.; Eanes, E. D.; Conn, K. M. Calcif. Tissue Int. 1980, 31, 247-51.
18. Nawrot, C. F.; Campbell, D. J.; Schroeder, J. K.; Van Valkenburg, M. Biochem. 1976, 15, 3445-9.
19. Butler, W. T. Meth. Enz. 1987, 145, 255-61.
20. Veis, A. Ciba Found. Symp. 1988, 136, 161-77.
21. Wheeler, A. P.; Rusenko, K. W.; Sikes, C. S. In Chemical Aspects of Regulation of Mineralization; Sikes, C. S.; Wheeler, A. P., Eds.; University of South Alabama Publication Services: Mobile, AL, 1988; p 9- 13.
22. Dubois, M.; Gilles, K. A.; Hamilton, J. K.; Rebers, P. A.; Smith, F. Ann. Chem. 1956, 28, 350-6.
23. Marsh, B. B. Biochem. Biophys. Acta 1959, 32, 351-61.
24. Wheeler, A. P.; Harrison, E. W. Comp. Biochem. Physiol. 1982, 71B, 629-36.
25. Lowry, O. H.; Rosebrough, N. J.; Farr, A. L.; Randall, R. J. J. Biol. Chem. 1951, 193, 265-75.
26. Miller, G. L. Ann. Chem. 1959, 31, 1964.
27. Veis, A.; Spector, A. R.; Zamoscianyk, H. Biochem. Biophys. Acta, 1972, 257, 404-13.
28. Whitaker, J. R.; Granum, P. E. Anal. Biochem. 1980, 109, 156-9.
29. Plantner, J. J.; Carlson, D. M. Meth. Enz. 1972, 28, 46-8.
30. Cornish-Bowden, A. Meth. Enz. 1983, 91, 60-75.
31. Weiner, S. Biochem. 1983, 22, 4139-45.
32. Krippner, R. D.; Nawrot, C. F. J. Dent. Res. 1977, 56, 873.

33. Lechner, J. H.; Veis, A.; Sabsay, B. In The Chemistry and Biology of Mineralized Connective Tissues; Veis, A., Ed.; Elsevier North Holland, Inc.: Amsterdam, 1981; p 395-8.
34. Veis, A.; Sabsay, B.; Wu, C. B. In Surface Reactive Peptides and Polymers: Discovery and Commercialization; Sikes, C. S.; Wheeler, A. P., Eds.; ACS Books: Washington, 1990; p .
35. Schultz, J.; Allison, H.; Grice, M. Biochem. 1962, 1, 694-8.
36. Tsung, C. M.; Fraenkel-Conrat, H. Biochem. 1965, 4, 793-801.
37. Inglis, A. S. Meth. Enz. 1983, 91, 324-32.
38. Hardie, D. G.; Coggins, J. R. In Multidomain Proteins-Structure and Function; Hardie, D. G.; Coggins, J. R., Eds.; Elsevier Science Publishers: Amsterdam, 1986; p 333-44.
39. Rusenko, K. W. Ph.D. Dissertation, Clemson University, Clemson, SC, 1988.
40. Hayashi, R. Meth. Enz. 1977, 47, 84-93.
41. Sikes, C. S.; Wheeler, A. P. In Chemical Aspects of Regulation of Mineralization; Sikes, C. S.; Wheeler, A. P., Eds.; University of South Alabama Publication Services: Mobile, AL, 1988, p 15-20.
42. Veis, D. J.; Albinger, T. M.; Clohisy, J.; Rahima, M.; Sabsay, B.; Veis, A. J. Exp. Zool. 1986, 240, 35-46.
43. Waller, T. R. Philos. Trans R. Soc. London Ser. B 1972, 284, 345-65.
44. Rahima, M.; Veis, A. Calcif. Tissue Int. 1988, 42, 104-12.

RECEIVED August 27, 1990

# Chapter 9

# Invertebrate Calcium Concretions
## Novel Biomineralization Systems

Harold Silverman, Jody M. Myers, and Thomas H. Dietz

Department of Zoology and Physiology, Louisiana State University, Baton Rouge, LA 70803

Molecules influencing natural biomineralization processes have been instructive models for choosing and designing chemical agents for industrial mineralization processes and, perhaps more importantly, in industrial strategies for the inhibition of mineralization. Vertebrate bones, teeth and invertebrate shells have been the most extensively studied. This report documents a naturally occurring and widespread biomineralization phenomenon, the invertebrate calcium concretions. These concretions may lead to the discovery of additional novel peptides, peptide complexes or other polymers that may have important implications to the initiation and inhibition of mineralization processes. We review the invertebrate concretions presently known and the extent of our chemical knowledge of these concretions, with an emphasis on the concretion forming system in bivalves, where a novel phosphate-binding protein-complex has been identified.

Many promising molecules for industrial use, either as inhibitors or initiators of mineralization processes, mimic molecules initially discovered in, and isolated from, biological mineralization systems (e.g.,1,2). Biomineralization systems that have been extensively studied include vertebrate bone and teeth, and invertebrate shells and tests. These systems produce biominerals largely in an extracellular environment. The mineral in all of these cases is associated with a small amount of organic material (as low as about 0.1% by weight) (e.g., 3,4). The organic material is composed of a variety of peptides, glycopeptides, and some non-polar molecules. While limited in terms of quantity, these components are thought to act in concert to regulate the mineralization process *in vivo*. There are specific examples of polyanionic proteins that are thought to be both inhibitory and/ or stimulatory to mineralization (5-10). Some of these molecules can be phosphorylated, a process common to molecular regulation (e.g., 11,12). While there are data with reference to these molecules there is at present no complete understanding how these molecules control mineralization (3,13,14). Many of

0097–6156/91/0444–0125$06.00/0

the molecules have been, at least partially, characterized and several have unique structures that have potential industrial applications. As an example, peptides containing polycarboxylic acids are of interest for their apparent ability to inhibit mineralization (2). More recently, some of the proteolipids have become of interest for their ability to initiate or pattern the initial mineralization process. Indeed, there is evidence that the initial mineralization events in bone and cartilage may be intravesicular being specifically associated with particular membrane components and their enzymatic activity (e.g., 15,16).

## Calcium Concretions

As reviewed by Brown in 1982 (17), calcified concretions, corpuscles, granules, bodies, and a host of other named mineralized products are produced in virtually all of the invertebrate Phyla and in the vertebrates. At the time of Brown's review, perhaps the best studied concretions in terms of chemical composition were those found in cestodes (e.g., 18) and those that were end-products of cellular degradation process or tertiary lysosomes (19 for review). While many of these concretions have a similar morphology and a size ranging from 0.5 to 3 $\mu$m (although large concretions are 100-200 $\mu$m in diameter), the chemical composition of the concretions are different from species to species and even between tissues in the same animal. Most are calcium-based minerals with either carbonate, phosphate, or oxalate serving as the primary anion. They can be rather complex, with other divalents replacing calcium to some extent, and some have more than one predominant anion. The mode of concretion production among tissues is remarkably different; some concretions being produced extracellularly and others being produced intracellularly. All of these differing concretion systems provide largely unstudied models of biomineralization which could yield biological polymers of interest. Perhaps more importantly, the concretion systems allow for the study of intracellular mineralization. This process has not been as extensively studied by comparison with extracellular mineralization, and may provide new clues to the molecular events associated with biomineralization in general. Alternatively, novel mechanistic events, molecules, or both may be present which could at some point be exploited for their chemical properties.

The chemical composition of only a few of the many concretion systems have been studied in detail with reference to their chemical compositions. Our understanding of even the more well-studied concretion systems is largely limited to their inorganic composition. Examples of well-studied systems include the concretions produced in the hepatopancreas of snails, the concretions produced in Malpighian tubules in association with molting in face fly larvae, and concretions found in the gills of freshwater unionid mussels. The tertiary lysosome system found in the digestive and/or excretory organs of some molluscs and crustaceans also has been characterized, but will not be covered here. Taking the three examples cited above, and comparing chemical differences and similarities gives some insight into the diversity of chemical structure and function for these calcium mineralized structures.

## Gastropod Concretions

There are at least two different types of concretions in the gastropods which have two relatively unique and independent functions. The last major review of this system was written by Fornie and Chetail (20). The following is intended as an overview of what is known about the chemical composition of these system and is not meant to be an exhaustive review. One type of concretion is produced in cells of the hepatopancreas of the animal. These concretions are thought to be produced intracellularly in association with Golgi and endoplasmic reticulum (21,22). They are composed largely of calcium pyrophosphate mineral (23,24) and can contain a variety of divalent cations substituted for calcium. In animals from metal-polluted sites the polluting divalent is preferentially accumulated into these intracellularly produced concretions or granules (25). Thus, these concretions are produced by cells with access to the external environment and appear to be used for the binding and eventual elimination of toxic divalents by secretion of the concretion into the digestive tract (22,25). The organic content of these granules is minimal according to C, H, and N analysis, and has not been analyzed in further detail. Carbonate mineral does not appear to be a major component of these concretions (23).

In contrast, there are connective tissue cells (amoebocytes) in gastropods which produce carbonate-based concretions (20,26,27). The divalents present are mainly calcium and magnesium and the concretions do not tend to accumulate significant quantities of other divalent cations. In two species, *Pomacea paludosa* and *Pila virens* the calcium carbonate is largely in the form of amorphous crystal while some vaterite crystals also are observed (27,28). These granules apparently are produced intracellularly by connective tissue cells of gastropods (e.g., 28,29). These granules are used to supply calcium to reproductive organs for passage to the egg capsule (26,27). Many others have suggested that they also are used for damaged shell regeneration (e.g., 30,31). Their function in this role is not as clear and has been disputed by some investigators (32,33). The egg capsule is particularly rich in calcium, and capsule formation appears to be related to a solubilization of granules during egg production and laying. Thus, in the same animal there are at least two widely divergent concretion systems both in terms of biological function and chemical properties. Neither of these concretion systems has had any extensive analyses for organic constituents. According to histochemical data, the calcium storage concretions of *P. paludosa* contain acid mucopolysaccharides, and amino acid analysis of concretion digests suggest a peptide component rich in aspartic and glutamic acid (28).

**Insect Concretions**

Several organs in various insects have the capacity to produce concretions. Again the object here is not to provide an exhaustive review but to give an introduction to the better studied insect concretion systems. The most common location for concretion accumulations are the Malpighian tubules (34,35). These concretions have different functions in different insects depending on the ecological physiology of the organism. The composition of the concretions also appears to differ among insects even though they are associated with the same organ, the Malpighian tubules. In the face fly larvae the granules are formed from calcium extracted from the larval food. Calcium is stored in granules of calcium phosphate with small quantities of magnesium and carbonate (14%) also present (35,36). The mineral phase is amorphous calcium phosphate, perhaps orthophosphate, although the data are not conclusive (37). The organic matrix of these concretions is a minor component (<1% by weight) and is probably composed mainly of, as yet unidentified, peptides since amino acids make up the bulk of the organic material (37). Like the organic matrix of most mineralized structures, glutamic and aspartic acid are the most abundant amino acids. There also is a small amount of carbohydrate (<0.2% by weight) present. Face flies harden their cuticle, in part, using mineralized salts and the disappearance of the Malpighian tubule granules is correlated with an increased mineralization of the cuticle during pupation (35). The amorphous nature of the concretions may allow for their ease of mobilization (3), while the use of phosphate as the primary anion may allow for better physiological control of mobilization than if the anion were carbonate. Insects which do not use mineral to harden their cuticle also contain granules in their Malpighian tubules. Accordingly, the function of these granules will probably turn out to be multi-faceted. For example, such concretions may be used as osmotic regulators. Beetles receiving less than usual water allotments secrete more granules to the lumen of their tubules suggesting an osmoregulatory role (34). Their position near the lumen of the Malpighian tubule suggests these granules also may play a role in binding and elimination of toxic metals, a function similar to that seen in the hepatopancreas of snails. Additionally, Herbst and Bradley (38) have described a "lime gland" consisting of a portion of the Malpighian tubules in the alkali fly *Ephydra hians*, which lives in hypersaline alkaline lakes. In this case, the inorganic component of the granules is almost exclusively calcium carbonate as judged by EDAX and confirmed by analysis of liberated $CO_2$. Studies on the organic composition of these granules are not available.

**Unionid Gill Concretions**

The concretions found in the gills of freshwater unionid mussels have been extensively studied by our laboratory. The concretions found in unionid gills are located extracellularly in connective tissue. The concretions make up 19-53% of the dry weight of the gills depending on the species (39-41). All North American and European unionids examined to date (Table I) have such gill calcium concretions.

Table I. Calcium concretion content in several freshwater unionid mussels (dry concretion as % of dry gill weight).

| Species | Concretion Content mean ± SEM | Reference |
|---------|-------------------------------|-----------|
| *Anodonta anatina* | 53 ± 1 | 40 |
| *Anodonta cygnea* | 32 ± 1 | 40 |
| *Anodonta grandis* | 56 ± 1 | 48 |
| *Carunculina parva texasensis* | 19 ± 1 | 48 |
| *Elliptio crassidens* | 28 ± 2 | 48 |
| *Ligumia subrostrata* | 25 ± 1 | 48 |
| *Unio pictorum* | 20 ± 1 | 40 |

These concretions are composed of an inorganic and an organic matrix. The organic matrix is relatively unique among mineralized organic matrices in that it comprises 20-25% of the concretion dry weight (42). These concretions are composed of an amorphous calcium phosphate with less than 3% carbonate and no measurable oxalate present. The phosphate is present in an amorphous orthophosphate form based on infrared spectrometry and X-ray crystallography data (T. Aoba and H. Silverman, unpublished observations). The divalent cation content of these concretions is highly regulated *in vivo* with calcium accounting for 25% and manganese 8% of concretion dry weight. The high manganese content in the concretions is confirmed for European unionids (40,41). Iron is present at 0.5% and magnesium at 0.1% by dry weight. These percentages mimic neither blood nor pondwater ionic content. These percentages do not change even after manipulation designed to elevate blood levels of divalent cations to more than double their normal value (43). Zinc, cadmium, and manganese have all been tested using multiple experimental designs and methods (see Table II, for examples). Holwerda and his colleagues have also shown that the gill concretions have little affinity for accumulating cadmium in similar but more long term experiments (40). These data taken together indicate that the concretions are not preferential sites of binding for any heavy metals.

If concretions are isolated and *in vitro* experiments performed, concretions are easily loaded with divalent cations further suggesting the tight *in vivo* regulation of divalent cation content. In addition, the concretions are not mobilized even during conditions of hypoxia which causes blood calcium to rise and pH to drop (42,44). Lastly, with regard to concretion regulation, as mentioned above infrared spectroscopy data indicate that the concretion mineral phase is an amorphous calcium orthophosphate. There is no evidence for hydroxyapatite formation in concretions isolated across sucrose gradients. If the concretions are isolated using NaOH to dissolve gill tissue, then evidence of hydroxyapatite appears. The thermodynamically favored tendency of amorphous orthophosphates to undergo hydrolysis to form hydroxyapatite in the presence of water is well known (45-47). The fact that there is no evidence of even a small amount of hydroxyapatite is indicative of the tight regulation of the environment of concretions *in vivo*.

Table II. Loading of heavy metals into the blood and concretions of *Anodonta grandis* by exposure to metals in the pondwater (Ref. 43).

| Metal | mg/l Pondwater concentration | Mean ± SEM | |
|---|---|---|---|
| | | Control | Metal-treated |
| Blood | | mmol/l | |
| Zn | 65 | 0.004 ± 0.000 | 0.013 ± 0.002 |
| Cd | 50 | undetected | 0.012 ± 0.002 |
| Mn | 100 | 0.12 ± 0.01 | 0.28 ± 0.05 |
| Concretion | | % dry weight | |
| Zn | 65 | 0.044 ± 0.030 | 0.085 ± 0.020 |
| Cd | 50 | undetected | 0.050 ± 0.025 |
| Mn | 100 | 8.65 ± 0.25 | 8.98 ± 0.50 |

In comparison to the concretions already described, the physiological role for unionid gill concretions is a relatively long-term calcium store which makes their function most similar to the function of face fly larvae concretions and the calcium carbonate concretions of gastropods. These concretions are produced intracellularly by a connective tissue cell and then are "secreted" to the extracellular space for storage (39). During the reproductive season, the gill of females is altered such that the gill serves as a brooding chamber for embryos. Briefly, the water channels of the gill are converted from water flow chambers into chambers which are isolated from external water flow. These chambers receive eggs that will develop into a fully shelled larval stage (glochidia) before they are released from the gill. In some of the large species, as many as 600,000 embryos can develop in a single female during a breeding season (48). Concomitant with the formation of glochidial shells, calcium is released from the concretions.

Demonstration that the calcium released from the concretions is used to construct glochidial shells comes from a series of radio-label experiments (49). The first experiment involves giving a female mussel a single injection of Ca-45 months before the animal becomes reproductive. Using the proper controls, any label that appears in the glochidia from such animals must have come from a maternal source (Table III). In addition, Ca-45 was given to animals with glochidia already developing, either by injection of label into the maternal blood stream or by placing label in the pondwater.

Table III. Ca-45 measured in various components of *Anodonta grandis* >30 days after being injected with radiolabel (Ref. 49)

| Tissue | As percent of CPM/mg dry weight of medial gill concretions | |
|---|---|---|
| | Non-reproductive | Reproductive |
| Whole gill | 47 ± 6 | 57 ± 14 |
| Gonad | 7 ± 2 | 7 ± 1 |
| Kidney | 11 ± 3 | 11 ± 3 |
| Concretions | | |
| Medial | 100 | 100 |
| Lateral | 115 | none |
| Glochidial shells | none | 19 ± 7 |

The results of this experiment (Table IV) indicate that virtually all of the calcium in the glochidial shells is of maternal origin. These data, together with the disappearance of calcium concretions from the gill during reproduction indicates that the concretions serve as a source of calcium for the brooded embryos. These data also indicate the strict regulation of processes involved in concretion production and, in this case, mobilization. Concretions are not mobilized even under harsh hypoxic conditions, yet they are mobilized on cue during reproduction. They rapidly reappear shortly after reproduction.

Table IV. Ca-45 measured in various components of *Ligumia subrostrata* after 4 h exposure to injected label or by incubating the animal in Ca-45 in the bath (Ref. 49)

| Tissue | | Exposure method | |
|---|---|---|---|
| | | Injection | Bath |
| Blood | (cpm/ml) | 14,100 ± 1,600 | 2,200 ± 400 |
| Soft tissue | (cpm/mg dry) | 697,000 ± 7,500 | 96,100 ± 14,000 |
| Shell | (cpm/mg dry) | 16,200 ± 800 | 80,200 ± 5,000 |
| Gill | (cpm/mg dry) | 402,800 ± 46,600 | 65,300 ± 9,500 |
| Glochidia | (cpm/mg dry) | 42,600 ± 4,100 | 8,900 ± 1,200 |

One report has not been able to replicate our observations with reference to concretion disappearance during reproduction (41). This is somewhat puzzling because even in the absence of maternal ion transfer, the amount of morphological transformation and stretching required by the gill to brood the large number of embryos almost certainly dictates that the massive accumulations of concretions and the rigid calcified chitinous rods of the gill must give way to accommodate the developing glochidial shells, particularly in large *Anodonta* species (49,50).

**Organic Constituents.** The mechanisms by which the concretions in the gill are tightly regulated *in vivo* are unknown. The cell producing the concretions has been identified, and we have some success maintaining the cells in culture. This should allow us to more easily study the organelles involved in the production of the concretions and the process of cellular secretion. However, even understanding these events will not completely explain all of the regulation described above. The concretions of the mussel are unique in that they contain a larger fraction of organic material than other mineralized tissues and even other mineralized concretions. The organic material present has been determined using various procedures and the results obtained in our laboratory roughly match those obtained by T. Aoba (personal communication) on similar material (25% vs 18%, respectively, and substantially different from the <1% for other mineralized structures). These values also are confirmed to some extent by our ability to cut large quantities of this material using a microtome without tearing the sections. This suggests our material is "softer" than the material others are working with (compare micrographs in Silverman 39,42,48,49 with those of Watabe *et al.*, 28). The function of the extensive organic matrix is not known and we are just beginning to analyze its composition. Histochemistry and EDAX analysis suggested that the concretions contain some polysaccharides and a proteinaceous component. Both methods indicated that there was little sulfur present either as sulfur amino acids or as sulfated saccharide moieties (51). Amino acid analysis of an acid hydrolyzate of whole concretions indicated that glutamic and aspartic acid were the predominant amino acids (51).

**Calmodulin.** Further analysis of the organic fraction of the concretions was accomplished using SDS-PAGE gels to analyze the peptide fractions. Initially we found that two peptide fractions could be isolated by EDTA dissolution techniques. Following exposure to EDTA, an EDTA-soluble fraction can be separated from an EDTA insoluble fraction (51). Both fractions were suspended in SDS/mercaptoethanol and separated using SDS-PAGE. In initial experiments we concentrated on low molecular weight peptides 10-97,000 kDa. Within this range there were a relatively small number of peptides (10-15) detected by both Coomasie blue and "Stains-all" suggesting that the concretion preparation was relatively pure. The EDTA-soluble fraction and the insoluble fraction were complimentary and an electrophoretic lane using whole concretion material contained the bands of the two fractions. To determine whether any of the peptides were calcium-binding peptides, the peptides were transferred to nitrocellulose paper and exposed to Ca-45. After washing, autoradiographs were prepared to check for specific Ca-45 binding (52). One band in the EDTA soluble fraction bound the Ca-45. This band had an apparent molecular weight of 17 kDa. Further examination reveals that this band co-migrates with authentic vertebrate calmodulin. Antibodies to authentic vertebrate calmodulin cross-react with the 17 kDa concretion protein. Conversely, a polyclonal antibody prepared against concretion proteins (53) and which reacts only with the 17 kDa concretion protein also reacts with authentic vertebrate calmodulin. These tests demonstrated that calcium binding still occurred after heat treatment and SDS/mercaptoethanol denaturation, suggesting an E-F hand calcium-binding protein. The cross-reactivity of antibodies to authentic vertebrate calmodulin combined with the co-migration suggests that a clam calmodulin is associated with the concretions. Immunocytochemistry indicates that calmodulin is associated with the concretions *in situ* but that more calmodulin is associated with concretion forming cells than with the concretions themselves.

Since calmodulin is a molecule that normally functions as a regulatory molecule inside the cell at calcium concentrations of 0.1-1 $\mu$m (54), and blood calcium concentration in mussels is 4 mM, it is unlikely that this molecule is an important regulatory molecule once concretions are secreted from the cell. The concretion-associated calmodulin is a portion of the EDTA-solubilized fraction and may be a residue left from its role in intracellular concretion formation. This view is supported by the intracellular localization of the calmodulin in intracellular granules of concretion-forming cells (51).

**Non-Polar Components.** While solubilizing the concretions for SDS- PAGE analysis it was determined that some portion of the concretion core never solubilized under mercaptoethanol or SDS treatment. Some of this core was likely non-polar components associated with the organic matrix of the concretions. Treatment of isolated concretions with 3:1 chloroform/methanol for an hour resulted in an extraction of roughly 6% of the concretion's dry weight (55). This remaining unsolubilized material was dissolved in SDS/mercaptoethanol plus EDTA and then re-extracted in 3:1 chloroform/methanol. Following these procedures, the extracts were separated by silica gel thin layer chromatography (TLC). The solvent system contained n-hexane, ethyl ether, acetic acid and water (92:8:1:1). Seven individual components were identified with only four of those identified coming from the initial extraction procedure. These components clearly are important constituents of the organic matrix, although none of these hydrophobic components has been identified to date. The function of these components will have to await further evaluation, but the association of mineral initiation events with hydrophobic materials is documented in bone (16), bacteria (56) and in model *in vitro* mineralization systems (57-60). In addition, Lautie *et al.* (41) suggest that procaryotes in the gills of unionids may have a role in $Mn^{4+}$ and $Fe^{2+}$ generation through coupled oxidation/reduction reactions. While this hypothesis remains equivocal, a hydrophobic environment likely would be an important component in such a system. Perhaps more importantly, the ability of the concretion to maintain an amorphous orthophosphate mineral phase with no evidence of hydroxyapatite formation suggests a shielding of the mineral, once secreted, from connective tissue water in a hydrophobic environment.

**Phosphate-binding Proteins.** As mentioned above, Silverman *et al.* (51) noted an array of 10-15 proteins associated with the concretions with molecular weights between 12-97 kDa. Myers (55) extended this analysis using Laemmli (61) gradient gels (6-18%) and expanded the range from 8-300 kDa thereby accounting for an addition 15-20 proteins.

In an attempt to determine if any of the matrix proteins were phosphorylated, gills strips excised and placed in organ culture were exposed to $^{32}PO_4$. In cultured strips $^{32}PO_4$ is rapidly and specifically incorporated into the calcium concretions. Label is incorporated both into the mineral phase and into the organic fraction of the concretions. Following such procedures about 0.8% of the label associated with the concretions is in the organic fraction. Autoradiograms of SDS-PAGE gels of isolated protein demonstrated a single major protein band showing $^{32}PO_4$ binding. The protein has an apparent molecular weight of 200 kDa (see schematic Figure 1A). Radiogram intensity was proportional to the amount of protein loaded onto the gel. Additional bands with very limited $^{32}PO_4$ binding also were observed. To determine whether or not the 200 kDa band was an aggregate of peptides associated by hydrophobic interactions, 8 M urea was added to the extraction medium and 4 M urea was included in the gel. Autoradiograms of urea-gels containing urea-extracted proteins contain two intense bands at 18 and 21 kDa with no band at 200 kDa (Figure 1B). If urea was not included in the gel matrix when urea-extracted proteins were electrophoresed then autoradiogram bands appear at 18, 21, and 200 kDa (Figure 1C). These data suggest that the 200 kDa band may represent an assemblage of the 21 and 18 kDa peptides held together by strong non- covalent interactions that are largely resistant to treatment with SDS/mercaptoethanol and totally disrupted by SDS/mercaptoethanol/8 M urea. The absence of urea in the gel matrix appears to allow the 21 and 18 kDa peptides to partially reassemble into the 200 kDa complex. Determination of the covalent nature of the phosphate-binding was shown by treatment with alkaline phosphatase (62). Covalent binding was indicated by the marked reduction in autoradiogram band intensity in lanes loaded with enzyme- treated proteins. Immunologic data confirms the relationship of the 21 and 18 kDa peptides to the 200 kDa peptide complex. Polyclonal antibodies made to the 200 kDa protein band cross-react with the 18 and 21 kDa peptides on immunoblots. While the exact function for this phosphoprotein complex is not yet known, there are conservatively at least 10 phosphate-binding sites per 200 kDa complex.

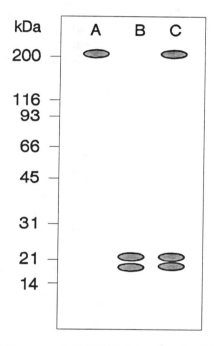

Figure 1. Autoradiogram of SDS-PAGE gels with $^{32}PO_4$ labelled concretion protein (A, no urea; B, urea in extract and gel; C, urea in extract but not gel).

## Conclusion

We have reviewed three invertebrate mineralizing systems that are not as well-known as skeletal systems of invertebrates and vertebrates. The systems all show some capacity for intracellular mineralization. Some may be very instructive with reference to matrix-granule initiation of mineralization in the skeletal systems. The freshwater mussel concretions have a uniquely high quantity of organic matrix which may contain novel regulating molecules. We have already identified an interesting phosphate-binding peptide complex whose function remains to be resolved. We are exploring prospects for resolving questions related to intracellular mineralization in a cell culture system. The fundamental role the concretions play in the biology of the mussel is now apparent. The entirely different and multiple roles played by concretions in the insects is just beginning to be understood. The same organ in different groups of insects appears to make concretions based on different mineral composition and form depending on the functional role of the concretion. The tight regulation of concretions within the life cycle of certain gastropods, freshwater unionids, and some insects suggests that there are neurohormonal pathways to be explored and perhaps novel mechanisms for demineralization. Should this be the case, then obvious pharmacological or industrial uses will derive from the study of such unique systems from the organismal, physiological, biochemical, to the inorganic physical chemistry level. Such systems are experimentally approachable at all these levels and may potentially yield important information on the general mechanisms of biomineralization.

Lastly, the concretion system is but one example of mineralizing or metal-binding protein systems available for study in the invertebrate kingdom. Other examples, include the tertiary lysosome systems of invertebrates, phosphoprotein carriers (63,64), and mineral aggregates of much smaller size (nm range) (65). All of these systems are of interest in the overall understanding of biomineralization.

## Acknowledgments

This work was supported by National Science Foundation grants DCB88-02320 and DCB87-01504.

## Literature Cited

1. Reddy, M.M.; Nancollas, G.H. Desalination 1973, 12, 61-73.
2. Sikes, C.S.; Wheeler, A.P. CHEMTECH 1988, 18, 620-6.
3. Mann, S. Struct. Biol. 1983, 54, 127-74.
4. Weiner, S.; Traub, W; Lowenstam, H. In Biomineralization and Biological Metal Accumulation. Westbroek, P.; DeJong, E.W., Eds.; Reidel: Dordrectht, Holland, 1983, p 205.
5. Wheeler, A.P.; Sikes, C.S. In Chemical Perspectives in Biomineralization: Chemical and Biochemical Perspectives. Mann, S.; Webb, J.; Williams, R.J.P., Eds. VCH Publishers: Weinheim, W. Germany, 1989, p 95.

6. Aoba, T.; Moreno, E.C.; Hay, D.I. Calcif. Tiss. Int. 1984, 36, 651-8.
7. Aoba, T.; Fukae, M.; Tanabe, T.; Shimizu, M.; Moreno, E.C. Calcif. Tiss. Int. 1987, 41, 281-9.
8. Hay, D.I.,; Carlson, E.R.; Schluckebier, S.K.; Moreno, E.C.; Schlesinger, D.H. Calcif. Tiss. Int. 1987, 40, 126-32.
9. Addadi, L.; Moradian, J.; Shay, E. Maroudas, N.G.; Weiner, W. Proc. Natl. Acad. Sci. USA 1987, 84, 2732-6.
10. Addadi, L; Weiner, S. Proc. Natl. Acad. Sci. USA 1985, 82, 4110-4.
11. Butler, W.T. Meth. Enzymol. 1987, 145, 255-61.
12. Wheeler, A.P.; Rusenko, K.W.; George, J.W.; Sikes, C.S. Comp. Biochem. Physiol. 1987, 87B, 953-60.
13. Lowenstam, H.A.; Weiner, S. In Biomineralization and Biological Metal Accumulation. Westbroek, P.; DeJong, E.W., Eds.; Reidel: Dordrecht, Holland, 1983, p 123.
14. Weiner, S. CRC Crit. Rev. Biochem. 1986, 20, 365-408.
15. Anderson, H.C. J. Cell. Biol. 1969, 41, 59-72.
16. Boyan, B.D.; Schwartz, Z.; Swain, L.D. In Surface Reactive Peptides and Polymers: Discovery and Commercialization. Sikes, C. S.; Wheeler, A. P., Eds.; ACS Books: Washington, D.C., 1990.
17. Brown, B.E. Biol. Rev. 1981, 57, 621-67.
18. Von Brand, T.; Scott, D.B.; Nylen, M.U.; Pugh, M.H. Exp. Parasit. 1965, 16, 382-91.
19. George, S.G. In Physiological Mechanisms of Marine Pollutant Toxicity, Vernberg, W.B.; Calabrese, A.; Thurberg, F.P.; Vernberg, F.J., Eds. Academic Press: New York, 1982, p 3.
20. Fornie, J.; Chetail, M. Malacologia 1982, 22, 265-84.
21. Simkiss, K. Symp. Soc. Exp. Biol. 1976, 30, 423-44.
22. Mason, A.Z.; Simkiss, K. Exp. Cell Res. 1982, 139, 383-91.
23. Howard, B.; Mitchell, P.C.E.; Ritchie, A.; Simkiss, K.; Taylor, M. Biochem. J. 1981, 194, 507-11.
24. Taylor, M.; Greaves, G.N.; Simkiss, K. In Biomineralization and Biological Metal Accumulation. Westbroek, P.; DeJong, E.W., Eds.; Reidel: Dordrecht, Holland, 1983, p 373.
25. Simkiss, K.; Taylor, M.; Mason, A.Z. Marine Biol. Lett. 1982, 3, 187-201.
26. Fournie, J.; Chetail, M. Malacologia 1982, 22, 285-91.
27. Meenakshi, V.R.; Blackwelder, P.L.; Watabe, N. Calcif. Tiss. Res. 1974, 16, 283-91.
28. Watabe, N.; Meenakshi, V.R.; Blackwelder, P.L.; Kurtz, E.M.; Dunkelberger, D.G. In The Mechanisms of Mineralization in the Invertebrates and Plants; Watabe, N.; Wilbur, K.M., Eds. University of South Carolina Press: Columbia, 1976, p 283.
29. Sminia, T.; DeWith, N.D.; Bos, J.L.; Van Niewmengen, M.E.; Witter, M.P.; Wondergem, J. Neth. J. Zool. 1977, 27, 195-208.
30. Abolins-Krogis, A. In Studies on Structure, Physiology, and Ecology of Molluscs. Fretter, V., Ed.; Academic Press: New York, 1968, p 75.
31. Abolins-Krogis, A. Cell Tiss Res. 1976, 172, 497-505.

32. Marsh, M.E.; Summerhall, R.D.; Sass, R.L. J. Ultrastruct. Res. 1981, 77, 46-59.
33. Timmermans, L.P.M. 1973. Malacologia 1973, 14, 53-61.
34. Hanrahan, S.A.; Nicolson, S.W. Int. J. Insect Morphol. 1987, 16, 99-119.
35. Grodowitz, M.J.; Broce, A.B. Ann. Ent. Soc. Am. 1983, 76, 418-24.
36. Darlington, M.V.; Meyer, H.J.; Graf, G.; Freeman, T.F. J. Insect. Physiol. 1983, 29, 157-62.
37. Grodowitz, M.J.; Broce, A.B.; Kramer, K.J. Insect Biochem. 1987, 17, 335-45.
38. Herbst, D.B.; Bradley, T.J. J. Exp. Biol. 1989, 145, 63-78.
39. Silverman, H. In Origin, Evolution, and Modern Aspects of Biomineralization in Plants and Animals: Proc. Vth Internatl. Symp. Biomineralization, Crick R., Ed.; Plenum Press: New York, 1989, p 361.
40. Pynnonen, K.; Holwerda, D.A.; Zandee, D.I. Aquatic Toxicol. 1987, 10, 101-14.
41. Lautie, N.; Carru, A.-M.; Truchet, M. Malacologia 1988, 29, 405-17.
42. Silverman, H.; Steffens, W.L.; Dietz, T.H. J. Exp. Zool. 1983, 227, 177-89.
43. Silverman, H.; McNeil, J.W.; Dietz, T.H. Can. J. Zool. 1987, 65, 828-32.
44. Malley, D.F.; Huebner, J.D.; Donkersloot, K. Arch. Environ. Contam. Toxicol. 1988, 17, 479-91.
45. Eanes, E.D.; Posner, A.S. Trans. NY. Acad. Sci. 1965, 28, 233-41.
46. Boskey, A.L.; Posner, A.S. J. Phys. Chem. 1976, 80, 40-5.
47. Tung, M.S.; Brown, W.E. Calcif. Tiss. Int. 1983, 35, 783-90.
48. Silverman, H.; Steffens, W.L.; Dietz, T.H. J. Exp. Zool. 1985, 236, 135-44.
49. Silverman, H.; Kays, W.T.; Dietz, T.H. J. Exp. Zool. 1987, 242, 137-46.
50. Richard, P.E. MS Thesis, Louisiana State University, 1987.
51. Silverman, H.; Sibley, L.D.; Steffens, W.L. J. Exp. Zool. 1988, 247, 224-31.
52. Maruyama, K.; Mikawa, T.; Ebashi, S. J. Biochem. (Tokyo) 1984, 95, 511-19.
53. Steffens, W.L.; Silverman, H.; Dietz, T.H. Can. J. Zool. 1985, 63, 348-54.
54. Carafoli, E. Ann. Rev. Biochem. 1987, 56, 395-433.
55. Myers, J.M. MS Thesis, Louisiana State University, 1989.
56. Boyan, B.D.; Boskey, A.L. Calcif. Tiss. Int. 1984, 36, 214-8.
57. Eanes, E.D.; Costa, J.L. Calcif. Tiss. Int. 1983, 35, 250-57.
58. Eanes, E.D.; Hailer, A.W.; Costa, J.L. Calcif. Tiss. Int. 1984, 36, 421-30.
59. Eanes, E.D.; Hailer, A.W. Calcif. Tiss. Int. 1987, 40, 43-8.
60. Mann, S.; Heywood, B. R.; Rajam, S. In Surface Reactive Peptides and Polymers: Discovery and Commercialization. Sikes, C. S.; Wheeler, A. P., Eds.; ACS Books: Washington, D. C., 1990, p .
61. Laemmli, U.K. Nature 1970, 227, 680-85.
62. Turner, B.M.; Davies, S. FEBS. Lett. 1986, 197, 41-4.
63. Marsh, M.E.; Sass, R.L. Biochemistry 1984, 23, 1448-56.
64. Marsh, M.E.; Sass, R.L. J. Exp. Zool. 1985, 234, 237-42.
65. Sparkes, S.; Greenaway, P. J. Exp. Biol. 1984, 113, 43-54.

RECEIVED August 27, 1990

# Chapter 10

# Mineral Regulating Proteins from Fossil Planktonic Foraminifera

Lisa L. Robbins[1] and Julie E. Donachy[2]

[1]Department of Geology, University of South Florida, Tampa, FL 33620–5200
[2]Department of Biological Sciences, University of South Alabama, Mobile, AL 36688

Although the protozoan planktonic foraminifera are one of the major $CaCO_3$ producers in the world's ocean, little is known about the primary structure of their skeletal organic matrix proteins which may play an important role in biomineralization. $CaCO_3$ nucleation assays of test matrix proteins from two species of planktonic foraminifera, *Orbulina universa* and *Pulleniatina obliquiloculata*, demonstrate the involvement of at least one major protein in the regulation of mineralization by inhibiting calcium carbonate crystal nucleation. Amino acid compositional data indicate that this protein is composed of approximately 31% aspartic acid, 31% serine and 15% glycine. Additionally, approximately 14.5% of this protein consists of hydrophobic amino acid residues. The serine residues in these fossil proteins are nonphosphorylated. Preliminary data from Edman degradation of the protein reveals that the N-terminus is comprised of a sequence of at least 9 aspartic acids in *P. obliquiloculata* and 5 in *O. universa*. These data indicate that the protein is composed of a polyanionic domain at the N-terminus and a hydrophobic domain is postulated to exist based on compositional data.

Planktonic foraminifera constitute a principal source of oceanic $CaCO_3$, producing 47% of the total pelagic sediment of the world's ocean ([1]). These protists can be broadly classified into two taxonomic groups, spinose and non-spinose. Spinose genera are highly porous and thinly calcified as compared to the non-spinose genera. A comparison of proteins between these two groups may shed light onto different mechanisms of biomineralization.

0097–6156/91/0444–0139$06.00/0

Chamber morphogenesis, in planktonic foraminfera, generally involves the secretion of a primary organic membrane (POM) which serves as a template for deposition of the initial layer of $CaCO_3$ (2,3). During chamber addition an alternation of organic laminae (OL) and $CaCO_3$ occurs resulting in the typical foraminiferal bilamellar test (4). Another major organic membrane associated with the test, the Inner Organic Lining (IOL), covers the internal surface of the $CaCO_3$ chamber but is separate from the protoplasm of the protozoan. It is not known whether these organic membranes are similar in composition, although Spero (5) has noted a common origin. The proteins of the organic membranes are believed to regulate the mineralization process, reflecting the intricate differences in test microstructure (6).

The regulation of $CaCO_3$ crystallization by organic matrix proteins from a number of carbonate systems has been studied, including mollusc shell matrix (7-13) and sea urchin spicule matrix (14). These proteins consist of polyanionic and hydrophobic domains which are believed to have specific functions in crystal growth regulation (15). A number of studies have shown that matrix proteins from biominerals are N-terminally blocked to Edman degradation, thereby hindering structural characterization (7,16, Donachy and Sikes, unpublished data). However, proteins from biominerals are attracting interest as nature's approach to regulating formation of crystalline phases (17,18), a process having broad commercial applications. A type of biomineral that is important globally, but has not received much attention is the skeleton, or test, of the foraminifera; unicellular, marine protozoans.

This paper presents a structural and chemical analysis of matrix proteins, including N-terminal sequences, from two species of planktonic foraminifera, *Orbulina universa* and *Pulleniatina obliquiloculata*, and discusses possible functions in $CaCO_3$ crystallization regulation.

## Materials and Methods

**Collection of samples and isolation of proteins.** Box core OCE205-2 obtained at Lat. 26° 09.25′N, Long. 77° 44.40′W and box core ERDC-Bx 92 obtained at Lat. 2° 13′S, Long. 156° 59′E were washed with distilled water through a 250 micron sieve and freeze dried. Approximately 500 mg of tests of each species, *O. universa* and *P. obliquiloculata*, were picked by a vacuum apparatus and cleaned by repeated brief intervals of sonication in distilled water.

Samples were freeze dried and subsequently decalcified in 30 ml of 10% EDTA (pH 8.0). Following centrifugation at 25,000 x g, the supernatant was dialyzed against water (Spectrapor nominal molecular weight cut off 6,000- 8,000 daltons). The dialysate was centrifuged (40,000 x g) and the supernatant freeze dried. This fraction represents the soluble matrix (SM).

**Gel elution.** The SM was applied to a 12% SDS polyacrylamide gel according to the methods of Laemmli (19) and stained with silver stain (BioRad). Two major bands were observed at approximately 65,000 and 54,000 daltons and were designated FP1 and FP2 respectively (20). An earlier study indicated FP1 to be comprised of a number of protein components (21) and therefore this band was cut out of the gel and electroeluted for further purification by reverse phase HPLC. Pieces of gel were macerated and eluted 3 hours in 50mM ammonium bicarbonate, 0.1% SDS buffer using a 3,000 MWCO membrane. The FP1 eluate was then freeze dried.

**Reverse phase HPLC (RPHPLC).** FP1 was reconstituted in 100 $\mu$l of water and 50 $\mu$l injected onto a C-18 Column (Vydac, Separations Group). Buffer B was 0.1% TFA in 100% acetonitrile and A was 0.1% TFA in water. The system was run at 25° C at 1 ml/min using a linear gradient of 0 to 100% B over 60 min. The eluant was monitored at 280 nm. Samples were collected from major protein fractions and analyzed for amino acids.

**Amino acids.** Fractions from reverse phase HPLC were dried in Kimax tubes (Corning) and prepared for amino acid analysis (PICO-TAG, Waters). Samples were treated with 6N HCL (Pierce) *in vacuo* for 18 h at 110°C. The hydrolyzates were dried *in vacuo* and subsequently reconstituted in 20 $\mu$l of methanol-water-triethylamine (TEA) (2:2:1) and redried. The solution was then derivatized in methanol-water-TEA-phenylisothiocyanate (PITC) (7:1:1:1) for twenty minutes and dried *in vacuo*.

The sample was again reconstituted in $Na_2HPO_4$ and acetonitrile (pH 7.4) and injected onto a PICO-TAG column (Waters) equilibrated with buffer (A) comprising of 0.14M sodium acetate (pH 6.4) and 0.5 ml/1 TEA. Buffer B was composed of 60% acetonitrile. The flow rate was 1 ml/min and temperature was 37°C. Amino acids were monitored at 254 nm.

**CaCO3 nucleation assay.** Solutions supersaturated with respect to $CaCO_3$ were prepared by adding separately 0.3 ml of 1.0 M $CaCl_2$ dihydrate and 0.6 ml of 0.4 M $NaHCO_3$ into 29.1 ml of artificial seawater (0.5 M NaCl, 0.011 M KCl) resulting in initial concentrations of 10.0 mM Ca and 8.0 mM dissolved inorganic carbon (DIC). The reaction vessel was a 50 ml, round bottom flask fitted with a pH electrode and closed to the atmosphere. The vessels were partially submersed in a water bath at 20°C. The reaction was monitored continuously by a strip chart recorder. After an induction period characterized by a stable pH, during which crystal nuclei form, there is a drop in pH. Inhibition of nucleation by foraminifer organic matrix was indicated by an increased induction period as compared to induction in the absence of matrix proteins.

**Phosphate analysis.** The extent of phosphorylation of proteins was measured spectrophotometrically upon formation of the phosphomolybdate complex. Soluble reactive phosphate was released from protein samples by $H_2SO_4$/persulfate digestion with autoclaving (22).

**Amino acid sequence determination.** Single peaks from gel eluates, separated by RPHPLC were used for sequence analysis. The $NH_2$-terminal amino acid sequence of foraminiferal matrix protein (FP1) was determined by automated

Edman degradation using the Applied Biosystems model 477A pulsed liquid protein/peptide sequencer with an online Model 120A PTH amino acid analyzer. Lyophilized matrix proteins were dissolved in distilled water and approximately 5 to 6 $\mu$g samples were sequenced.

## Results

The FP1 fraction of the organic matrix from *Orbulina universa* and *Pulleniatina obliquiloculata* demonstrated considerable inhibitory activity in the CaCO$_3$ crystal nucleation assay. The induction time of crystal nucleation increased from control values of 6 minutes to over 100 minutes at a dosage of approximately 0.1 to 0.2 $\mu$g/ml solution. This value is lower than values obtained for a highly phosphorylated matrix protein from the oyster, *Crassostrea virginica*, which increased nucleation time to over 230 minutes at 0.1 $\mu$g/ml solution (Table I). When this oyster matrix protein is dephosphorylated it shows inhibitory activity similar to the foraminifera matrix protein. Phosphate analysis indicated the presence of less than 1% bound phosphate in the matrix proteins from *O. universa*.

Table I. The effects of matrix proteins on CaCO$_3$ nucleation. Values are means ± SD, (N).

| Protein | Dosage (mg/ml) | Induction period (min) |
| --- | --- | --- |
| Control | 0 | 6.17 ± 0.98 (6) |
| Foraminifera FP1 | 0.1 | 117.00 ± 17.00 (2) |
| Oyster SEII[a] | 0.1 | 56.67 ± 11.70 (3) |
| Oyster RP-1[b] | 0.1 | 234.00 ± 26.10 (5) |
| Oyster RP-1 dephosphorylated[c] | 0.1 | 106.00 ± 11.50 (3) |

[a] Soluble matrix fraction retained by Sephacryl S-300 column chromatography (23).
[b] Most hydrophilic reverse phase fraction of SEII (23).
[c] Dephosphorylated using insoluble alkaline phosphatase for 24 h at 37 °C.

The 65,000 dalton band (FP1) observed upon gel electrophoresis consists of a heterogeneous assemblage of proteins as demonstrated by reverse phase HPLC (Figure 1). Amino acid analyses of fraction FP1-3, one of the major fractions, demonstrated high amounts of Asx (aspartic acid and asparagine), serine and glycine (Table II), although hydrophobic amino acids comprised up to 14.5% of the total.

Results from automated Edman degradation of the FP1-3 fraction of matrix material from *P. obliquiloculata* and *O. universa* indicated an NH$_2$-terminal domain of polyaspartic acid in each of these foraminifera. This domain extended at least 9 amino acids into the protein in *P. obliquiloculata* and 5 amino acids

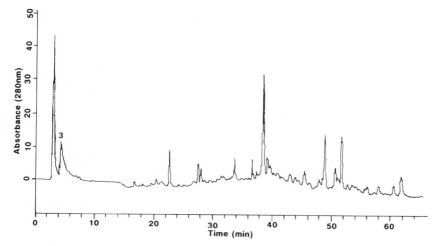

Figure 1. Reverse phase chromatography of FP1 gel extract from <u>Orbulina universa</u>. Fifty micrograms of FP1 were injected onto a C-18 column. Buffer "B" consisted of 0.1% TFA in 100% acetonitrile and buffer "A" consisted of 0.1% TFA in distilled water. The column was developed with a linear gradient of 0 to 100% B over 60 min. Fraction 3 was collected and subjected to further study.

in *O. universa*. After these domains of aspartic acid no other amino acids sequenced in this 65,000 dalton fraction indicating some form of blockage to further Edman degradation.

Table II.  Amino Acid Composition of FP1-3 of *Orbulina universa*.

| Amino Acid | Mol percent |
|---|---|
| Asx | 31.3 |
| Glx | 0.8 |
| Ser | 30.6 |
| Gly | 15.1 |
| Arg | .1 |
| Thr | Tr |
| Ala | 2.5 |
| Pro | N.D. |
| Tyr | 4.7 |
| Val | 4.0 |
| Met | 0.8 |
| Ile | 2.8 |
| Leu | 2.6 |
| Phe | 0.3 |
| Lys | 4.4 |

Tr = trace amounts
N.D. = not determined

## Discussion

A major protein fraction from the tests of *O. universa* and *P. obliquiloculata* demonstrates a similarity to those matrix proteins found in other mineralizing systems. Specifically they are high in aspartic acid, serine and glycine but also contain significant numbers of hydrophobic amino acids. These characteristics are common among matrix proteins isolated from diverse phyla representing both $CaCO_3$ and $CaPO_4$ biominerals as demonstrated in the echinoderms (24), protozoans (25,21), mollusks (7-9,26), and mammals (27,28).

*In vitro* $CaCO_3$ crystallization assays indicate that foraminifera proteins are involved in mineralization regulation as demonstrated by their inhibition of $CaCO_3$ crystal nucleation. The level of inhibition is slightly lower than that found in oysters (8,12,13) and may be a reflection of the level of phosphorylation. Using synthetic matrix analogues (29, Sikes et al., unpublished data) have demonstrated increased inhibitory activity in those peptides containing phosphoserine as compared to those without phosphate. Oyster shell matrix is approximately 15% by weight phosphate which may explain the observed increase in inhibitory activity as compared to the essentially non-phosphorylated foraminifer matrix proteins. It is not known if the low phosphate levels observed for the foraminifera reflect phosphate levels of matrix in living individuals or if this is a function of these proteins being extracted from core top material which may be three to four thousand years old. Loss of O-linked phosphate was observed in shell matrix of oyster dredged from dead shell beds (Wheeler, personal communication) suggesting that diagenetic loss may account for the lack of phosphate observed in matrix from the fossil foraminifera.

Previous attempts to elucidate matrix protein sequences have led to mixed results. From results obtained using mild acid hydrolysis in which the soluble matrix of several molluscs were hydrolyzed in 0.25 M acetic acid at 108 °C for 48 h, Weiner and Hood (30) reported a significant part of the soluble matrix to be comprised of $(Asp-Y)_n$ where Y is predominately serine or glycine, thus accounting for the high levels of Asp, Ser, and Gly common to matrix proteins in general. Using automated Edman degradation Butler et al. (27) showed two closely related $NH_2$-termini from the highly phosphorylated fraction of rat phosphodentin to consist of Asp-Asp-Asp-Asn and Asp-Asp-Pro-Asn. A genetically derived sequence of a prominent spicule matrix protein (SM50) has been reported by Sucov et al. (24). Each of these protein sequences contain negatively charged sites capable of interacting with crystal surfaces which may function in the mineralization process.

Preliminary sequence data indicate that FP1-3 is composed of polyaspartic acid at the amino terminus. In the case of *P. obliquiloculata*, the $NH_2$-terminal domain consists of at least 9 aspartic acids. A sequence of 5 aspartic acids was observed for *O. universa*. The significance of this difference is unclear at this time but may reflect a species difference. These negatively charged domains may be involved in crystal surface binding by these proteins. Sequencing is blocked at this point which may be the result of any one of a number of naturally occurring modifications to amino acids or changes which occur as a result of treatments used in the purification process. Imide structures, which are five membered rings that may occur when Asp is followed by Gly or Asp, have been demonstrated in natural proteins such as collagen (31) and in synthetic matrix analogues and polyaspartic acid (Donachy and Sikes, unpublished data). The high levels of Asp and Gly present in foraminifera matrix proteins suggest the strong possibility of the occurrence of these structures which would block the protein to further Edman degradation.

Other structures which may result in blockage to Edman degradation include pyroglutamic acid which has been identified at the N-terminus of phosphoprotein from human saliva (32) among other proteins. Also, the presence of carbohydrates, a known component of foraminifera organic matrix (20,25), may block Edman degradation (33,34). These possibilities are currently being addressed.

The foraminifera soluble matrix protein consists of a high percentage of hydrophobic amino acids (14.5%). Whether these are present in a discrete domain is as yet undetermined. This may well be the case as has been described in other mineralization regulating proteins (13,35-37). The presence of hydrophobic domains in association with polyanionic regions of statherin, a salivary protein, was necessary for complete inhibition of crystal nucleation (39). Similar results were observed for inhibition of calcium phosphate formation by the amelogenins from teeth (38). Additionally, Sikes and Wheeler (15) demonstrated enhanced inhibitory activity when a hydrophobic domain of polyalanine was added to the polyanionic peptide $Asp_{15}$. It is suggested that the negatively charged part of mineralization regulatory molecules adsorbs onto the crystal surfaces, blocking further growth and the hydrophobic domain creates a region around the crystal nuclei which blocks the diffusion of lattice ions to the crystal surface. It may be that this type of polyanionic-hydrophobic structure is a particularly effective arrangement for regulation of crystallization by water-soluble polymers (40).

At this point, it is not known if the POM, OL, and IOL are each comprised of different proteins with specific mineralization regulatory functions. Alternatively, it may be that these organic matrices are basically composed of a similar suite of proteins which may inhibit, promote, or otherwise modify calcification based on their specific structural configuration.

## Acknowledgments

This research was funded by NSF/EPSCoR and NSF OCE-8817343. We are grateful for fruitful discussions with S. Sikes and A. P. Wheeler. Box cores from Bahama cruise R/V Oceanus (1988) were kindly provided by Dr. W. Curry (NSF).

## Literature Cited

1. Berger, W. H. In Treatise on Chemical Oceanography, Riley, J. P.; Chester, R. Eds.; Academic Press: New York, 1976, Vol. 5, pp 265- 388.
2. Be', A. W. H.; Hemleben, Ch.; Anderson, O. R.; Spindler, M. J. Foram. Res. 1979, 10, 117-28.
3. Hemleben, C.; Anderson, O.R.; Berthold, W.U.; Spindler, M. In Biomineralization in Lower Plants and Animals: The Systematics

Association, Leadbeater, B. S. C.; Ridings, R., Eds.; Oxford; Spec. vol. no. 30, p 237- 49.

4. Towe, K. M. Proc. II. Plankt. Conf. 1971, p 1213-18.
5. Spero, H. J. Mar. Biol. 1988, 99, 9-20.
6. Towe, K. M.; Cifelli, R. J. Paleontol. 1967m 41, 742-62.
7. Weiner, S. Biochem. 1983, 22, 4139-45.
8. Wheeler, A. P.; Sikes, C. S. Amer. Zool. 1984, 24, 933-44.
9. Wheeler, A. P.; Sikes, C. S. In Chemical Perspectives on Biomineralizaiton. Mann, S.; Webb, J.; Williams, R. J. P., Eds.; VCH Publishers: Weinheim, W. Germany, 1989; 95-131.
10. Sikes, C. S.; Wheeler, A. P. In Biomineralization and Biological Metal Accumulation. Westbroek, P.; deJong, E. W. Eds.; Reidel Publishing Co.: Dordrecht, 1983; p 285-289.
11. Sikes, C. S.; Wheeler, A. P. Biol. Bull. 1986, 170, 494-505.
12. Wheeler, A. P.; George, J. W.; Evans, C. A. Science 1981, 212, 1397-8.
13. Wheeler, A. P.; Rusenko, K. W.; Swift, D. M.; Sikes, C. S. Mar. Biol. 1988, 98, 71-80.
14. Benson, S.C.; Benson, N.; Wilt, F. J. Cell. Biol., 1986, 120, 1878-86.
15. Sikes, C. S.; Wheeler, A. P. In Chemical Aspects of Regulation of Mineralization. Sikes, C. S.; Wheeler, A. P. Eds.; University of South Alabama Publication Services: Mobile, AL, 1988; p 15-20.
16. Veis, A.; Sabay, B.; Wu, C. B. In Surface Reactive Peptides and Polymers: Discovery and Commercialization. Sikes, C. S.; Wheeler, A. P. Eds.; ACS Books, Washington, D.C., 1990.
17. Sikes, C. S.; Wheeler, A. P. U.S. Patent 4 534 881, 1985.
18. Mann, S. Nature, 1988, 332, 119-24.
19. Laemmli, U.K. Nature 1970, 227, 680-85.
20. Robbins, L. L. Ph.D. Dissertation, University of Miami, Miami, 1987.
21. Robbins, L.L.; Brew, K. Geochim. Cosmochim. Acta 1990, 54.
22. Eisenreich, S.J.; Bannerman, R. T; Armstrong, D. E. Environ. Lett. 1975, 9, 43-53.
23. Rusenko, K. W.; Donachy, J. E.; Wheeler, A. P. In Surface Reactive Peptides and Polymers: Discovery and Commercialization. Sikes, C. S.; Wheeler, A. P., Eds.; ACS Books, Washington, D.C., 1990.
24. Sucov, H. M.; Benson, S.; Robinson, J. J.; Britten, R. J.; Wilt, F.; Davidson, E. H. Develop. Biol. 1987, 120, 507-19.
25. Weiner, S.; Erez, J. J. Foram. Res. 1984, 14, 206-12.
26. Weiner, S. Calcif. Tissue. Int. 1979, 29, 163-7.
27. Butler, W.T.; Bhown, M.; DiMuzio, M. T.; Cothran; W. C.; Linde, A. Arch. Biochem. Biophys. 1983, 225, 178-86.

28. Wasi, S.; Hofmann, T.; Sodek, J.; Fisher, L.; Tenebaum, H. C.; Termine, J. D. In The Chemistry and Biology of Mineralized Tissues. Butler, W. T., Ed.; Ebsco Media: Birmingham, AL, 1985; p 434.

29. Sikes, C. S.; Yeung, M. L. ; Wheeler, A.P. In Surface Reactive Peptides and Polymers: Discovery to Commercialization. Sikes, C. S.; Wheeler, A. P. Eds; ACS Books, Washington, D.C., 1990.

30. Weiner, S.; Hood, L. Science 1975, 190, 987-9.

31. Bornstein, P. Biochem. 1970, 9, 2408-21.

32. Grey, A.A.; Wong, R.; Bennick, A. J. Biol. Chem. 1979, 254, 4809.

33. Komatsu, S.K.; DeVries, A. L.; Feeney, R. E. J. Biol. Chem. 1970, 245, 2909.

34. Morris, H. R.; Thompson, M. R.; Osuga, D. T.; Ahmed, A. I.; Chan, S. M.; Vandenheede, J. R.; Feeney, R. E. J. Biol. Chem. 1978, 253, 5155-62.

35. Rusenko, K. W. Ph.D. Dissertation, Clemson University, Clemson, 1988.

36. Schlesinger, D. H.; Hay, D. I. Int. J. Peptide Res. 1986, 27, 373-9.

37. Wong, R. S. C.; Hofmann, T.; Bennick, A. J. Biol. Chem. 1979, 254, 4800-8.

38. Hay, D.I.; Moreno, E. C.; Schlesinger, D. H. Inorg. Persp. Biol. Med. 1979, 2, 271-85.

39. Aoba, T.; Moreno, E. C. In Surface Reactive Peptides and Polymers: Discovery and Commercialization. Sikes, C. S.; Wheeler, A. P., Eds.; ACS Books, Washington, D.C., 1990.

40. Sikes, C. S.; Wheeler, A. P. U.S. Patent 4 868 287, 1989.

RECEIVED August 27, 1990

# Chapter 11

# Phosphorylated and Nonphosphorylated Carboxylic Acids

## Influence of Group Substitutions and Comparison of Compounds to Phosphocitrate with Respect to Inhibition of Calcium Salt Crystallization

**J. D. Sallis, M. R. Brown, and N. M. Parker**

**Department of Biochemistry, University of Tasmania, Hobart, Tasmania 7001, Australia**

Multinegatively charged molecules are recognized inhibitors of hydroxyapatite (HA) crystallization and phosphocitrate (PC) with its unique character is no exception. To gain insight into PC's inhibitory properties, a range of phosphorylated and non-phosphorylated carboxylic acids was studied. A useful synthetic strategy for some phosphorylated compounds was to couple 1,2-phenylene phosphochloridate to ester protected carboxylates followed by hydrogenation, base hydrolysis and chromatography. More elaborate strategies were devised for hydrocarbon chain lengthening, the incorporation of a sulfate, amino or carboxyl group to the parent compound. Their inhibitory influence on HA and/or CaOx crystallization was compared to PC. Data indicate that the group arrangement inherent in PC presents as the most ideal for preventing HA formation.

Biological calcification with its several distinct crystalline salt forms can often be initiated from changing environmental circumstances leading to the pathological precipitation of the salt. In recognition of this aspect, considerable research effort has been applied to seek natural and synthetic compounds which might control or even regress unwanted calcification. The most successful compounds reported thus far have associated phosphate and/or carboxylate moieties. Interest, for example, in the role of pyrophosphate (P-O-P) as a urinary inhibitor of stone formation led to extensive investigations of the potential of the bisphosphonate (P-C-P) class of compounds (1). The latter group are among the most powerful of the inhibitors of hydroxyapatite. Their usefulness clinically however poses some problems as secondary undesirable responses have been noted and the compounds are not degraded by enzymes (2). In respect to carboxylic acids alone, they too have elicited attention for their ability to inhibit although it is primarily the

0097–6156/91/0444–0149$06.00/0

polycarboxylic acids which influence. Citric acid is recognized as a natural urinary inhibitor of calcium salt crystallization (3) but selected glycosaminoglycans appear more influential, displaying strong inhibitory action against calcium oxalate crystallization (4-6).

Over the past decade, our studies have centered around the inhibitory potential of phosphocitrate (PC), a compound containing both a phosphate and carboxylate group (7). With a reported natural occurrence in animal mitochondria (8), this molecule also has been verified as a very powerful inhibitor of hydroxyapatite crystallization (9). Its inhibitory action for the most part seems to derive from its multinegative charge/size ratio which together with its stereochemical character allows it to bind to a crystalline lattice preventing growth and aggregation.

Our previous studies (9) have highlighted some of the important chemical features which appear to confer inhibitory properties on a molecule. In the present studies, we have developed synthetic strategies for some additional compounds which have extended the range of compounds now examined, thus enabling consolidation and a reappraisal of the basic requirements. The structure-inhibitory activity relationships then of a range of phosphorylated and non-phosphorylated carboxylic compounds have been compared to the response evoked by PC. Consideration also has been given to the influence of group substitutions within some of the molecules.

## Methods

Inhibitor studies: (a) hydroxyapatite (HA): The inhibitory activity of compounds toward apatite crystallization was determined by comparing the time at which the amorphous calcium phosphate (ACP) phase transformed into the HA crystalline state. The method was as previously described (9) and based on a report by Meyer and Eanes (10) in which a solution of calcium phosphate was induced to spontaneously precipitate at pH 7.4 and temperature 25 °C. Protons ejected by the transformation of ACP to HA were then neutralized and quantitated by base titration. The time (induction time = $I_t$) at which transformation starts was determined (see Fig. 1) so that the time difference between solutions with an inhibitor compound present to that seen in its absence could be expressed as $\Delta I_t$.

(b) Calcium Oxalate: The rate of depletion of aqueous calcium from a metastable solution of a mixture of calcium chloride and sodium oxalate after seeding with mature calcium oxalate crystals was assessed as previously described (11).

Test compounds: Unless otherwise indicated, compounds were obtained from commercial sources. In respect to those compounds for which a synthesis was required, outline strategies are given below. Ultimate purification was generally accomplished by ion exchange chromatography and characterization was pursued by a variety of spectroscopic and chemical techniques.

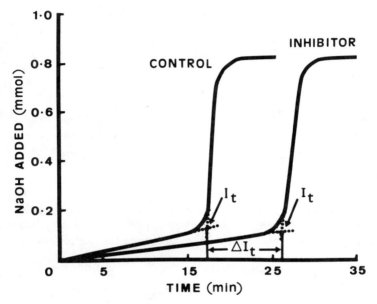

Figure 1. Measurement of induction time.

Phosphocitrate (PC). Phosphorylation of triethylcitrate followed by hydrogenation, base hydrolysis and chromatographic purification was effected as previously described to yield PC (7,12). A similar strategy was used to derive phosphomalate (PM), phosphomethylglutarate (PMG) and phosphoisocitrate (PIC).

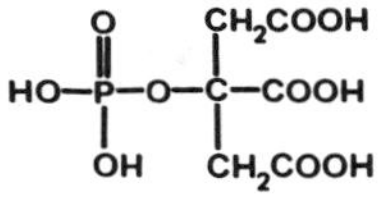

PC

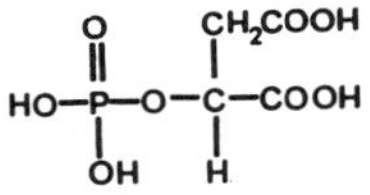

PM

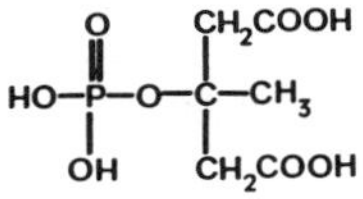

PMG

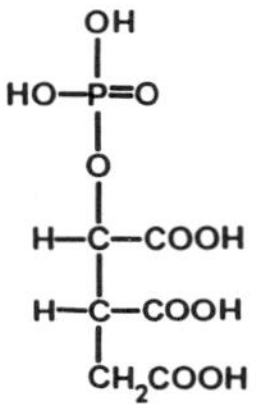

PIC

Tetracarballylate (TETCA: propane 1,2,2'3-tetracarboxylic acid). The intermediate, 2-cyano trimethyltricarballylate was initially prepared by dialkylation of methyl cyanoacetate with methyl bromoacetate in the presence of base (13).

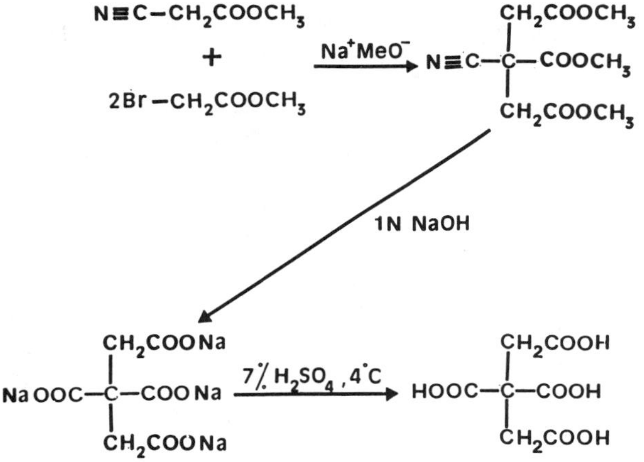

TETCA

Subsequent hydrolysis yielded the desired product, 2,4-Dimethylene tetracarballylate (DMTETCA: pentane 1,3,3',5 tetracarboxylic acid). Reaction of methyl

3-bromopropionate with methyl cyanoacetate in the presence of base yielded the trimethyl ester of pentane 3′ cyano 1,3,5 tricarboxylic acid. Hydrolysis of this product produced pentane 1,3,3′,5 tetracarboxylic acid or as trivialized here, 2,4 dimethylene tetracarballylate.

Phospho-2-amino tricarballylate (PAT). Trimethyl 2-nitro tricarballylate was synthesized as described by Kaji and Zen (14). Catalytic hydrogenation of this product yielded the intermediate, trimethyl 2-amino tricarballylate. Phosphorylation of this product was achieved by 2-cyanoethyl phosphate in the presence of dicyclohexyl carbodiimide (DCCD) and pyridine. The final product was obtained by base hydrolysis in the presence of calcium.

Sulfo-2-amino tricarballylate (SAT): We have previously reported a method for synthesis of this compound (15). Initially, 2-amino tricarballylate was prepared from diethyl 1,3-acetonedicarboxylate (16). Condensation of this reactant with pyridine-sulfur trioxide yielded SAT.

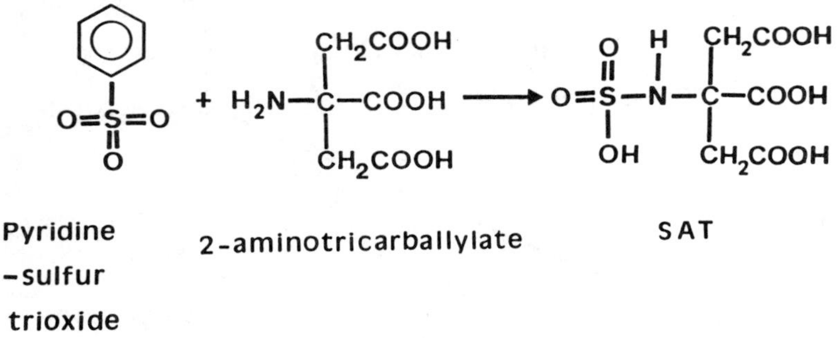

Pyridine            2-aminotricarballylate              SAT
-sulfur

trioxide

2-Methylidine aminophosphonate tricarballylate (MAPT: 2-methylidine aminophosphonate propane 1,2,3-tricarboxylic acid). Using as starting product 2-cyanotrimethyl tricarballylate, a series of reactions involving the isolation of an imidoaldehyde hydrochloride tin (IV) complex was employed in a similar manner to that described by Gancarz and Wieckzorke (17). Subsequent phosphorylation followed by hydrolysis led to the crude compound.

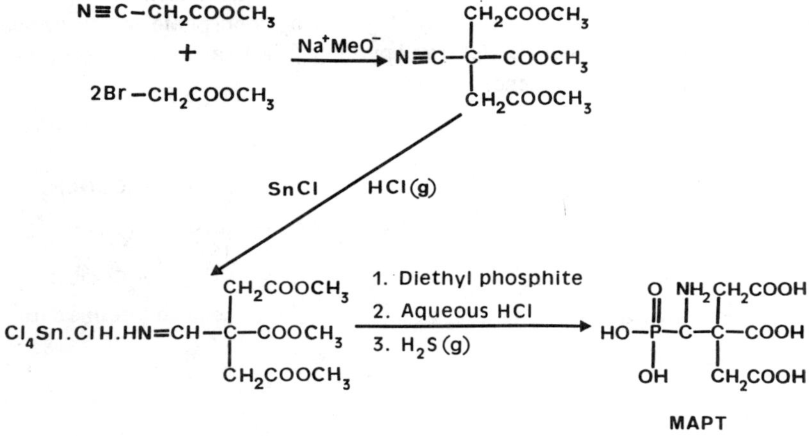

MAPT

## Results and Discussion

Table I describes differences in induction time for series A, carboxylated, non-phosphorylated compounds and series B, phosphorylated, non-carboxylated compounds. In general, the carboxylated compounds are very poor inhibitors, only those compounds with three or more carboxyl groups having any influence at all. Citrate requires at least a concentration in excess of $100\mu M$ to exhibit any inhibitory properties. TETCA on the other hand with four carboxyls is much stronger. It was interesting to note that extension of the side chains of TETCA did reduce the inhibitory power of the parent compound. Delocalization of the net negative charge over the larger extended molecule, thus in effect providing a smaller charge to size ratio could be one explanation.

Table I. Influence of Some Carboxylic Acids and Some Phosphorylated, Non-carboxylated Compounds on the Transformation of Amorphous Calcium Phosphate to Hydroxyapatite.

| Series A | Inhibitor Conc. $(\mu M)$ | $\Delta I_t$ (min) |
|---|---|---|
| Formate | 100 | zero |
| Oxalate, Succinate, Malate | " | zero |
| Tricarballylate | " | zero |
| Citrate | " | 2 |
| Tetracarballyate | " | 34 |
| 2,4 Dimethylene Tetracarballylate | " | 7 |
| Series B | | |
| Pyrophosphate | 50 | 15 |
| Adenosine 5'-diphosphate | " | 12 |
| Adenosine 5'-triphosphate | " | 18 |
| Imidobisphosphonate | " | 17 |
| Hydroxy ethane bisphosphonate | " | 26 |
| Tripolyphosphate | " | 19 |

By comparison, compounds of series B containing a phosphate moiety were significantly more inhibitory and at a lower concentration. While an increase in the number of phosphate groups in general seems beneficial (e.g. ATP>ADP), the configuration of groups can overide such an advantage. HEBP, the most active compound in the series probably gained its increase in potency over pyrophosphate through the attachment of the OH group at C1.

Figure 2 reveals however that much more inhibitory power is displayed by compounds possessing both phosphate and carboxylate moieties in comparison

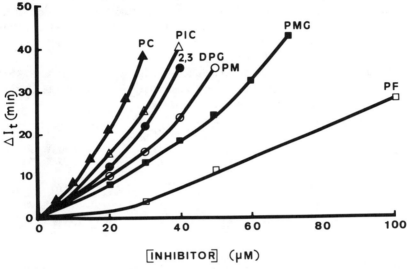

Figure 2. A comparison of the inhibitory influence of some phosphorylated, carboxylic acids on the induction time required for the transformation of amorphous calcium phosphate to hydroxyapatite.

to compounds with only one of the moieties. The number and positioning of these negatively charged groups then assumes major importance. Phosphono-formate (PF) and phosphomalate for example, are stronger inhibitors than their non-phosphorylated counterparts. Phosphomalate and phosphomethyl glutarate both display moderate inhibitory activity, suggesting that the inclusion of two carboxylates instead of one is more beneficial. Clearly however, phosphocitrate with three carboxylates is much more impressive. Of interest in this series, phosphoisocitrate is not quite as powerful as phosphocitrate suggesting that a C3 position of the $PO_4$ is important.

Modification of the basic phosphocitrate molecule by the inclusion of other groups does significantly alter inhibitory potential as shown in Fig. 3. An amino grouping appeared to lessen activity, probably as a result of the positive charge of this moiety. Although PAT revealed itself to be a very powerful inhibitor, other characteristics inherent in the molecule do not make it suitable as a potential drug. Under acid conditions it is very labile and additionally, the compound is also subject to enzyme deactivation. Replacement of the phosphate moiety in PAT with a sulfo-group (SAT) renders the compound less inhibitory. This effect could be explained by the fact that when $SO_4$ is incorporated into the molecule, there is a loss of a chelating group compared to phosphate. In addition, the moiety change also confers other interesting properties on the molecule. We have previously reported, that SAT is enzyme resistant and possesses good absorptive properties (18). MAPT, the other N-compound tested, displayed only relatively weak activity.

Whilst most of the research has focused on controlling hydroxyapatite formation, some limited observations have also been made regarding the influence of the more powerful compounds on calcium oxalate crystallization. As can be seen in Fig. 4 phosphocitrate is not as effective an inhibitor as HEBP suggesting that underlying differences in configuration are important for binding to the particular type of crystal lattice formed. Of interest, TETCA at $100\mu M$ proved almost as potent as phosphocitrate in this system.

Overall, the data emerging from these studies allow for some generalizations and conclusions to be drawn concerning the nature of groups required for inhibition and the importance of their relationship to one another. Inhibitors of calcium salt crystallizations are generally thought to act primarily by occupying defect sites on various faces of the crystals (19). Through negative charge for example, compounds bind tightly to hydroxyapatite. Although charge is important, size also influences as it is clear that commonly a larger molecule can exert a weaker association because of its respective spatial group orientation. TETCA for example with four negative groups tightly arranged around the central C possesses stronger inhibitory powers than the extension product, DMTETCA. Comparatively however, in respect to hydroxyapatite inhibition, TETCA is not as strong as phosphocitrate which probably reflects the fact that with phosphate present, five chelation sites are available as compared to four with TETCA. Also of course, the properties of hydroxyl (e.g. P-OH, C-OH, S-OH) or carboxyl groups would

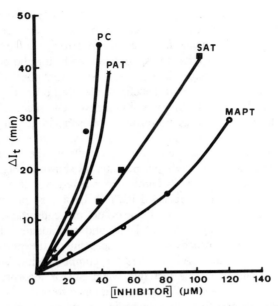

Figure 3. Comparative influence of group substituents on the time required to transform amorphous calcium phosphate into hydroxyapatite.

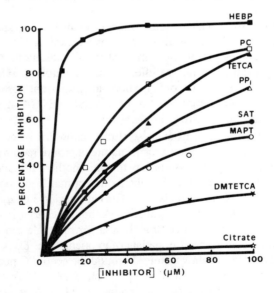

Figure 4. Dose-response relationship of some selected hydroxyapatite inhibitors to restrict calcium oxalate crystallization.

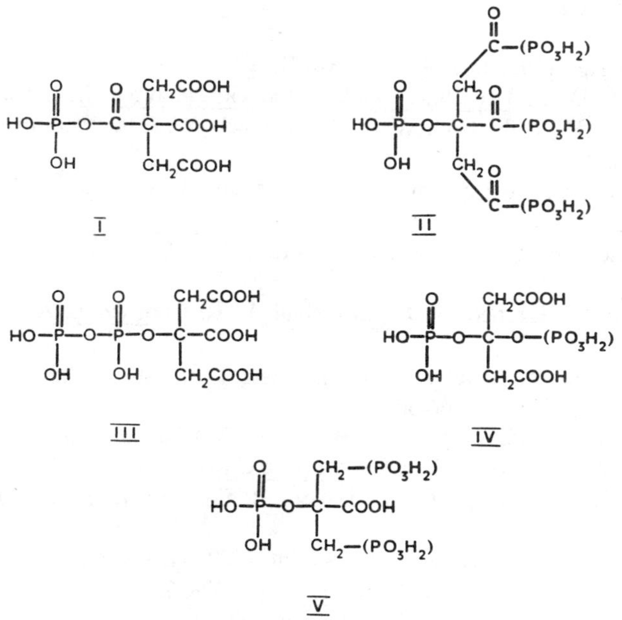

Figure 5. Structures of some compounds predicted to exert a strong inhibitory influence on calcium salt crystallization.

be subject to change under the influence of other groups. Substitution of groups at C3 appears to play a key role in the molecule's inhibitory properties. PC, TETCA, MAPT and 2,4 DMTETCA all have a central C atom from which chelating groups radiate tetrahedrally. The removal of one chelating arm (e.g. PM vs PC) or the substitution of one chelating functionality with that of a lower chelating ability or charge, leads to a significant depletion of inhibitory activity. In terms of contribution to inhibitory potential, the following order appears to emerge: $O\text{-}PO_3H_2 > \text{-}CH_2\text{-}PO_3H_2 > \text{-}NH\text{-}PO_3H_2 > \text{-}O\text{-}SO_3H$ (predicted) $> \text{-}NH\text{-}SO_3H > \text{-}COOH > \text{-}OH$.

On the basis of the information available then, it might be predicted that the following compounds shown in Fig. 5 would also show similar or even more powerful inhibitory action.

To date, these compounds have not been synthesized and of course if they were to become useful compounds with a clinical application they would need to possess the characteristics of being non-toxic and of being eventually cleared from systems. Alternatively, a search for natural compounds with a similar arrangement of chemical groups as suggested might provide a greater understanding on how biological mineralization can be controllable.

**Literature Cited**

1. Fleisch, H. Kidney Int. 1978, 13, 361-71.
2. Sallis, J. D. In Urolithiasis and related clinical research; Schwille, P. O.; Smith, L. H.; Robertson, W. G.; Vahlensieck, W. Eds.; Plenum: New York, 1985; pp 803-9.
3. Bisaz, S.; Felix, R.; Neuman, W. F.; Fleisch, H. Min. Electr. Metab. 1978, 1, 74-83.
4. Robertson, W. G.; Peacock, M.; Nordin, B. E. C. Clin. Chim. Acta. 1973, 43, 31-7.
5. Ryall, R. L.; Harnett, R. M.; Marshall, V. R. Clin. Chim. Acta. 1981, 112, 349-56.
6. Kok, D. J.; Papapoulos, S. E.; Blomen, L. J. M. J.; Bijvoet, O. L. M. Kidney Int. 1988, 34, 346-50.
7. Williams, G.; Sallis, J. D. Anal. Biochem. 1980, 102, 365-73.
8. Williams, G.; Sallis, J. D. In Urolithiasis, Clinical and Basic Research. Smith, L. H.; Robertson, W. G.; Finlayson, B. Eds.; Plenum: New York, 1981; pp 569-77.
9. Williams, G.; Sallis, J. D. Calc. Tiss. Int. 1982, 34, 169-77.
10. Meyer, J. L.; Eanes, E. D. Calc. Tiss. Res. 1978, 25, 59-68.
11. Meyer, J. L.; Smith, L. H. Invest. Urol. 1975, 13, 31-5.
12. Tew, W. P.; Mahle, C.; Benavides, J.; Howard, J. E.; Lehninger, A. L. Biochemistry 1980, 19, 1983-8.
13. Cope, A. C.; Holmes, H. L.; House, H. O. Organic Reactions 1957, 9, 107-331.
14. Kaji, E.; Zen, S. Bull. Chem. Soc. Jap. 1973, 46, 337-8.
15. Brown, M. R.; Sallis, J. D. Anal. Biochem. 1983, 132, 115-23.
16. Dornow, A.; Rombusch, K. Chem. Ber. 1955, 88, 1334-41.
17. Gancarz, M.; Wieckzorke, C. Synthesis 1977, 9, 625.
18. Brown, M. R.; Sallis, J. D. In Urolithiasis and Related Clinical Research; Schwille, P.O.; Smith, L.H.; Robertson, W.G.; Vahlensieck, W., Eds.; Plenum: New York, 1985; pp 891-4.

RECEIVED August 27, 1990

# Chapter 12

# Synthesis of *O*-Phosphoserine-Containing Peptides by Solution- and Solid-Phase Methods

**J. W. Perich[1]**

**Department of Organic Chemistry, University of Melbourne, Parkville, Victoria 3052, Australia**

The preparation of Ser(P̲)-containing peptides is described by the use of either (a) a 'global phosphite-triester phosphorylation' approach which uses dialkyl N̲,N̲- dialkylphosphoramidites or by (b) a more general approach which utilizes protected Boc-Ser(PO₃R₂)-OH derivatives in the Boc mode of solution or solid phase peptide synthesis. In the first approach, a protected Ser-containing peptide is synthesized and phosphorylated using dibenzyl or di-t̲-butyl N̲,N̲-diethylphosphoramidite/1H̲-tetrazole followed by oxidation of the resultant 'phosphite-triester' intermediate. The cleavage of benzyl phosphate groups is effected by hydrogenolysis while t̲-butyl phosphate groups are cleaved by mild acidolytic treatment. In the second and more efficient approach, various Boc-Ser(PO₃R₂)-OH derivatives are prepared from Boc-Ser-OH by a high yielding three step route using either the phosphorotriester or 'phosphite-triester' phosphorylation procedures. The benzyl derivative, Boc-Ser(PO₃Bzl₂)-OH, has been used for the synthesis of Ser(PO₃Bzl₂)-containing peptides followed by a final palladium-catalyzed deprotection step. However, this derivative has restricted use in peptide synthesis due to the susceptibility of benzyl phosphate groups to acidolytic treatments used for Boc-cleavage. The phenyl derivative, Boc-Ser(PO₃Ph₂)-OH, has been used for the synthesis of several complex Ser(PO₃Ph₂)-containing peptides with complete cleavage of the phenyl groups effected by platinum oxide-mediated hydrogenation. This derivative has also been used in solid phase methodology for the synthesis of various simple Ser(P̲)-containing peptides.

[1]Current address: Biochemistry & Molecular Biology Unit, School of Dental Science, Faculty of Medicine & Dentistry, The University of Melbourne, 711 Elizabeth Street, Parkville, Victoria 3052, Australia

0097–6156/91/0444–0161$06.00/0

From the early studies of bovine milk caseins, it was found that the phosphorus content of the milk caseins was in the form of phosphorylated serine and that these O-phosphoseryl [Ser(P)] residues readily bound calcium and were responsible for maintaining the structural integrity of the casein micelle by calcium-phosphate bridging (1). A striking feature of the milk caseins was that the majority of the Ser(P) residues occur in clusters and the tetra- phosphorylated sequence -Glu-Ser(P)-Ile(Leu)-Ser(P)-Ser(P)-Ser(P)-Glu-Glu- was a common sequence in both bovine $\alpha_{s1}$- and A-casein, and human $\beta$-casein. Due to the important structural roles played by Ser(P) residues in milk phosphoproteins and the difficulty in obtaining peptide fractions from these natural sources, this led early workers to turn to chemical methods for the synthesis of model Ser(P)-containing peptides for use in biochemical and structure-function studies.

The chemical synthesis of Ser(P)-peptides was first investigated by Folsch (2) between 1957 and 1965 and these studies led to the preparation of twenty-one single Ser(P)-containing di- and tri-peptides, and the dipeptide Ser(P)-Ser(P), by the (a) phosphorylation of the protected seryl peptide with either diphenyl or dibenzyl phosphorochloridate followed by (b) removal of the phosphate protecting groups by catalytic hydrogenation. In a modification of this approach, Theodoropoulos and coworkers (3) prepared several simple Ser(P)-containing peptides by the phosphorylation of the protected Ser-peptide with di-4- nitrobenzyl phosphorochloridate/imidazole followed by palladium-catalysed hydrogenolytic deprotection. The synthesis of Ser(P)-peptides by the 'global' phosphorylation approach has been reviewed by Folsch (2) and synthetic Ser(P)-peptides prepared up to 1984 have been documented by Frank (4).

However, despite these early studies, there was no further development in the chemical methods for the synthesis of Ser(P)-peptides between 1965 and 1985. In 1987 though, three groups used the diphenyl phosphorochloridate 'global phosphorylation' approach for the repeat synthesis of Ser(P)-Ser(P) (5), and the synthesis of the cAMP-dependent protein kinase peptides, Arg-Arg-Ala-Ser(P)-Val-Ala (6) and Ac-Arg(Leu)-Ala-Ser(P)-OMe (7).

In 1981, Perich and coworkers (8) reported this global phosphorylation approach was unsuitable for the phosphorylation of some Ser-peptides due to rearrangement of the seryl residue on phosphorylation, this problem being particularly prominent for multiple Ser-containing peptides (9,10). From $^{31}$P NMR spectroscopy studies using Ac-Ser-NHMe and Ac-Gly-Ser-Gly-NHMe as model substrates, it was observed that rapid phosphorylation of the seryl hydroxyl group was followed by cleavage of the diphenyl phosphate moiety from the Ser(PO$_3$P$_2$)-residue (Figure 1) (Perich and Johns, Aust. J. Chem., Part IV, in press). In the case of the Ac-Ser-NHMe phosphorylation, the subsequent isolation of serine N-methylamide diphenyl phosphate indicates that the amino group of the seryl residue plays a role in the dephosphorylation process. These findings indicate that the successful outcome of a phosphorylation is dependent on (a) the peptide sequence about the seryl residue and (b) the nature of the phosphate substituent, and may provide a possible explanation for the unsuccessful phosphorylation of several Ser-peptides reported in the literature (3,7).

In consideration of the above synthetic difficulties, it was recognized that the efficient synthesis of complex Ser(P)-peptides and Ser(P)-cluster peptides would require the development of improved chemical methods. Since 1982, two synthetic approaches were developed which involved (a) the global 'phosphite-triester' phosphorylation of seryl-containing peptides using the highly reactive dibenzyl or di-t-butyl N,N-diethylphosphoramidite (11, also Perich and Johns, Aust. J. Chem., Part V, in press) or (b) the use of protected N-(t-butoxycarbonyl)- O-(diaralkylphosphoro)serine [Boc-Ser(PO₃R₂)-OH] derivatives (12-16) in the Boc mode of peptide synthesis (Perich, Alewood and Johns, Aust. J. Chem., Part VII, in press).

## Synthesis of Ser(P)-Peptides by 'Global' Phosphite-Triester Phosphorylation

In the first approach, the phosphorylation procedure involves the initial phosphitylation of the hydroxy compound with (i) dibenzyl N,N-diethylphosphoramidite (17)/1H-tetrazole and subsequent *in situ* oxidation of the resultant dibenzyl phosphite-triester with either MCPBA or t-butylhydroperoxide (18). The high reactivity and efficiency of this 'phosphite-triester' phosphorylation approach has led to its use in the phosphorylation of alcohols (19), protected serine derivatives (11,19) protected tyrosine derivatives (11,19), the protected seryl-peptides Boc-Glu(OBu⁺)-Ser-Leu-OBu⁺ (11) and Boc-Asp(OBzl)-Ala-Ser-Gly-Glu(OBzl)-OBzl (20), and the protected threonyl-peptide Boc-Thr-Leu-Arg(Mbs)-OMe (21).

By the use of this approach, the phosphite-triester phosphorylation of Boc-Glu(OBu⁺)-Ser-Leu-OBu⁺ followed by palladium-catalysed hydrogenolytic cleavage of the benzyl phosphate groups from the Ser(PO₃Bzl₂)-tripeptide gave the Ser(P)-peptide Glu-Ser(P)-Leu in 96% yield (11). In a later study, this approach was used for the preparation of the three peptides, Ac-Ser(P)-NHMe, Ac-Ser(P)-Ser(P)-NHMe and Ac-Ser(P)-Ser(P)-Ser-(P)-NHMe (Figure 2) in 90, 60 and 15% yields respectively (Perich and Johns, Aust. J. Chem., Part VI, in press). However, the decrease in yield with increasing number of seryl residues indicated that the efficiency of this procedure decreased with increasing complexity of the Ser-peptide and suggests that this approach may be limited to the preparation of simple serine-containing peptides.

Apart from dibenzyl N,N-diethylphosphoramidite, the 4- halobenzyl derivatives, di-4-fluorobenzyl N,N-diisopropylphosphoramidite, di-4-chlorobenzyl N,N-diisopropylphosphoramidite, and di-4-bromobenzyl N,N-diethylphosphoramidite have also been prepared and used for the phosphorylation of alcohols and protected serine derivatives (Perich and Johns, Aust. J. Chem., Part X, in press). Later, Bannwarth and Kung (22) extended this system to the use of diallyl N,N-diisopropylphosphoramidite for the phosphorylation of a protected serine,

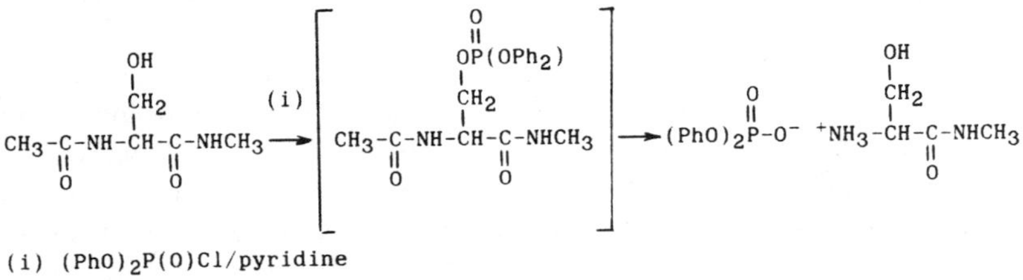

Figure 1. Product obtained from the phosphorylation of Ac-Ser-NHMe with diphenyl phosphorochloridate in pyridine.

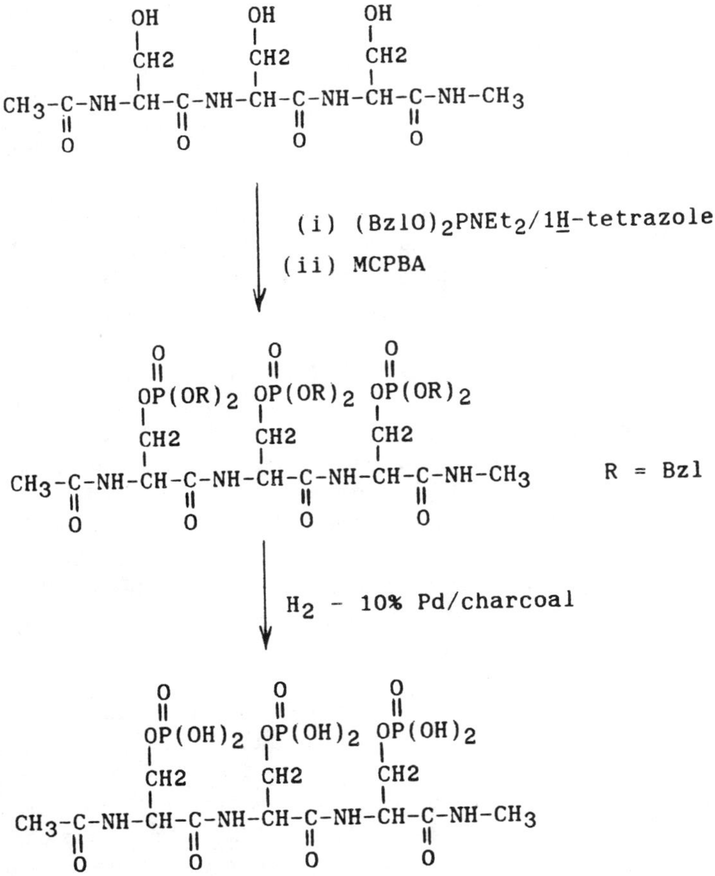

Figure 2. 'Phosphite-triester' phosphorylation of Ac-Ser-Ser-Ser-NHMe using dibenzyl N,N-diethylphosphoramidite.

threonine and tyrosine derivative and the protected Ser-peptide, Z-Asp(O$^\pm$Bu)-Ala-Ser-Glu(O$^\pm$Bu)$_2$.

In addition to the synthesis of Glu-Ser(P)-Leu via benzyl phosphate protection, this Ser(P)-tripeptide was also prepared by the use of t-butyl phosphate protection. This synthesis involved the initial phosphorylation of Boc-Glu(OBu$^\pm$)-Ser-Leu-OBu$^\pm$ with di-t-butyl N,N-diethylphosphoramidite/1H-tetrazole - m-chloroperoxybenzoic acid (23) followed by acidolytic cleavage of the t-butyl groups from Boc-Glu(OBu$^\pm$)-Ser(PO$_3$$^2$Bu$_2$)-Leu-OBu$^\pm$ with 40% TFA/AcOH (11) (Figure 3). A feature in the use of t-butyl phosphate groups is that these protecting groups are particularly sensitive to acidolytic cleavage, both t-butyl groups being cleaved from di-t-butyl isobutyl phosphate within 3 min using 4 M HCl/dioxane and within 4 h using 1 M HCl/dioxane.

## Synthesis of Ser(P)-Peptides using Boc-Ser(PO$_3$R$_2$)-OH derivatives

In the second and more useful synthetic approach, the solution-phase synthesis of Ser(P)-peptides is accomplished by the incorporation of protected Boc-Ser(PO$_3$R$_2$)-OH derivatives into peptide synthesis followed by hydrogenolytic deprotection of the protected Ser(PO$_3$Bzl$_2$)- or Ser(PO$_3$Ph$_2$)-peptides (16,24-26). This procedure is comprised of three stages, (a) the efficient preparation of Boc-Ser(PO$_3$R$_2$)-OH derivatives, (b) incorporation of these derivatives in the Boc mode of peptide synthesis and (c) the final deprotection of the assembled Ser(PO$_3$R$_2$)-containing peptide. A major feature of this synthetic approach is that it permits a phosphorylated serine residue to be incorporated at any desired position within a peptide, and is also suitable for the synthesis of multiple Ser(P)-containing peptides and mixed Ser(P)/Ser-containing peptides.

## Synthesis of Ser(P)-peptides through benzyl and 4-bromobenzyl phosphate protection

The synthesis of Boc-Ser(PO$_3$Bzl$_2$)-OH is accomplished in good overall yield by a three-step procedure which involves (a) initial protection of the carboxyl group of Boc-Ser-OH as its 4-nitrobenzyl ester, (b) phosphorylation of the serine hydroxyl group by the use of either the phosphorotriester or 'phosphite-triester' phosphorylation procedures and (c) removal of the 4-nitrobenzyl group from the carboxyl terminus by sodium dithionite reduction (13,24) (Figure 4).

The crucial phosphorylation step in this synthesis is accomplished by the use of either dibenzyl phosphorochloridate/pyridine (13) or dibenzyl N,N-diethylphosphoramidite/1H-tetrazole - MCPBA (18). While only 50-60% yields of Boc-Ser(PO$_3$Bzl$_2$)-ONBzl were initially obtained by the use of dibenzyl phosphorochloridate in pyridine, this yield was increased to 94% by the use of a modified procedure in which the phosphorylation is conducted at low temperature (-40 °C) and includes a delayed second addition of dibenzyl phosphorochloridate. The development of these conditions resulted from $^{31}$P NMR studies which showed

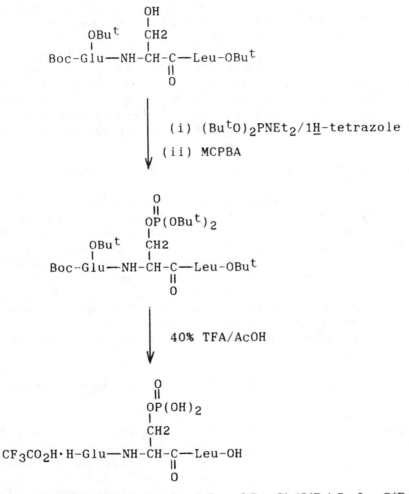

Figure 3. 'Phosphite-triester' phosphorylation of Boc-Glu(O$^{\pm}$Bu)-Ser-Leu-O$^{\pm}$Bu using di-t-butyl N,N-diethylphosphoramidite.

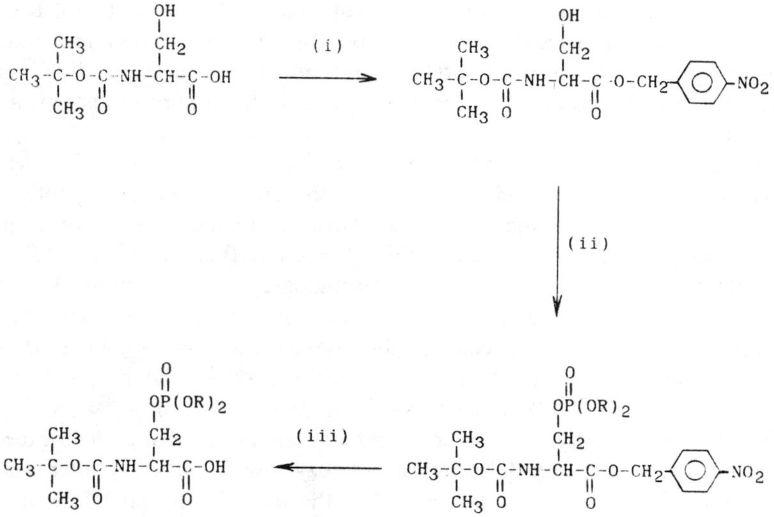

(i) NBzl-Br, Et₃N (6 h, 60°C);

(ii) (BzlO)₂P(O)Cl/pyridine (6 h, -40°C) or

(BzlO)₂PNEt₂/1H-tetrazole then MCPBA/CH₂Cl₂;

(iii) Na₂S₂O₄, pH 8.6 (1 h, 50°C)

Figure 4. Synthesis of the protected Boc-Ser(PO₃Bzl₂)-OH derivative.

that for dibenzyl phosphorochloridate in pyridine, the phosphoro-N-pyridinium intermediate is the active phosphorylation species and undergoes rapid pyridine-mediated debenzylation at temperatures above -20 °C (13).

In the case of phosphite-triester phosphorylation, the phosphorylation is effected in near-quantitative yield by initial phosphitylation of Boc-Ser-ONBzl with dibenzyl N,N-diethylphosphoramidite/1H-tetrazole followed by *in situ* oxidation of the resultant phosphite-triester with m-chloroperoxybenzoic acid (MCPBA). The advantages of this procedure is that both the phosphitylation and oxidation reactions are rapid, the phosphitylation reaction is conducted at 20 °C and the phosphorylation procedure gives the dibenzyl phosphorotriester in a near-quantitative yield.

By the use of (a) Boc-Ser(PO$_3$Bzl$_2$)-OH in peptide synthesis, (b) 4 $\underline{M}$ HCl/dioxane (30 min) for the cleavage of the Boc group from the dipeptide, Boc-Ser(PO$_3$Bzl$_2$)-Leu-OBzl and (c) mixed anhydride coupling of the resultant dipeptide hydrochloride with Boc-Glu(OBzl)-OH, Boc-Glu(OBzl)-Ser(PO$_3$Bzl$_2$)-Leu-OBzl was obtained in a yield of 33% after chromatographic purification. A limitation in the use of Boc-Ser(PO$_3$Bzl$_2$)- OH in peptide synthesis is that the benzyl phosphate group is subject to acidolytic cleavage during the routine use of 4 $\underline{M}$ HCl/dioxane or 50% TFA/CH$_2$Cl$_2$ for the successive removal of the N-terminal Boc group from Boc-protected Ser(PO$_3$Bzl$_2$)-containing peptides. From a $^{31}$P NMR study using dibenzyl *iso*butyl phosphate as a test compound, it was determined that there was approximately 35% and 50% benzyl loss respectively after a 30 min treatment with the above solutions. Also, the use of 1 $\underline{M}$ HCl/formic acid, 1 $\underline{M}$ HCl/acetic acid or formic acid caused approximately 90, 10 and 1% benzyl loss respectively after a 60 min treatment.

In peptide synthesis Perich and Johns (24) reported the yield of Boc-Glu(OBzl)-Ser(PO$_3$Bzl$_2$)-Leu-OBzl increased from 33 to 81% by changing from 4 $\underline{M}$ HCl/dioxane (30 min) to formic acid (60 min) for the cleavage of the Boc group from the dipeptide, Boc-Ser(PO$_3$Bzl$_2$)-Leu-OBzl and subsequent mixed anhydride coupling of the resultant dipeptide with Boc-Glu(OBzl)-OH. However, the additional isolation of the mono benzyl phosphate tripeptide, Boc-Glu(OBzl)-Ser(PO$_3$Bzl,H)-Leu-OBzl, in 7% yield indicated that minor benzyl loss from the Ser(PO$_3$Bzl$_2$)- dipeptide had occurred during Boc removal. In later work, this low-level benzyl loss was established to occur during the evaporative removal of formic acid at 40 °C. Thus, by the use of a minimal volume of formic acid and the rapid removal of formic acid under high vacuum at 20 °C (< 5 min), benzyl loss was minimized and the isolated yield of the Ser(PO$_3$Bzl$_2$)- tripeptide increased to 94%.

Peptide deprotection was readily accomplished by the palladium-catalysed hydrogenolysis of the Ser(PO$_3$Bzl$_2$)-tripeptide in formic acid, the zwitterionic Ser(P)-tripeptide Glu-Ser(P)-Leu being obtained in quantitative yield. The benzyl phosphate groups are also susceptible to acidolytic cleavage, both benzyl phosphate groups being observed by $^{31}$P NMR spectroscopy to be cleaved by a prolonged treatment with 4 $\underline{M}$ HCl/dioxane (172 h) or a short treatment with 2.8 M HBr/ AcOH ($<$ 15 min).

**Synthesis of Glu-Ser(P̲)-Leu Using Ppoc Amine Protection.** To overcome acidolytic benzyl loss from Ser(PO₃Bzl₂)-residues during peptide synthesis, Perich (16) prepared the 2- phenyl̲isopropyloxycarbonyl (Ppoc) protected derivative, Ppoc-Ser(PO₃Bzl₂)-OH, in consideration that the Ppoc group is cleaved by 2% TFA/ CH₂Cl₂ during peptide synthesis. The Ppoc-protected derivative was prepared by a simple three-step synthetic procedure (as for Figure 4) which featured the phosphorylation of Ppoc-Ser-ONBzl by either dibenzyl phosphorochloridate or dibenzyl N̲,N̲-diethylphosphoramidite followed by sodium dithionite reduction of the 4-nitrobenzyl group from Ppoc-Ser(PO₃Bzl₂)-ONBzl.

By the use of Ppoc-Ser(PO₃Bzl₂)-OH in peptide synthesis and 0.5 $\underline{M}$ HCl/ dioxane (or 4% TFA/CH₂Cl₂) for the cleavage of the Ppoc group from the Ppoc-Ser(PO₃Bzl₂)-Leu-OBzl, the mixed anhydride coupling of the resultant dipeptide hydrochloride (or dipeptide trifluoroacetate) with Boc-Glu(OBzl)-OH gave the tripeptide Boc-Glu(OBzl)-Ser(PO₃Bzl₂)-Leu-OBzl in 94% yield. As above, hydrogenolytic deprotection of this tripeptide in formic acid gave Glu-Ser(P̲)-Leu in quantitative yield.

While this synthesis demonstrates the application of Ppoc-Ser(PO₃Bzl₂)-OH in peptide synthesis, it must be remembered that for the use of this derivative in general peptide synthesis, subsequent extension of the Ser(PO₃Bzl₂)-peptide necessitates the incorporation of the following N̲-protected amino acids as their Ppoc- or Bpoc-derivatives.

**Synthesis of Glu-Ser(P̲)-Leu Using 4-Bromobenzyl Phosphate Protection.** In a second approach to minimize benzyl phosphate loss during peptide synthesis, Perich and Johns (Aust. J. Chem., Part X, in press) used the 4-bromobenzyl group for phosphate protection since this group offered greater acid stability than either the 4-fluorobenzyl or 4-chlorobenzyl groups. In comparison to the benzyl group, the 4-bromobenzyl phosphate group exhibited a four-fold increase in stability in either 1 $\underline{M}$ HCl/acetic acid or formic acid solutions. However the 4-bromobenzyl group offered no increased resistance to acidolytic debenzylation in either 4 $\underline{M}$ HCl/dioxane or 50% TFA/CH₂Cl₂ solutions.

The preparation of Boc-Ser(PO₃BrBzl₂)-OH was accomplished in 40% yield by a one-step procedure which involved initial *in situ* t̲-butyldimethylsilyl protection of Boc-Ser-OH followed by phosphorylation with di-4-bromobenzyl N̲,N̲-diethylphosphoramidite/1H-tetrazole, oxidation of the resultant phosphite-triester with t̲-butylhydroperoxide and final mild acidolysis of the t̲-butyldimethylsilyl group with aqueous acetic acid. By the use of Boc-Ser(PO₃BrBzl₂)-OH in peptide synthesis and formic acid for the cleavage of the Boc group from Boc-Ser(PO₃BrBzl₂)-Leu-OBzl, the mixed anhydride coupling of the resultant dipeptide formate with Boc-Glu(OBzl)-OH gave the protected tripeptide, Boc-Glu(OBzl)-Ser(PO₃BrBzl₂)-Leu-OBzl in 94% yield. As with benzyl phosphate protection, the 4-bromobenzyl groups were readily cleaved by palladium-catalysed hydrogenolysis and gave Glu-Ser($\underline{P}$)-Leu in quantitative yield.

**Synthesis of Ser($\underline{P}$)-Peptides Through Benzyl Phosphate Protection Using a Convergent Approach.** In two recent studies, the synthesis of two protected Ser(PO$_3$Bzl$_2$)-peptides has been accomplished by the use of a convergent approach which involves the use of 1 $\underline{M}$ HCl/AcOH for the removal of the Boc group. In the first of these studies, Bannwarth and Trzeciak ([20]) used 1 $\underline{M}$ HCl/AcOH (5 min, 20 °C) for the removal of the Boc group from Boc-Asp(OBzl)-Ala-Ser-Gly-Glu(OBzl)-OBzl and the subsequent mixed anhydride coupling of the resultant pentapeptide hydrochloride with Z-Trp-Ala-Gly-Gly-OH gave the protected Ser(PO$_3$Bzl$_2$)-nonapeptide Z-Trp-Ala-Gly-Gly-Asp(OBzl)-Ala-Ser-Gly-Glu(OBzl)-OBzl in 69% yield. Subsequent palladium-catalysed hydrogenolysis of the Ser(PO$_3$Bzl$_2$)-nonapeptide in aqueous acetic acid gave the Ser($\underline{P}$)-nonapeptide in good yield after DEAE-Sephadex purification.

In the second study, De Bont and coworkers ([21]) used 1 $\underline{M}$ HCl/AcOH (7 min) for the removal of the Boc group from Boc-Thr(PO$_3$Bzl$_2$)-Leu-Arg(Mbs)OMe and subsequent active ester coupling of the resultant tripeptide hydrochloride with Boc-Lys(Z)-Arg(Mbs)-OBt gave the protected pentapeptide, Boc-Lys(Z)-Arg(Mbs)-Thr(PO$_3$Bzl$_2$)-Leu-Arg(Mbs)-OMe in 60% yield. In this case, the Thr(PO$_3$Bzl$_2$)-pentapeptide was deprotected by initial TFMSA/thioanisole/$\underline{m}$-resol/TFA treatment followed by saponific cleavage of the methyl ester. Although this convergent approach can be successfully utilized for the synthesis of particular Ser($\underline{P}$)-peptides, this approach is not recommended for the synthesis of peptides containing multiple Ser($\underline{P}$)- or Thr($\underline{P}$)- residues.

**Synthesis of Ser($\underline{P}$)-peptides through phenyl phosphate protection**

The synthesis of protected Boc-Ser(PO$_3$Ph$_2$)-OH is accomplished in good overall yield by a three-step procedure which involves (a) initial protection of the carboxyl group of Boc-Ser-OH as its 4-nitrobenzyl ester, (b) phosphorylation of the serine hydroxyl group by the use of diphenyl phosphorochloridate/pyridine and (c) hydrogenolytic removal of the 4-nitrobenzyl group from the carboxyl terminus ([14]) (as for Figure 4).

By the use of Boc-Ser(PO$_3$Ph$_2$)-OH in peptide synthesis and 4 $\underline{M}$ HCl/dioxane for the removal of the Boc group from the Boc-dipeptide, the protected tripeptide Boc-Glu(OBzl)-Ser(PO$_3$Ph$_2$)-Leu-OBzl was prepared in an overall yield of 87% (respective coupling yields of 93 and 94% respectively). While the platinum catalysed hydrogenolysis of phenyl phosphate groups from Ser(PO$_3$Ph$_2$)-peptides has generally been reported in the literature to be incomplete, Perich and Johns ([14]) reported that rapid and complete phenyl cleavage could be effected by the use of molar equivalents of platinum dioxide per phenyl-phosphate group and 50% TFA/AcOH as hydrogenation solvent. The development of these conditions arose from the observation that (a) benzyl or phenyl cleavage was facilitated by the introduction of trifluoroacetic acid to the hydrogenation solution and (b) that Jung and Engel (27) reported that phosphorus-oxygen cleavage was the first step in the hydrogenolytic removal of the phenyl group. By the use of these conditions, the hydrogenation of H-Glu-Ser($PO_3Ph_2$)-Leu-OH proceeded rapidly and with complete phenyl cleavage to give H-Glu-Ser(P)-Leu-OH.TFA in high yield.

In later work, Perich and Johns used this procedure for the efficient synthesis of the multiple Ser(P)-containing peptides, Ser(P)-Ser(P)-Ser(P)-NHMe (25) and Ac-Glu-Ser(P)-Leu-Ser(P)-Ser(P)-Ser(P)-Glu-Glu-NHMe (16,26) in high yields. Also, this approach has been used for the synthesis of a variety of Ser(P)-containing peptides (see Table 1) and the synthesis of Ser(P)-Ser(P)-Ser(P)-Glu-Glu-NHMe is outlined in Figure 5.

An advantage in the use of phenyl phosphate groups is that the above protected Ser($PO_3Ph_2$)-peptides are readily soluble in common organic solvents (such as ethyl acetate and dichloromethane) and this facilitates peptide synthesis. A major feature of this synthetic approach is that due to high purity of protected peptides obtained by this synthetic approach and the cleanness of the deprotection procedure, the obtained Ser(P)-peptides are obtained in high purity and do not require chromatographic purification.

As shown in Table I, the approach is also suitable for the synthesis of mixed Ser/Ser(P)-peptides in which either Boc-Ser($PO_3Ph_2$)-OH or Boc-Ser(Bzl)-OH is incorporated at the specific site during peptide synthesis. This method was used for the high-yielding synthesis of the two Ser(P)-peptides, Ser-Ser(P)-Ser(P)-NHMe and Ser(P)-Ser-Ser(P)-NHMe (Perich and Johns, unpublished data). In these cases, simultaneous removal of the seryl benzyl ether is effected during the platinum hydrogenation step.

The characterization of the synthetic Ser(P)-peptides is routinely performed by the use of $^{13}C$ NMR spectroscopy, $^{31}P$ NMR spectroscopy and FAB mass spectrometry. A distinct feature for the $^{13}C$ NMR spectra of Ser(P)-containing peptides is that the $C\alpha$ and $C\beta$ carbons of the Ser(P)-residue are observed as characteristic phosphorus-coupled doublet signals with coupling constants ranging from 4.4 to 8.8 Hz.

In the case of FAB mass spectrometry, this soft ionization technique generally gives a high intensity pseudo-molecular ion in the FAB mass spectrum of Boc-Ser($PO_3R_2$)-OH derivatives (28,29) and Ser(P)-peptides (30). However, the interpretation of FAB mass spectra for Ser(P)-peptides can often be difficult due to the high molecular weight mass spectral region being complicated by extensive metal complexation of the molecular ion. In the case of multiple Ser(P)-peptides, metal-complex species are generally observed which comprise to (a) sodium, (b) potassium, (c) mixed sodium-potassium and (d) platinum (i.e., ½ Pt) complex ions. Although these metal ions are present in the peptide in only trace amounts, the observation of these high intensity metal-complex ions in the FAB mass spectrum indicate that multiple Ser(P)-cluster peptides possess a structural conformation which permits extensive metal complexation.

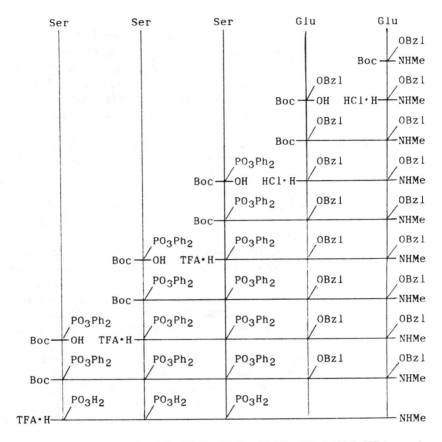

Figure 5. Synthesis of Ser(P)-Ser(P)-Ser(P)-Glu-Glu-NHMe.TFA.

Table I

| | |
|---|---|
| Ser (P̲)-Leu | Ref. (24) |
| Glu-Ser(P̲)-Leu | Ref. (24) |
| Ac-Glu-Ser(P̲)-Leu-NHMe | (Perich and Johns*) |
| Ac-Gly-Ser(P̲)-Gly-NHMe | (Perich and Johns*) |
| Ser(P̲)-NHMe | Ref. (25) |
| Ser(P̲)-Ser(P̲)-NHMe | Ref. (25) |
| Ser(P̲)-Ser(P̲)-Ser(P̲)-MHMe | Ref. (25) |
| Ser(P̲)-Ser-Ser(P̲)-NHMe | (Perich and Johns*) |
| Ser-Ser(P̲)-Ser(P̲)-NHMe | (Perich and Johns*) |
| Ac-Ser-Ser(P̲)-Ser(P̲)-NHMe | (Perich*) |
| Ac-Ser(P̲)-NHMe | (Perich and Johns*) |
| Ac-Ser(P̲)-Ser(P̲)-NHMe | (Perich and Johns*) |
| Ac-Ser(P̲)-Ser(P̲)-Ser(P̲)-NHMe | (Perich and Johns*) |
| Asp-Ser(P̲)-Ser(P̲)-Glu-Glu-NHMe | (Perich and Reynolds*) |
| Ser(P̲)-Ser(P̲-)Ser(P̲)-Glu-Glu-NHMe | (Perich and Reynolds*) |
| Ile-Ser(P̲)-Ser(P̲)-Ser(P̲)-Glu-Glu-NHMe | (Perich and Reynolds*) |
| Ser(P̲)-Ile-Ser(P̲)-Ser(P̲)-Ser(P̲)-Glu-Glu-NHMe | (Perich and Reynolds*) |
| Glu-Ser(P̲)-Ile-Ser(P̲)-Ser(P̲)-Ser(P̲)-Glu-Glu-NHMe | (Perich and Reynolds*) |
| Ac-Glu-Ser(P̲)-Ile-Ser(P̲)-Ser(P̲)-Ser(P̲)-Glu-Glu-NHMe | (Perich and Reynolds*) |
| Ac-Glu-Ser(P̲)-Leu-Ser(P̲)-Ser(P̲)-Ser(P̲)-Glu-Glu-NHMe | Ref. (26) |
| Ile-Val-Pro-Asn-Ser(P̲)-Val-Glu-Glu-NHMe | (Perich and Reynolds*) |

*unpublished data

## Synthesis of Ser(P̲)-Peptides By Solid-Phase Methods

In consideration that solid-phase peptide synthesis has been widely used for the synthesis of complex high molecular weight peptides, there has been interest since 1985 in the use of this method for the synthesis of large and complex Ser(P̲)-containing peptides. Although this approach is not as well developed as the solution phase synthesis of Ser(P̲)-peptides, the synthesis of some Ser(P̲)-peptides has been reported by the use of either the 'global' phosphorylation or the 'synthon incorporation' method.

**Solid phase synthesis of Ser(P̲)-peptides by 'Global Phosphorylation'.** In 1988, Perich and John (11) reported the synthesis of Glu-Ser(P̲)-Leu by the 'global' phosphorylation of Boc-Glu(OBu$^\pm$)-Ser-Leu bound to a polystyrene resin support. The resin-bound seryl peptide was prepared by the use of Fmoc-Ser(OBu$^\pm$)-OH in peptide synthesis followed by acidolytic cleavage of the seryl t̲-butyl ether with

95% TFA/CH$_2$Cl$_2$. The phosphorylation of the seryl residue was effected by initial phosphitylation using dibenzyl or di-t-butyl N,N- diethylphosphoramidite/ 1H-tetrazole followed by oxidation of the resultant phosphite-triester with m-chloroperoxybenzoic acid. In the case of the protected Ser(PO$_3$Bu$^+_2$)-tripeptide, deprotection was effected by initial removal of the Boc and t-butyl groups with 95% TFA/CH$_2$Cl$_2$ followed by cleavage of the Ser(P)-tripeptide from the polystyrene resin by high pressure hydrogenolysis (palladium acetate).

In 1989, Otvos and coworkers (31) reported the synthesis of Gly-Gly-Ser(P)-Pro-Val-Glu-Lys and its N-acetylated derivative by the phosphorylation of a seryl peptide bound to a 4-alkoxybenzylalcohol-resin. The seryl peptide-resin was prepared by the Fmoc mode of peptide synthesis (with the Fmoc amino acids and Fmoc-Ser-OH incorporated as their pentafluorophenyl esters) and was phosphorylated by dibenzyl phosphorochloridate in pyridine/toluene. Peptide deprotection was effected by a 1 h treatment with TFA/CH$_2$Cl$_2$/anisole (62:30:8) and the crude Ser(P)-heptapeptide was purified by C$_{18}$ reverse-phase HPLC. In addition, they reported the use of this approach for the synthesis of the two-mono- and the di-phosphorylated forms of a tridecapeptide (shown below) in crude yields of 25 to 27%.

Lys-Ser(P)-Pro-Val-Pro-Lys-Ser(P)-Pro-Val-Glu-Glu-Lys-Gly

While this approach was used for the synthesis of five Ser(P)-peptides in low to moderate yields, a limitation of the synthetic approach is that the dibenzyl phosphorochloridate phosphorylation of the seryl peptide is incomplete and the phosphorylation proceeds with the formation of a major yellow byproduct.

**Solid Phase Synthesis of Ser(P)-peptides Using Boc-Ser(PO$_3$Ph$_2$)-OH.** As with solution phase peptide synthesis, the protected Boc-Ser(PO$_3$R$_2$)-OH derivatives are also amenable with the Boc mode of solid phase peptide synthesis. For example, Perich and coworkers (32) demonstrated the synthesis of Glu-Ser(P)-Leu by (a) the use of Boc-Ser(PO$_3$Ph$_2$)-OH in a Merrifield solid phase synthetic protocol, (b) hydrogenolytic cleavage (palladium acetate) of the Ser(PO$_3$Ph$_2$)-peptide from the polystyrene resin support followed by (c) platinum reduction of the phenyl phosphate groups (27). Final C$_{18}$ reverse-phase HPLC purification of the crude peptide gave Glu-Ser(P)-Leu in high yield.

In 1989, Arendt and coworkers (33) prepared Leu-Arg-Arg-Ser(P)-Leu-Gly by the use of a similar synthetic approach except that they employed a HF step for the cleavage of the Ser(PO$_3$Ph$_2$)-peptide from the resin support. While it has been reported that HF causes the decomposition of both -Ser(PO$_3$R$_2$)-(R = Et, Me, Bzl) and Ser(PO$_3$H$_2$)-residues, the above authors found that in the case of the HF-stable phenyl phosphate groups, there is no detectable HF-mediated decomposition of the Ser(PO$_3$Ph$_2$)-residue during the peptide-resin cleavage step.

## Conclusion

In view of the recent developments in the chemical methods for the 'global phosphorylation' and the 'synthon incorporation' approaches, it is now evident that both these methods will find wide application in the solution and solid phase synthesis of Ser(P)-containing peptides. However, of the two methods, the greater synthetic flexibility in the use of Boc-Ser($PO_3R_2$)-OH derivatives suggests that the 'synthon incorporation' approach has the greater potential for the preparation of large and complex Ser(P)-containing peptides which contain either single or multiple Ser(P)-residues.

## Literature Cited

1. Farrell, H.M. Jr. In Fundamentals of Dairy Chemistry; Wong, N. P.; Jenness, R.; Keeney, M.; Marth, E. H., Eds.; Van Nostrand Reinhold Company: New York, 1988; Chapt 9.
2. Folsch, G. Svensk Kemisk Tidskrift 1967, 79, 38.
3. Theodoropoulos, D.; Souchleris, I. Biochem. 1964, 3, 145.
4. Frank, A. W. CRC Crit. Rev. Biochem., 1984, 16, 51.
5. Schlesinger, D. H.; Buku, A.; Wyssbrod, H. R.; Hay, D. I. Int. J. Peptide Protein Res. 1987, 30, 257.
6. Grehn, L.; Fransson, B.; Ragnarsson, U. J. Chem. Soc. Perkin Trans. I 1987, 529.
7. Johnson, T. B.; Coward, J. K. J. Org. Chem. 1987, 52, 1771.
8. Alewood, P. F.; Johns, R. B.; Perich, J. W. Proceedings of the 7th American Peptide Symposium, 1981, p 65.
9. Perich, J. W.; Alewood, P. F.; Johns, R. B. Aust. J. Chem. 1987, 40, 257.
10. Perich, J. W.; Langford, N. J.; Johns, R. B. Aust. J. Chem. 1987, 40, 1389.
11. Perich, J. W.; Johns, R. B. Tetrahedron Lett. 1988, 29, 2369.
12. Alewood, P. F.; Perich, J. W.; Johns, R. B. Synth. Commun. 1982, 12, 821.
13. Alewood, P. F.; Perich, J. W.; Johns, R. B. Aust. J. Chem. 1984, 37, 429.
14. Perich, J. W.; Alewood, P. F.; Johns, R. B. Tetrahedron Lett. 1986, 27, 1373.
15. Perich, J. W.; Alewood, P. F.; Johns, R. B. Synthesis, 1986, 572.
16. Perich, J. W. Ph.D. Thesis, The University of Melbourne, Melbourne, 1986.
17. Smirnova, L. I.; Malenkovskaya, M. A.; Predvoditelev, D. A.; Nifant'ev, E. E. Zh. Org. Khim., 1980, 16, 1170.
18. Perich, J. W.; Johns, R. B. Tetrahedron Lett., 1987, 28, 101.
19. Perich, J. W.; Johns, R. B. J. Org. Chem. 1989, 54, 1750.
20. Bannwarth, W.; Trzeciak, A. Helv. Chim. Acta 1987, 70, 175.
21. de Bont, H. B. A.; Liskamp, R. M. J.: O'Brian, C. A.; Erkelens, C.; Veeneman, G. H.; van Boom, J. H.; Int. J. Peptide Protein Res. 1989, 33, 115.
22. Bannwarth, W.; Kung, E. Tetrahedron Lett., 1989, 30, 4219.

23. Perich, J. W.; Johns, R. B. Synthesis 1988, 142.

24. Alewood, P. F.; Perich, J. W.; Johns, R. B. Tetrahedron Lett., 1984, 25, 987.

25. Perich, J. W.; Johns, R. B. J. Org. Chem. 1988, 53, 4103.

26. Perich, J. W.; Johns, R. B. Chem. Comm., 1988, 664.

27. Jung, E.; Engel, R. J. Org. Chem., 1975, 40, 244.

28. Perich, J. W.; Johns, R. B.; Liepa, I.; Chaffee, A. L. Org. Mass Spectrom., 1988, 23, 680.

29. Perich, J. W.; Johns, R. B.; Liepa, I.; Chaffee, A. L. Org. Mass Spectrom., 1988, 23, 797.

30. Johns, R. B.; Alewood, P. F.; Perich, J. W.; Chaffee, A. L.; MacLeod, J. K. Tetrahedron Lett., 1986, 27, 4791.

31. Otvos Jr., L.; Elekes, I.; Lee, V. M-Y. Int. J. Peptide Protein Res., 1989, 34, 129.

32. Perich, J. W.; Valerio, R. M.; Johns, R. B. Tetrahedron Lett., 1986, 27, 1377.

33. Arendt, A.; Palczewski, K.; Moore, W. T.; Caprioli, R. M.; McDowell, J. H.; Hargrave, P. A. Int. J. Peptide Protein Res. 1989, 33, 468.

RECEIVED August 27, 1990

# Chapter 13

# Factors Contributing to Dental Calculus Formation and Prevention

**Donald J. White and Ed R. Cox**

**The Procter and Gamble Company, Sharon Woods Technical Center, Cincinnati, OH 45241**

Low and high molecular weight saliva species have been suggested as promoters and inhibitors of dental calculus formation. This dual role simulates the reactivity of similar macromolecules in controlling bone and tooth formation. Historically, agents found effective for dental calculus prevention have included molecules effective at inhibiting calcium phosphate mineralization. The activity of inhibitors in calculus prevention is surprising considering their relative inhibitory activity in comparison to salivary species. In this study, constant composition mineralization experiments demonstrated that inhibitors are limited by overgrowth phenomena and that continuous control of biomineralization requires reservoir levels of inhibitor in solution. It is proposed that a major limitation of salivary macromolecules in controlling plaque mineralization involves their slow diffusivity into plaque prohibiting them from controlling localized increases in supersaturation.

Over 85% of adults form mineralized deposits, called calculus, on their teeth. Dental calculus (also called tartar) is a petrified plaque deposit containing roughly 70-80% calcium phosphate salts. Calculus formation begins with the deposition of a macromolecular pellicle upon the teeth and this is followed by the deposition of microorganisms and the formation of a biofilm. Calculus forms in appositional layers and, in people prone to heavy tartar formation, mineralization of a layer of calculus can occur in less than a week. Typically, the course of plaque mineralization involves nucleation of precursor mineral phases of dicalcium phosphate dihydrate (DCPD) and octacalcium phosphate (OCP) along with amorphous calcium phosphate in extracellular spaces between microorganisms. This is followed by crystal growth and phase transformations to more stable hydroxyapatite mineral phases (1,2).

Natural polymers are thought to play important roles in calculus mineralization and prevention. Microscopic analysis reveals that the initial nucleation of mineral

0097–6156/91/0444–0177$06.00/0

within plaque is associated with extracellular matrix composed of precipitated proteolipids and peptides. Once crystals are formed, the actions of these natural polymers are apparently reversed and salivary macromolecules, in particular phosphorylated peptides known as PRP's, rapidly adsorb to crystal surfaces inhibiting further mineralization (3). In a series of experiments, we have recently shown that calculus deposits exhibit mineralization and dissolution kinetics an order of magnitude slower than synthetic minerals of comparable surface area and composition (4).

Despite the inhibitory properties of saliva macromolecules, and their essentially constant excretion into the saliva, the nucleation and mineralization of discrete calculus layers occurs quite rapidly (1,2). One approach to tartar control has been the topical application of compounds known to interfere with the crystal growth and phase transformation of calcium phosphate minerals and in fact commercial products today are prepared with crystal growth inhibitors such as soluble pyrophosphate, diphosphonates and zinc salts (5). Like saliva macromolecules, tartar control inhibitors do not completely control calculus mineralization. Their clinical efficacy ranges on average from 30-50% leaving considerable room for improvement (6). The development of improved inhibitors for tartar control requires increased understanding of various factors contributing to and limiting their performance. Previously, we have speculated that overgrowth of minerals and nucleation and rapid mineralization at localized high supersaturations could contribute to ineffective action of both natural and synthetic inhibitors.

Traditional methods for the evaluation of biological inhibitors and tartar control substances have involved conventional mineralization techniques where changes in solution composition can affect phase transformations and concomitant growth and dissolution of complex calcium phosphate phases compromising accurate determinations of inhibitor action (5). Furthermore, the decreasing thermodynamic driving force during mineralization prohibits the measurements of overgrowth, inhibitor breakdown, secondary nucleation and other effects which might contribute to benefits/limitations of inhibitor actives.

The constant composition (CC) technique provides important advantages for the assessment of mineralization kinetics of calcium phosphates. By virtue of its control over solution thermodynamics, the CC method enables the controlled study of inhibitor influences on the mineralization of calculus precursor phases such as OCP and DCPD which play critical roles in the evolution of calculus. In addition, the ability to grow minerals for extended periods provides the opportunity to quantitatively examine processes contributing to limitations of inhibitor agents, including overgrowth and secondary nucleation effects.

The purpose of the present study was to compare the activity of synthetic inhibitors and natural saliva species in the CC system and to continue our studies on the limitations of tartar control inhibitors by examining the effects of solution supersaturation on long term inhibitory activity of model tartar control inhibitors.

The results are examined in terms of their broad implications toward the study of inhibitor effects on biological systems.

## Methods and Materials

**Solution/Solid Preparation and Analysis.** Analytical reagent grade chemicals and distilled/deionized water were used to prepare solutions and titrants, which were filtered (0.22 $\mu$m Millipore) to remove particulate impurities. Inhibitor solutions were prepared using reagent grade chemicals with the exception of disodium ethanehydroxy diphosphonic acid (EHDP) which was supplied by Norwich Eaton Pharmaceuticals.

Calcium phosphate seed materials were characterized by X-ray diffraction and IR spectroscopy and compared to literature values published by LeGeros (7). Specific surface areas were measured by single point nitrogen/helium adsorption. Synthetic hydroxyapatite (HAP) (Ca/P = 1.66:SSA = 28 $m^2$/g) was prepared by the method of Nancollas and Mohan (8) and stored at 37°C as an aqueous slurry (44 mg/ml). DCPD (Ca/P = 1.02/SSA = 1.8 $m^2$/g) was prepared by the method of Marshall and Nancollas (9) and also stored at 37°C as an aqueous slurry at pH 5.35 (50 mg/ml).

**Constant Composition Crystal Growth.** CC crystallization experiments followed the general protocols set forth in a prior publication (5). The supersaturated reaction solutions each contained a background of NaCl to adjust solution ionic strength to 0.15 M. The composition of reaction solutions included: 1.75 mM Ca, 2.00 mM P, pH 7.40 (HAP high supersaturation); 1.10 mM Ca, 0.66 mM P, pH 7.40 (HAP low supersaturation); 5.36 mM Ca, 5.36 mM P, pH 6.00 (DCPD low supersaturation); 8.40 mM Ca, 9.59 mM P, pH 5.60 (DCPD high supersaturation). Inhibitors were either added directly to metastable reaction solutions prior to seed addition or conversely seeds were pretreated in concentrated inhibitor solutions, centrifuged, water washed and recentrifuged, and subsequently added directly to reaction solutions. Following seed inoculation the mineralization rates were monitored at constant supersaturation by the controlled addition of lattice ion titrant solutions delivered from a pair of piston driven burettes in a modified Brinkman pH stat apparatus (5). The composition of lattice ion titrants and the thermodynamic supersaturation of reaction solutions were determined using an iterative computer speciation program which takes into account the formation of ion pairs and activity coefficient terms. Solution aliquots taken during mineralization were analyzed for calcium and phosphate to verify constancy of supersaturation.

## Results

The effect of saliva macromolecules on the in cell CC mineralization of one of the calculus precursor phases, dicalcium phosphate dihydrate (DCPD) is shown in Figure 1. ("In cell" refers to experiments where inhibitor is added into the background reaction solution and seed crystal materials are added to initiate growth). Saliva dialyzed with a 5,000 mw cutoff membrane showed excellent mineralization inhibition. Dose response experiments demonstrated that saliva macromolecules could exert a 50% inhibition of mineral growth when diluted 300 fold. Undialyzed saliva demonstrated superior inhibition to the dialyzed portion, illustrating the importance of lower molecular weight inhibitors. The whole saliva effected 50% inhibition when diluted 900 fold in the CC mineralization cell. Despite the excellent inhibition conferred by saliva inhibitors, Figure 1 illustrates that overgrowth occurred following a lag period for both dialyzed and undialyzed samples.

Similar effects are observed in experiments with materials used commercially for tartar control. Figure 2 shows the in cell mineralization rate of DCPD in the presence of pyrophosphate and EHDP at low and high supersaturations at equivalent inhibitor concentrations. As was the case with saliva, the inhibition of mineralization with the tartar control inhibitors proceeded during a lag phase following which mineralization recommenced at high levels. This type of behavior is indicative of overgrowth or secondary nucleation processes, which should be strongly dependent upon supersaturation. In agreement with this, increases in supersaturation resulted in both 1) decreases in lag periods and 2) diminished differences among inhibitors.

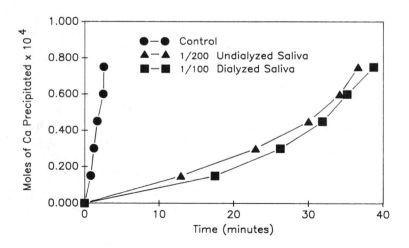

Figure 1. Saliva effects in cell on DCPD mineralization.

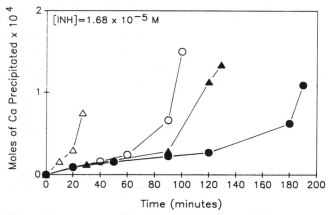

Figure 2. Effect of supersaturation ($\sigma$) on inhibitory effects of pyrophosphate (triangles) and EHDP (circles) on DCPD in cell mineralization. Concentration of inhibitor is constant as shown in all experiments. Filled symbols reflect experiments carried out at low supersaturation ($\sigma$ = 0.4 DCPD - see ref. 5); open symbols reflect experiments carried out at higher ($\sigma$ = 0.5) supersaturation.

Pretreatment of minerals with concentrations of inhibitor equivalent to those used in antitartar topicals resulted in similar crystallization profiles, with lag periods followed by accelerated mineralization. Changes in lag periods preceeding mineralization (calculated by drawing tangents to pre- and post- accelerated mineralization profiles) are shown in Table I.

Table I. HAP Mineralization Lag Periods - Effect of Inhibitors and Supersaturation

| Inhibitor | Conc. (mM) | Lag Period (Hrs) | |
| --- | --- | --- | --- |
| | | Low $\sigma$ | High $\sigma$ |
| EHDP | 26.8 | 16.4 | 0.67 |
| Pyrophosphate | 26.8 | 8.3 | 0.03 |
| Zinc Chloride | 26.8 | 5.6 | 0.33 |

A natural question which arises from these findings is what levels of mineralization inhibitors are necessary to maintain inhibition or re-inhibit overgrowth. To study this we examined the effect of pyrophosphate on in cell CC mineralization of DCPD as shown in Figure 3. The addition of $50\mu M$ pyrophosphate to reaction solution initially (point A) resulted in considerable inhibition. Following the characteristic lag period, mineralization recommenced and at point B (where the growth rate equaled an initial control rate) pyrophosphate was added at twice the original concentration eg. 100 $\mu$M. As shown, the reintroduction of inhibitor resulted in only temporary and modest reinhibition.

To study the fate of inhibitor during overgrowth periods the concentration of Zn ion was monitored during CC overgrowth both in solution and in solid precipitate as shown in Figure 4. The results showed that all inhibitor could be accounted for in the solid deposit following overgrowth, ie. the solution concentration had become depleted to levels ineffective to maintain inhibition. The fate of the Zn ions was not determined in this study but SEM analysis revealed large numbers of new crystals following overgrowth with varied morphologies from the original DCPD seed materials, suggesting that the Zn was not necessarily incorporated into the other crystals but that the surface area of new nucleated mineral overwhelmed the available inhibitor concentration.

## Discussion

Macromolecules are thought to have dual roles in promoting and retarding biological mineralization processes in normal formation of shells, bones and teeth (10). In cases of ectopic mineralization, there is reason to suspect that the actions of macromolecules may be similar, physical chemically, although the net result is undesired mineral deposition. In the latter case, such as in the case of dental calculus formation, promotion of nucleation by macromolecules confounds the normal protective actions of biological fluids against supersaturation, and subsequent crystal growth takes place in the presence of these same inhibitors.

Amongst biomineralization researchers considerable effort has been directed against studying the comparative inhibitory characteristics of biological and synthetic inhibitors. For example, in the case of salivary macromolecules, Moreno et al. (11,12) have carefully studied the surface active effects of a broad class of inhibitors. The focus on comparative studies among inhibitors is interesting when one considers that most biological inhibitors and certainly all synthetic agents designed to control mineralization are present (or applied) at concentrations far in excess of those necessary to typically control mineralization. A focus of our research has thus included attempts to determine how and why both biological and synthetic inhibitors fail. The constant composition method is ideal for studying long term mineralization reactions since the controlled supersaturation enables direct studies of overgrowth and secondary mucleation phenomena.

The results of the present study using the CC technique provide preliminary insight into a number of physical chemcial factors which must be of interest to researchers studying the effect of low molecular weight molecules or macromolecules on biomineralization. First, results of extended CC studies show that both natural saliva macromolecules and synthetic tartar control agents were overgrown following lag periods. Clearly, this must be a consideration in any evaluations of inhibitor activity (indeed dental calculus is an ectopic example of overgrowth).

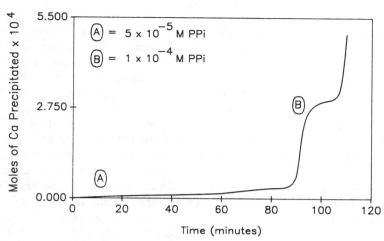

Figure 3. In cell DCPD crystallization affected by solution pyrophosphate added at points A and B in crystallization reaction.

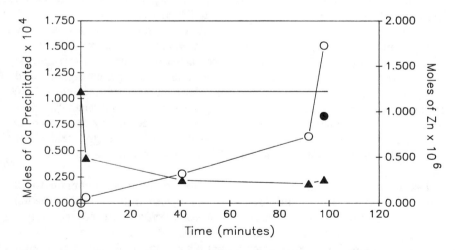

Figure 4. Change in Zn levels during in cell crystallization of DCPD. Horizontal line reflects original Zn level in reaction cell. Filled triangles represent analyzed Zn levels in solution (corrected for dilution) during crystallization and overgrowth. Filled circle represents Zn level in solid material at the end of reaction. Open circles represent calcium mineralized during reaction.

Second, the results of the saliva inhibitory activity clearly point to the discrepancy often encountered in the concentrations used for test conditions vs the real situation (this frequently occurs following separation techniques for biological inhibitors - these are only infrequently tested at *in vivo* levels). In calculus, localized crystallization may occur in regions of high supersaturation overcoming the effects of macromolecular inhibitors which may be slow in diffusing into the plaque biofilm.

Third, the differences in lag periods which we observed at various supersaturations, in particular the decrease in differences between inhibitors, points out the importance of testing all ranges of thermodynamic driving force to be expected in the biological mineralization media. Predicting accurate supersaturation conditions for *in vivo* modeling can be difficult, due to analytical limitations, however researchers can easily examine ranges underneath the barrier for spontaneous precipitation. It is at this limit that the maximum in localized supersaturation could occur within any concentrating system. In the case of dental calculus, localized increases in supersaturation occur upon microbial death where high internal phosphate concentrations in organisms are exposed to the calcium rich saliva environment.

Fourth, the results of the experiment described in Figure 3, where overgrown material was only marginally inhibited by two fold increased inhibitor levels, point to the importance of examining the reactivity of biominerals both before and following overgrowth. Analysis post-overgrowth is particularly important since this is indicative of the reactivity of most natural deposits (4).

Lastly, the fate of inhibitors during mineralization and overgrowth is obviously a key parameter to all biomineralization researchers which must be more fundamentally explored. In the present case, increased numbers of crystals and altered morphologies resulted in depleted Zn levels in solution permitting continued overgrowth. In calculus research, for example, a number of researchers have speculated that protease activity may destroy protein inhibitors on mineral surfaces on a localized basis thereby promoting mineralization. Accordingly, attempts have been made to correlate protease levels with calculus incidence (13), despite our lack of physical chemical understanding of whether the proteins can in fact be overgrown.

Dental calculus formation involves a number of processes common to all biomineralization reactions - the development of supersaturation, initial nucleation and crystallization in the presence of macromolecules. Our efforts to find technologies to control calculus deposition have led us into research into the limitations of inhibitory agents which is applicable to low molecular weight tartar control agents or higher molecular weight macromolecules present in biological mineralizing media. A fundamental understanding of the physical chemical factors related to crystallization and inhibition coupled with further insights into the nature and type of macromolecules (and low molecular weight solutes) present during biomineraliztion will lead us to more complete understanding of the controlling aspects of all biomineral formation, including calculus.

## Literature Cited

1. Mandel, I.D. <u>Comp. Cont. Ed. Dent. Suppl</u> 1987, <u>8</u>, 5235.
2. Schroeder, H. E. <u>Formation and Inhibition of Dental Calculus</u>; Hans Huber:Berne, 1969.
3. Moreno, E. C.; Aoba, T.; Gaffar, A. In <u>Recent Advances in the Study of Dental Calculus</u>; ten Cate, J. M., Ed.; IRL: Oxford, 1989; p 129.
4. McClanahan, S. F.; Cox, E. R.; Boehm, L. A.; White, D. J. <u>J. Dent. Research</u> 1989, <u>68A</u>, 380 (#1587).
5. White, D. J.; Bowman, W. D.; Nancollas, G. H. In <u>Recent Advances in the Study of Dental Calculus</u>; ten Cate, J. M., Ed.; IRL: Oxford, 1989; p 175.
6. Stookey, G. K.; Jackson, R. D.; Beiswanger, B. B.; Stookey, K. R. In <u>Recent Advances in the Study of Dental Calculus</u>; ten Cate, J. M., Ed.; IRL: Oxford, 1989; p 235.
7. LeGeros, R. Z.; LeGeros, J. P. In <u>Phosphate Minerals</u>; Nriagu, J. D.; Moore, P. B., Ed.; Springer-Verlag: New York, 1984; p 351.
8. Nancollas, G. H.; Mohan, M. S. <u>Arch. Oral Biol.</u> 1970, <u>15</u>, 731.
9. Marshall, R. W.; Nancollas, G. H. <u>J. Phys. Chem.</u> 1969, <u>78</u>, 3838.
10. Wheeler, A.P.; Rusenko, K.W.; Sikes, C.S. In <u>Chemical Aspects of Regulation of Mineralization</u>; Sikes, C.S.; Wheeler, A.P., Eds.; Univ. of South Alabama Publication Services: Mobile, AL, 1988; p. 9.
    UNX, 48.
11. Moreno, E.C.; Kresak, M.; Hay, D.I. <u>J. Biol. Chem.</u> 1982, <u>257</u>, 2981.
12. Moreno, E.C.; Kresak, M.; Hay, D.I. <u>Calc. Tissue Int.</u> 1984, <u>36</u>, 48.
13. Watanabe, T.; Morita, M. In <u>Recent Advances in the Study of Dental Calculus</u>; ten Cate, J. M., Ed.; IRL: Oxford, 1989; p 65.

RECEIVED August 27, 1990

# Chapter 14

# Commercial Production of Bovine Somatotropin in *Escherichia coli*

**James F. Kane, Steven M. Balaban, and Gregg Bogosian**

**Animal Sciences Division, Monsanto Agricultural Company, Chesterfield, MO 63198**

The gene for bovine somatotropin (BST) has been cloned behind the *Escherichia coli* tryptophan promoter in a pBR322 based plasmid designated pBGH1. Plasmid pBGH1 was transformed into *E. coli* K-12 strain W3110G. When indole acrylic acid was added to a culture of W3110G[pBGH1] growing in a fermenter, high levels of BST accumulated inside the cells. The BST was present as a denatured component of insoluble proteinaceous aggregates called inclusion bodies. In this report we discuss some properties of the host during growth and product accumulation as well as some of the properties of these inclusion bodies. Additionally a number of engineering and process development hurdles, which had to be overcome to prepare this product for commercialization, are discussed.

Bovine somatotropin (BST) is a protein produced by the pituitary gland of the cow. This protein is composed of 190 or 191 amino acids, has either a phenylalanine or an alanine as the N terminal amino acid, and contains two disulfide bonds (Figure 1). As early as the 1930's it was known that when this protein was injected into cows these cows could produce more milk (1) and could do so more efficiently. Despite this attractive possibility for increasing the productivity and reducing the costs to the dairy farmer, the potential commercialization of BST remained untapped because the only source of this protein was bovine pituitary glands. The market projections for commercial use of BST is in the hundreds of millions of grams per year, making BST a "commodity protein." It was simply impossible to produce enough of this protein in a large enough scale to consider seriously BST as a commercial product.

With the advent of techniques in recombinant DNA, it became reasonable to reconsider the production of the massive quantities of BST that would be required in commercialization of this product. The technological challenges were substantial. One had to clone the gene responsible for the synthesis of BST and place it in a vector that would provide for the efficient expression of the cloned

0097–6156/91/0444–0186$06.00/0

```
-1    1
Met-Phe-Pro-Ala-Met-Ser-Leu-Ser-Gly-Leu-Phe-Ala-Asn-Ala-
14
Val-Leu-Arg-Ala-Gln-His-Leu-His-Gln-Leu-Ala-Ala-Asp-Thr-
28
Phe-Lys-Glu-Phe-Glu-Arg-Thr-Tyr-Ile-Pro-Glu-Gly-Gln-Arg-
42
Tyr-Ser-Ile-Gln-Asn-Thr-Gln-Val-Ala-Phe-Cys-Phe-Ser-Glu-
56
Thr-Ile-Pro-Ala-Pro-Thr-Gly-Lys-Asn-Glu-Ala-Gln-Gln-Lys-
70
Ser-Asp-Leu-Glu-Leu-Leu-Arg-Ile-Ser-Leu-Leu-Leu-Ile-Gln-
84
Ser-Trp-Leu-Gly-Pro-Leu-Gln-Phe-Leu-Ser-Arg-Val-Phe-Thr-
98
Asn-Ser-Leu-Val-Phe-Gly-Thr-Ser-Asp-Arg-Val-Tyr-Glu-Lys-
112
Leu-Lys-Asp-Leu-Glu-Glu-Gly-Ile-Leu-Ala-Leu-MET-Arg-Glu-
126
Leu-Glu-Asp-Gly-Thr-Pro-Arg-Ala-Gly-Gln-Ile-Leu-Lys-Gln-
140
Thr-Tyr-Asp-Lys-Phe-Asp-Thr-Asn-Met-Arg-Ser-Asp-Asp-Ala-
156
Leu-Leu-Lys-Asn-Tyr-Gly-Leu-Leu-Ser-Cys-Phe-Arg-Lys-Asp-
168
Leu-His-Lys-Thr-Glu-Thr-Tyr-Leu-Arg-Val-Met-Lys-Cys-Arg-
182
Arg-Phe-Gly-Glu-Ala-Ser-Cys-Ala-Phe
```

Figure 1. Amino acid sequence of bovine somatotropin. The methionine residues in the sequence are highlighted. The first methionine residue is designated "-1" in order to maintain the conventional numbering system for the amino acid residues in BST.

gene. This was accomplished by scientists at Genentech, a biotechnology company based in San Francisco. These researchers cloned the BST gene using messenger RNA isolated from bovine pituitary glands. They constructed a vector, the details of which have been published, and placed this vector into the microorganism *Escherichia coli* (2). The major advantage of *E. coli*, as far as being a host for recombinant DNA experiments is concerned, is the voluminous data base on its biochemistry, physiology and genetics. While it is true that we know more about *E. coli* than any other organism, it is also true that data on biochemistry and physiology were obtained from small scale low density shake flask cultures. Thus, *E. coli* has been the workhorse of molecular genetics, but it has less of a history in large scale fermentations. This has some interesting ramifications on developing a process to produce this protein. For commercialization it was necessary to grow *E. coli* to high cell densities, greater than 25-30 grams of dry weight per liter, while generating 25 to 30% of the cellular protein as BST. Although *E. coli* is considered a facultative anaerobic microorganism, that is, it can grow with or without oxygen, the levels of energy required of high cell mass and product accumulation necessitates an aerobic environment. In addition to the engineering challenges to make a highly aerobic environment for growth and product formation in a large scale high density fermentation, there were the additional concerns of microbial physiology. Simply put, what are the effects of high rates of BST synthesis on the metabolism of *E. coli* in high cell density fermentations? There were some surprises during scale up which are discussed in this report.

A new phenomenon was observed when BST (and other eucaryotic proteins as well) was cloned into *E. coli*. The accumulated protein was found in the insoluble fraction (sedimented at relatively low speeds of 5,000 to 10,000g). The molecules of BST inside the cell aggregated and formed insoluble, highly refractile, inclusion bodies in the cytoplasm (3-5). The plus side of this effect was that the product was readily isolated from the fermentation broth by the relatively simple steps of cell breakage and centrifugation (6). At this point BST was at least 40% pure. The down side of these refractile bodies was that the molecules were present in a reduced and denatured state. The technological challenges were to dissolve the insoluble particles and to renature the protein thus allowing the two disulfide bonds to form in the proper manner. After renaturation, further purification steps were required to remove extraneous protein and lipopolysaccharides (endotoxins) from the *E. coli* host and make a bulk product that was greater than 95% pure. As a further challenge, this entire, rather complicated sequence of steps had to be accomplished at the lowest possible cost in order to keep the price reasonable for the diary farmer. While techniques for protein purification are highly advanced, most of these applications are only useful in analytical studies or with low quantity, high cost products. The BST market has just the opposite requirements: millions of grams of material needed at commodity (non-pharmaceutical) prices. This area of the process could not call upon existing technologies and new approaches had to be developed.

The last technological hurdle in reaching commercialization concerned the formulation of a biologically active BST protein into a stable, injectable material with a relatively long shelf life that was slowly released into the animal. The technology of slow release injectable proteins is in its infancy, so that there were innumerable technical challenges in this aspect of the project as well.

In retrospect, the cloning of the BST gene into *E. coli* was the easiest task of the entire project because there was a tremendous amount of data available to help in this endeavor. The majority of the process, however, involved operations for which there were no precedents. In this chapter we will address a few of those challenges.

## Genetics of BST Expression System

**Properties of pBGH1**. The plasmid used in the production of BST (2) is shown in Figure 2. Basically, the vector is pBR322 (7) containing the gene that encodes BST as well as flanking control elements. The expression of the cloned gene is controlled by an *E. coli* tryptophan promoter which is located on a 289 base pair (bp) *Eco*RI fragment (Figure 3). Although not described in the Genentech article, at the 3′ end of the gene in the plasmid pBGH1, there is a 293 base pair *Hind*III fragment which contains tandem *lac UV5* promoters that function in the termination of transcription initiated at the tryptophan promoter (8). The nucleotide sequence of this region has been published (9).

**Control of Transcription**. The tryptophan promoter has been used in a number of expression vectors (10). Transcription initiation depends upon the interaction between a repressor complex, composed of L-tryptophan and the aporepressor protein (the product of the *trp R* gene), and the tryptophan operator sequence (11) just upstream from the BST gene. When the repressor complex binds to this region of the DNA, the RNA polymerase is unable to bind to the promoter and transcription is prevented (repression). When tryptophan dissociates from the aporepressor, there is no complex to interact with the DNA, and the RNA polymerase binds to the promoter, initiating transcription (derepression). In the fermentation process, therefore, it was necessary to reduce or disrupt the repressor complex at the appropriate time so that transcription of the BST gene could begin. This was achieved by the addition of the tryptophan analog, indole acrylic acid (IAA). When IAA is added to the medium, it is taken up by the cell, and it competes with endogenously produced tryptophan for binding to the aporepressor. IAA has approximately a 30-fold greater affinity for the aporepressor than tryptophan (12,13) but the complex that is formed has at least a 60-fold lower affinity for the tryptophan operator region compared to the normal tryptophan-aporepressor complex. Thus, the IAA- aporepressor complex forms more readily than the tryptophan-aporepressor complex but is unable to bind to the tryptophan operator. This allows RNA polymerase access to the promoter region and high rates of transcription are initiated.

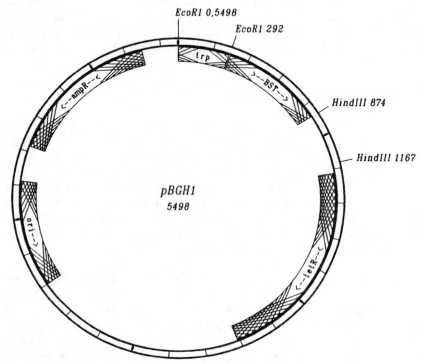

Figure 2. A schematic representation of the vector pBGH1. The locations of the restriction enzyme sites that flank the promoter, the structural gene for bovine somatotropin, and the transcription termination fragment are indicated.

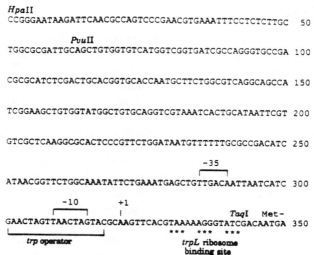

*HpaII*
CCGGGAATAAGATTCAACGCCAGTCCCGAACGTGAAATTTCCTCTCTTGC 50

*PvuII*
TGGCGCGATTGCAGCTGTGGTGTCATGGTCGGTGATCGCCAGGGTGCCGA 100

CGCGCATCTCGACTGCACGGTGCACCAATGCTTCTGGCGTCAGGCAGCCA 150

TCGGAAGCTGTGGTATGGCTGTGCAGGTCGTAAATCACTGCATAATTCGT 200

GTCGCTCAAGGCGCACTCCCGTTCTGGATAATGTTTTTTGCGCCGACATC 250

−35
ATAACGGTTCTGGCAAATATTCTGAAATGAGCTGTTGACAATTAATCATC 300

−10 +1 *TaqI* Met−
GAACTAGTTAACTAGTACGCAAGTTCACGTAAAAAGGGTATCGACAATGA 350
*** *** ***
trp operator trpL ribosome binding site

Figure 3. Nucleotide sequence of the *E. coli* tryptophan promoter. The critical elements of the promoter are highlighted. These include the startpoint of transcription (labeled '+1') and the RNA polymerase recognition elements labelled '-35' and '-10' (indicating their coordinates relative to the startpoint of transcription). The Trp repressor binding site (the *trp* operator), which overlaps the promoter, is highlighted below the nucleotide sequence. The *trpL* ribosome binding site is highlighted downstream from the startpoint of transcription. The asterisks indicate the homology between this ribosome binding site and the 3' end of 16S ribosomal RNA. The initial methionine residue of the *trp* leader peptide is indicated downstream from the ribosome binding site. Prior to this time, the nucleotide sequence upstream of the *E. coli* tryptophan promoter had been elucidated only as far as the 'C' residue at coordinate 205 of this figure (10). Our determination of the nucleotide sequence of the *E. coli* tryptophanon promoter on the plasmid pGM1 (23,24) has extended the known sequence by 204 base pairs, to the *HpaII* site indicated in the figure. For the construction of pBGH1, the tryptophan promoter was cloned from the plasmid pGM1 as a *PvuII-TaqI* fragment (using the restriction sites indicated on the figure) to which were added *EcoRI* linkers (25); other alterations were made in the vicinity of the ribosome binding site to yield the 289 base pair *EcoRI* fragment on pBGH1 (2,25). The nucleotide sequence of this same upstream portion of the tryptophan promoter, independently derived from the bacteriophage λRStrp570-91 (26), has recently been elucidated and is in exact agreement with that presented here (J.H. Zeilstra-Ryalls, B. Hagewood, and R. L. Somerville, personal communication).

## Host Physiology

*E. coli* is perhaps the best studied microorganism in molecular biology. The basic information on the genetics, biochemistry and physiology of this microorganism is unparalleled. However, fermentation of recombinant organisms is a relatively new field, and there is a paucity of data on the conditions required to obtain optimal growth and productivity of the cloned gene products.

**Production of BST.** In the case of the recombinant *E. coli* K-12 strain, W3110G[pBGH1], we have used the following scheme for the growth of the recombinant *E. coli*, and the accumulation of the BST protein. A production seed vial of the BST host-vector was thawed and added to a rich medium, such as Luria Broth (contains tryptone, yeast extract and NaCl), that contains the antibiotic tetracycline. The culture was grown for several hours until a density of approximately $10^9$ cells per ml was reached. This culture was transferred to a minimal salts, glucose production medium and was grown to a density of approximately $10^{10}$ cells per ml. At this point IAA was added to the fermentation vessel to begin high level synthesis of BST. Within an hour the product began to accumulate in the cell in the form of insoluble inclusion bodies. The results in Figure 4 illustrate the increase in the BST protein as a function of time after the addition of IAA. It is apparent from the intensity of the protein bands that BST is the major protein synthesized during this phase.

**Synthesis and Incorporation of Norleucine.** The microorganism was placed under a significant amount of metabolic stress during the production of the BST protein. In the relatively short time span of 6 to 10 hours, the recombinant microbe produced 5 to 6 grams of the BST protein per liter of medium. This represented approximately 65% of the total protein produced by the cell during this period, and resulted in a metabolic imbalance because the amino acids required for the synthesis of the BST were drastically different from the amino acid needs for normal *E. coli* proteins (14). Leucine, for example, represents approximately 15% of the total amino acid content of BST, and this is at least two-fold higher than the leucine content of native *E. coli* proteins. As a consequence of this imbalance there was a derepression of the amino acid biosynthetic pathway responsible for the synthesis of leucine. The leucine biosynthetic enzymes have broad substrate specificities (15), and this leads to the production of unusual amino acids such as norvaline and norleucine. The *in vivo* synthesis of the amino acid analog norleucine and its subsequent incorporation into protein in place of methionine (14) was an unexpected result in the high density fermentations used to make BST. In this specific case the pool level of the metabolite $\alpha$-ketobutyrate, a product of the enzyme threonine deaminase, accumulated and was a substrate for $\alpha$-isoproplymalate synthase, the first enzyme of the leucine biosynthetic pathway. This is an abnormal substrate for this enzyme and leads

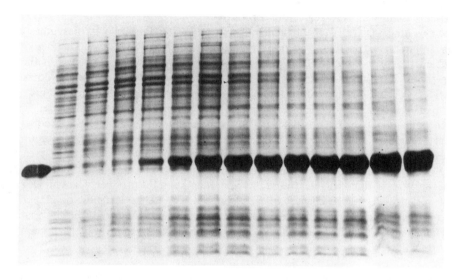

Figure 4. Polyacrylamide gel of samples from a typical 15 liter fermentation to produce bovine somatotropin. Samples were removed from the fermenter every hour beginning 2 hours before the addition of IAA. All of the samples were diluted to a constant turbidity and an aliquot was removed and the cells were lysed in a denaturing and reducing buffer. Equal volumes were loaded onto the gel. The gel was stained with Coomasie blue. The lanes from left to right represent: lane 1 - BST standard; lane 2 - 2 hours before IAA addition; lane 3 - 1 hour before IAA addition; lane 4 - time of IAA addition (IO); lanes 5 to 14 - 1 to 10 hours after the addition of IAA, respectively.

to the production of at least two straight chain aliphatic amino acids, norvaline (5-carbon) and norleucine (6-carbon). Norleucine, which is incorporated into protein in place of methionine (16), was found at all five methionine residues of BST (see Figure 1). Misincorporation at residue 123 was unique in that it affected the elution profile of the BST on a reverse phase HPLC column (17). This easy method of detecting misincorporation of norleucine was a tremendous asset that allowed us to correct this problem.

This study (14) was the first demonstration of the intracellular biosynthesis of an amino acid analog by a normal prototrophic strain of *E. coli*. The factors which pre-disposed *E. coli* to generate norleucine were all associated with the metabolic imbalances created by high rates of synthesis of the cloned gene product BST. Similar effects have been reported for other recombinant proteins that have a high percentage of leucine residues (18-20). It is important to realize that *E. coli* is the catalyst used to make BST. The quantity and quality of the product are dependent upon the availability of the required amino acids. Fermentation conditions or protein synthesis requirements that adversely affect the pool levels of endogenously produced amino acids could negatively impact the efficiency of the catalyst, thus reducing the quality and quantity of the product. The synthesis of norleucine occurred because of a derepressed level of the leucine biosynthetic pathway and high levels of $\alpha$-ketobutyrate. There is every reason to believe that other types of metabolic imbalances would lead to the synthesis of other amino acid analogs.

**Protein Solubility and Stability**. As mentioned above, BST accumulates in *E. coli* as an insoluble protein aggregate. This particular tendency is shared by other cloned gene products (3-5) and has some distinct advantages for the commercial production of this protein. First, the product is stabilized when it aggregates. In the absence of this aggregation BST is readily degraded. The evidence for this conclusion comes from several sources: (i) When the recombinant strain was grown in the production medium, the tryptophan promoter was not fully repressed since the microorganism had to make tryptophan for growth. This was verified by measuring the activity of the enzyme anthranilate synthase, the first enzyme specific to the tryptophan pathway. The specific activity of this enzyme in the production strain was equivalent to that of the plasmidless *E. coli* host growing in minimal salts glucose medium. This further means that the RNA polymerase has access to the tryptophan promoter on the plasmid. Despite the fact that this gene was being transcribed, essentially no BST accumulated in the cell. (ii) When the lacZ gene replaced the BST gene, $\beta$-galactosidase was produced during growth in the minimal salts medium. $\beta$-Galactosidase is a very stable enzyme and indicated the level of transcriptional activity from the tryptophan promoter located on the plasmids. (iii) The addition of IAA resulted in marked increases in the rates of transcription initiations. This led to a further increase in the rate of protein synthesis and BST began to accumulate. The BST probably aggregated and precipitated in the cell because of the inherent properties of this protein. It is highly non-polar and insoluble at the intracellular pH of *E. coli*. (iv) The inclusion bodies remained visible in the cell for several hours after the cells were resuspended in a medium lacking IAA (Figure 5).

At the completion of a typical production fermentation, a one liter aliquot of the culture was transferred into 9 liters of fresh pre-warmed minimal salts glucose medium that lacked IAA. After transfer, the percentage of cells containing inclusion bodies remained relatively constant for 6 hours. Subsequently, the percentage of inclusion body-containing cells dropped and this drop coincided with the increase in cell growth as evidenced by an increase in the optical density of the culture. Second, the presence of the refractile bodies is ready evidence of the accumulation of the product (Figure 6). This provides an on-line analysis of the progress of the fermentation and can be used to quantitate the amount of product made during the production phase. Third, the isolation of the BST from the cells is made easier by the presence of these inclusion bodies. By simply disrupting the cells and centrifuging the cell slurry, the protein product can be isolated in a relatively high state of purity (at least 40%) by inexpensive and readily scalable procedures. This means that fewer isolation steps are required to obtain the product in a highly purified state.

## Engineering and Process Development

Once a functional genetic system was developed for the synthesis of BST, it was necessary to address key engineering and process development questions in order to make this process a commercial reality. Since tons of purified recombinant protein were required, the large scale process was developed in discrete steps in both the fermentation and purification areas. Initially the process required many steps with a low overall yield of product. The initial expression level from the fermentation was only 1 gram per liter, and the fermentation raw material costs were unacceptable for a cost- effective large-scale production. Furthermore, the number and type of purification steps which worked for a low quantity, high value product were not conducive to a high-volume low-cost process needed for the market position required of BST. Sterile processing methodology was an absolute necessity and represented a further impediment to achieving cost targets.

All phases of research, from microbial physiology through sterile processing, had to be adapted to these large scale needs while at the same time the required chemical and analytical abilities had to be developed. These requirements had to be met so that the material made for clinical trials and toxicological studies would match the material to be produced at the large scale. In other words, we had to have cGMP (Current Good Manufacturing Practices) which meant that the process had to be defined and reproducible.

**Fermentation Process Development.** As mentioned above, the original fermentation protocol produced less than a gram of product per liter of fermentation

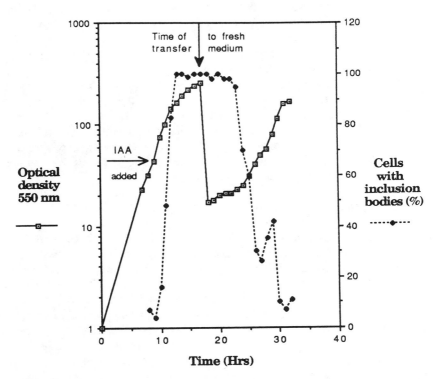

Figure 5. Persistence of inclusion bodies in recombinant cells after transfer to fresh medium. The host-vector for bovine somatotropin was grown in a 15 liter Biolafitte fermenter and product formation was initiated by the addition of IAA. After approximately 10 hours a 1 liter sample was transferred to fresh pre-warmed minimal salts glucose medium and maintained at 37 °C. Aliquots were removed at the indicated times and the optical density (at 500 nm) and presence of inclusion bodies were ascertained.

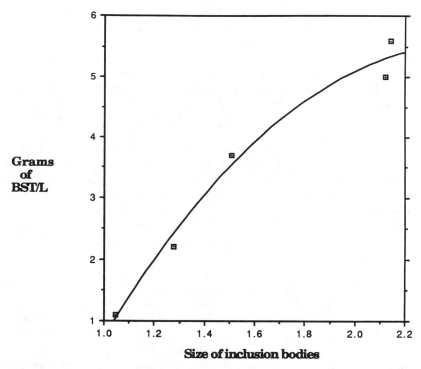

Figure 6. Relationship between bovine somatotropin produced and the size of the intracellular inclusion bodies. After the addition of IAA to a 15 liter fermentation vessel, samples were removed at 2 hour intervals. One sample was used to determine the concentration of bovine somatotropin by HPLC analysis. A second sample was examined with computer enhanced image analysis to measure the size of the inclusion bodies. The size was recorded in square microns.

broth. Furthermore, productivity was inconsistent. The cell density was low and the growth kinetics of the cells after induction could -not be reproduced run after run. A program to understand the physiology and growth requirements was successful in establishing a protocol that used a low cost chemically-defined production media to produce a high cell density prior to the addition of IAA. Product expression reached a level where at least 50% of the protein made after the addition of IAA was BST. As a result the fermentation costs were no longer a controlling part of the cost of goods. From an engineering perspective, however, a new problem emerged. *E. coli* required a highly aerobic environment for maximum productivity. This problem of course is exacerbated as the volume of the fermenter vessel increases.

Fermentation development progressed from the 15 liter to the 100 liter then to the 1500 liter size. Surprisingly good agreement was observed as the process was increased from the 15 liter scale. Continuous measurement of agitation power, heat transfer, and oxygen transfer rates permitted the design of a fermentation protocol that would be directly scalable to the larger plant-size fermenters. While additional work had indicated that even higher cell densities could be achieved, the cost of the increasing agitation power and heat transfer capabilities made this approach unattractive. Fermentation monitoring, both during and after the growth cycle, was developed to ensure compliance with the guidelines (Points to Consider) proposed by the FDA. A consistent fermentation protocol that produced a high level of expression of bovine somatotropin permitted the rapid development of a downstream purification process that used fewer processing steps and had a higher yield.

**Inclusion Body Isolation**. The bovine somatotropin was associated with insoluble aggregates called inclusion bodies. It was necessary to optimize the isolation of these bodies from the cell debris. The first step in this procedure was the disruption of the *E. coli* cells with an homogenizer. Since these particles were akin to grains of sand, special equipment was needed for this portion of the process. Cooperation between the process engineers and suppliers was developed so that designs such as the ceramic cell homogenization disruption valve could be constructed and tested on site. This interaction permitted the isolation of the maximum amount of product with the longest lasting, lowest maintenance equipment. As a result a consistent starting material could be generated for the next purification step.

**Refold Technology**. It was known early in the program that the BST protein in the refractile body was not folded in the correct configuration, and that the appropriate disulfide bonds were not formed in most of the protein monomers. A methodology had to be developed for first dissolving the protein from the refractile body in a suitable inexpensive solvent. Chaotropic agents such as guanidine HCl (21) could be used but the costs and disposal of such large amounts of this material were prohibitive. The dissolution of the particles with alternate chaotropic agents was not difficult but the refolding and oxidation to the native configuration was more doubtful. Technology was developed that allowed for these reactions to occur once the refractile bodies were solubilized.

**Purification.** The purification system evolved rapidly after the fermentation and refractile body isolation steps were optimized. This resulted in a preparation of BST at least 40% pure as the starting material for the purification area. These initial purification steps; that is, homogenization and centrifugation of the refractile bodies, permitted a reduction in the number and complexity of the subsequent operational steps. The scale-up from grams per batch to kilograms per batch quickly followed.

**Formulation.** The first product type was a daily dose formulation that gave an excellent lactation response in the treated animals (22). While this was a useful development tool, there was evidence that the market potential for such a product was too labor intensive to be commercially useful. Therefore, it was important to develop sustained release formulations and this was undertaken essentially during the development stages of the overall process. Many methodologies had promise but the oil suspension form was chosen because it offered an ease in the formulation processing coupled with simple administration by the user. Since the product was injected, the final processing steps, from the drying stage forward, had to occur under sterile conditions. Sterile rooms meeting cGMP conditions were constructed, validated and operated to produce the clinical and toxicological material.

**Conclusions**

All of the microbial genetics and analytical chemistry discussed herein were put in place to support the commercial process. The process parameters for the fermentation, purification and formulation were scaled up without significant difficulty to full size plant which was constructed in Austria. This overall process remains as a substantial achievement in the area of biotechnology, since no recombinant protein had been produced previously at this scale, under these operating conditions.

**Literature Cited**

1. Asimov, G. J.; Krouze, N. K. J. Dairy Sci. 1937, 20, 289-306.
2. Seeburg, P. H.; Sias, S.; Adelman, J.; de Boer, H. A.; Hayflick, J.; Jhurani, P.; Goeddel, D. V.; Heyneker, H. L. DNA 1983, 2, 37-45.
3. Marston, F. A. O. Biochem. J. 1986, 240, 1-12.
4. Kane, J. F.; Hartley, D. L. Trends in Biotech. 1988, 6, 95-101.
5. Krueger, J. K.; Kulke, M. H.; Schutt, C.; Stock, J. Biopharm. Manufact. 1989, 2 (Mar), 40-5.
6. Kane, J. F.; Bogosian, G. Biopharm. Manufact. 1987, 1 (Nov), 26-51.

7. Bolivar, F.; Rodriguez, R. L.; Greene, P. J.; Betlach, M. C.; Heyneker, H. L.; Boyer, H. W.; Crosa, J. H.; Falkow, S. Gene 1977, 2, 95-113.

8. Calcott, P. H.; Kane, J. F.; Krivi, G. G.; Bogosian, G. Dev. in Indust. Microbiol. 1988, 29, 257-66.

9. Bogosian, G.; Kane, J. F. Nucleic Acids Res. 1987, 15, 7185.

10. Somerville, R. L. Biotechnol. and Genetic Engineering Reviews 1988, 6, 1-41.

11. Yanofsky, C.; Crawford, I. P. In Escherichia coli and Salmonella typhimurium: Cellular and Molecular Biology; Ingraham, J. L.; Low, K. B.; Magasanik, B.; Schaechter, M.; Umbarger, H. E., Eds.; American Society for Microbiology; Washington, D.C, 1987; Vol. 2, p 1453-72.

12. Marmorstein, R. Q.; Joachimiak, A.; Sprinzl, M.; Sigler, P. G. J. Biol. Chem. 1987, 262, 4922-7.

13. Marmorstein, R. Q.; Sigler, P. B. J. Biol. Chem. 1989, 264, 9149-54.

14. Bogosian, G.; Violand, B. N.; Dorward-King, E. J.; Workman, W. E.; Jung, P. E.; Kane, J. F. J. Biol. Chem. 1989, 264, 531-9.

15. Kohlhaw, D. B.; Leary, T. R. In Metabolism of Amino Acids and Amines; Tabor, H.; Tabor, C. W., Eds.; Academic Press: New York; Methods in Enzymology Vol. 17A; 1970; p 771-7.

16. Cowie, D. B.; Cohen, G. N.; Bolton, E. T.; DeRobichon-Szulmajster, H. Biochim. Biophys. Acta 1959, 34, 39-46.

17. Violand, B. N.; Siegel N. R.; Bogosian, G.; Workman, W. E.; Kane, J. F. In Techniques in Protein Chemistry; Hugli, T. E., Ed.; Academic Press: New York; 1989; p 315-26.

18. Koide, H.; Yokoyama, S.; Kawai, G.; Ha, J.-M.; Oka, T.; Kawai, S.; Miyake, T.; Fuwa, T.; Miyazawa, T. Proc. Natl. Acad. Sci. USA 1988, 85, 6237-41.

19. Lu, H. S.; Tsai, L. B.; Kenney, W. C.; Lai, P.-H. Biochem. Biophys. Res. Commun. 1988, 156, 807-13.

20. Tsai, L. B.; Lu, H. S.; Kenney, W. C.; Curless, C. C.; Klein, M. L.; Lai, P.-H.; Fenton, D. M.; Altrock, B. W.; Mann, M. B. Biochem. Biophys. Res. Commun. 1988, 156, 733-9.

21. Langley, K. W.; Berg, T. F.; Strickland, T. W.; Fenton, D. M.; Boone, T. C.; Wypyck, J. Eur. J. Biochem. 1987, 163, 313-21.

22. Peel, C. J.; Bauman, D. E.; Gorewit, R. C.; Sniffen, C. J. J. Nutrition 1981, 111, 1662-71.

23. Miozzari, G. F.; Yanofsky, C. J. Bacteriol. 1978, 133, 1457-66.

24. Nichols, B. P.; Yanofsky, C. In Recombinant DNA, Part C; Wu, R.; Grossman, L.; Moldave, K., Eds.; Academic Press: New York; Methods in Enzymology Vol. 101; 1983; p 155-64.

25. De Boer, H. A.; Comstock, L. J.; Yansura, D. G.; Heyneker, H. L. In Promoters: Structure and Function; Rodriguez, R. L.; Chamberlin, M.J., Eds.; Preager: New York; 1982; p 462-81.

26. Bertrand, K. P.; Postle, K.; Wray, L. V. Jr.; Reznikoff, W. S. J. Bacteriol. 1984, 158, 910-9.

RECEIVED August 27, 1990

# Chapter 15

# Synthesis, Biological Activity, and Conformation of Cyclic Growth Hormone Releasing Factor Analogs

**Arthur M. Felix, Edgar P. Heimer, Thomas F. Mowles, David Fry, and Vincent Madison**

**Roche Research Center, Hoffmann–La Roche, Inc., Nutley, NJ 07110**

Growth hormone-releasing factor, GRF(1-44)-NH$_2$, has been shown to stimulate growth hormone release *in vitro* and *in vivo* in humans and a variety of animal species. Preliminary clinical studies have shown that GRF(1-44)-NH$_2$ can be used for the treatment of growth hormone deficient children. Structure-activity studies have shown that the shortened analog, GRF (1-29)-NH$_2$, retains the full intrinsic activity with only slightly reduced potency. Replacement analogs of GRF(1-29)-NH$_2$ have resulted in peptides with increased potency which are being used for performance enhancement applications in domestic livestock (pigs and lactating cows). Conformational analysis (molecular dynamics calculations based on NOE-derived distance constraints) has been carried out to correlate bioactivity with conformation. These studies have led to the design and synthesis of i-(i+4) cyclic analogs of GRF(1-29)-NH$_2$. A solid phase method for the synthesis of the cyclic peptides was developed using an orthogonal protection scheme. This included conventional N$\alpha$-Boc-benzyl protection and the incorporation of side-chain 9-fluorenylmethyl (-OFm) ester protection of Asp[8] and N$^\varepsilon$-Fmoc protection of Lys[12]. Following specific deprotection of Asp(OFm)[8] and Lys(Fmoc)[12] using piperidine, solid phase side chain to side chain lactamization was carried out most efficiently with the BOP reagent. The high biological activity of the resultant cyclo[8,12][Asp[8],Ala[15]]-GRF(1-29)-NH$_2$ system may be explained on the basis of retention of a preferred bioactive conformation which includes substantial $\alpha$-helicity in H$_2$O between residues 7-17 and 21-15. The $\alpha$-helical conformation of these cyclo[8,12] peptides was only minimally disrupted with either stereoisomeric or isosteric replacements and the corresponding peptides retained substantial biological

0097–6156/91/0444–0201$06.00/0

activity. The cyclic peptides were also more stable to dipeptidylpeptidase degradation than the linear analogs.

Peptides with growth hormone releasing activity were first isolated in 1982 from patients with pancreatic islet adenomas. The peptides were characterized independently by two groups (1,2). A total of three forms of homologous growth hormone releasing factor (GRF) peptides were isolated, found to contain 44, 40 and 37 amino acids and identified respectively as GRF(1-44)-NH₂, GRF(1-40)-OH and GRF(1-37)-OH. Of these, GRF(1-44)-NH₂ has been shown to be the primary structure on the basis of cloning and cDNA sequence analysis (3,4). GRF is produced in the hypothalamus and transported via the hypothalamic-hypophysial-portal system to the pituitary gland, where it binds to receptor sites on the somatotroph and stimulates the release and synthesis of growth hormone (GH). Growth hormone is released into the circulation and causes the production of somatomedin-C in target tissues which stimulates the events required for linear growth. GH release is modulated by somatostatin, which is responsible for the pulsatile manner in which GH is released, and by growth hormone and somatomedin-C feedback. Although there are primary structural differences of GRF among species, human GRF stimulates the secretion of GH in most animal species. The clinical utility of GRF for the treatment of growth hormone deficiency (of hypothalamic origin) has been reported (5,6) and the potential application of GRF analogs for performance enhancement and improved feed utilization in livestock has been established (7).

Structure-activity studies of GRF have shown that GRF(1-29)-NH₂ is the shortest fragment with full intrinsic activity and high *in vitro* and *in vivo* potency (8-11). Analogs of GRF(1-29)-NH₂ in which Gly[15] is replaced by Ala[15] have been shown to have increased potency (12,13) which may be related to enhanced α-helicity and maximization of amphiphilic structure. Conformational studies of [Ala[15]]-GRF(1-29)-NH₂ (circular dichroism and molecular dynamics calculations based on NOE-derived distance constraints) demonstrated the presence of helical segments (residues 9-15 and 21-26) in water (pH 3) (14). These observations prompted us to design a series of i-(i+4) cyclic analogs of [Ala[15]]-GRF(1-29)-NH₂ with side-chain to side-chain (lactamization) which can stabilize the α-helix in these regions and may lead to the retention of a preferred bioactive conformation.

## Methods

**Synthesis of Peptides.** The peptides were prepared by solid phase synthesis (15) using a Vega Model 1000 semi-automated synthesizer. The COOH-terminal residue, Boc-Arg(Tos)-OH, was linked to benzhydrylamine-resin (copolystyrene-1% divinylbenzene, 200-400 mesh) at a substitution level of 0.6 mequiv/g. The Boc/benzyl strategy was used for each coupling which was mediated with dicyclohexylcarbodiimide (DCC). Completion of coupling was determined by the qualitative ninhydrin method (16). In several cases double couplings were required to reach the end point. In special cases in which a third coupling was required, the hydroxybenzotriazole (HOBt) ester (17) and DCC were preactivated in DMF and diluted with toluene. The trifunctional amino acids were protected as follows: Boc-Arg(Tos)-OH, Boc-Asp(OcHex)-OH, Boc-Glu(OBzl)-OH, Boc-Lys(2ClZ)-OH, Boc-Ser(Bzl)-OH, Boc-Thr(Bzl)-OH and Boc Tyr(2,6-Cl₂Bzl)-OH. Boc-Gln-OH was coupled by the symmetric anhydride method in DMF. The cyclic peptides required the use of Boc-Asp(OFm)-OH (18) and Boc-Lys(Fmoc)-OH at the sites of lactamization. Following the selective removal of the side chain -OFm and Fmoc-protecting groups with 20% piperdine in DMF (20 min) solid phase cyclization (lactamization) was carried out by the procedure of Felix et al. (19) using benzotriazol-1-yl-oxy-tris-(dimethylamino) phosphonium hexafluorophosphate (BOP). Cleavage of peptide from resin was carried out by the single stage HF procedure using dithioethane as scavenger.

**Purification and Characterization of Peptides.** The peptides were purified by 2 stages of preparative high performance liquid chromatography (HPLC) on Synchropak RP-P and Nucleosil C₁₈ columns. Gradient elution was carried out with H₂0 (containing 0.025% - 0.1% TFA) and acetonitrile (containing 0.025% - 0.1% TFA). Fractions were collected at 1-minute intervals and evaluated by analytical HPLC using a Lichrosorb RP-8 column which was eluted with a linear gradient of 0.1 M NaClO₄ (pH 2.5)-acetonitrile. Appropriate fractions were pooled and lyophilized. The products were shown to be homogeneous (97.5%) by analytical HPLC and characterized by amino acid analysis (6M HCl, 110°, 72h), fast atom bombardment mass spectroscopy and sequence analysis. Circular dichroism (CD) spectra were recorded on a Jasco J-500A Spectropolarimeter at room temperature in 75% methanol (pH 6) and in water at pH 3. Two-dimensional n.m.r. studies were performed on Varian XL-400 and VXR-500 spectrometers in 75% CD₃OH/25% H₂0 at pH 6 and in 90% H₂0/10% D₂0 at pH 3 at 22.5°C. The CHARMM program (20) was used for constrained energy minimization and molecular dynamics optimization of the peptide structures.

**Biological Studies.** *In vitro* biological studies were carried out using a rat pituitary cell culture bioassay according to the procedure of Brazeau *et al.* (21). GRF and analogs were incubated for 4h at 37° and treatment media assayed by radioimmunoassay. *In vivo* studies were performed by subcutaneous administration in pigs using a 5 x 5 Latin square design as previously reported (22). Evaluation of potency was determined by GH area under the curve analysis.

**Results and Discussion**

Molecular dynamics calculations (based on NOE-derived distance constraints) in 75% methanol reveal that GRF(1-29)-NH₂ has a nearly ideal α-helical geometry between residues 4-29 and that it possesses stretches of α-helicity between residues 9-14 and 24-28 in water (23). Conformational studies of the potent analog, [Ala¹⁵]-

GRF(1-29)-$NH_2$, confirmed the enhancement of secondary structure predicted by Chou-Fasman calculations (24). Molecular dynamics calculations for [Ala$^{15}$]-GRF(1-29)-$NH_2$ in 75% methanol reveal an extended $\alpha$-helix between residues 3-29 and $\alpha$-helicity between residues 9-15 and 21-26 in water.

We have recently observed that cyclization (lactamization) of Asp$^8$ to Lys$^{12}$ in GRF results in a family of analogs which retain high *in vitro* (Table I) and *in vivo* biological activity (14).

Table I. Cyclic$^{8,12}$ Analogs of GRF(1-29)-$NH_2$

| GRF Analog | Relative Potency |
|---|---|
| GRF(1-44)-$NH_2$ | 1.00 |
| GRF(1-29)-$NH_2$ | 0.80 |
| [Ala$^{15}$]-GRF(1-29)-$NH_2$ | 4.10 |
| [D-Ala$^2$,Ala$^{15}$]-GRF(1-29)-$NH_2$ | 4.50 |
| [desNH$_2$Tyr$^1$,D-Ala$^2$,Ala$^{15}$]-GRF(1-29)-$NH_2$ | 4.59 |
| [N-MeTyr$^1$,D-Ala$^2$, Ala$^{15}$]-GRF(1-29)-$NH_2$ | 5.93 |
| Cyclo-(Asp$^8$-Lys$^{12}$)-[Ala$^{15}$]-GRF(1-29)-$NH_2$ | 0.77 |
| Cyclo-(Asp$^8$-Lys$^{12}$)-[D-Ala$^2$,Ala$^{15}$]-GRF(1-29)-$NH_2$ | 2.37 |
| Cyclo-(Asp$^8$-Lys$^{12}$)-[desNH$_2$Tyr$^1$,D-Ala$^2$,Ala$^{15}$]-GRF(1-29)-$NH_2$ | 2.47 |
| Cyclo-(Asp$^8$-Lys$^{12}$)-[N-MeTyr$^1$,D-Ala$^2$,Ala$^{15}$]-GRF(1-29)-$NH_2$ | 3.57 |
| Cyclo-(Lys$^8$-Asp$^{12}$)[D-Ala$^2$,Ala$^{15}$]-GRF(1-29)-$NH_2$ | 2.90 |
| Cyclo-(D-Asp$^8$-Lys$^{12}$)[D-Ala$^2$,Ala$^{15}$]-GRF(1-29)-$NH_2$ | 1.79 |
| Cyclo-(Asp$^8$-D-Lys$^{12}$)[D-Ala$^2$,Ala$^{15}$]-GRF(1-29)-$NH_2$ | 3.82 |
| Cyclo-(D-Asp$^8$-D-Lys$^{12}$)[D- Ala$^2$,Ala$^{15}$]-GRF(1-29)-$NH_2$ | 1.12 |

The solid phase side-chain to side-chain lactamization (19) was carried out efficiently using the Boc/benzyl strategy, selective deprotection of Asp$^8$(OFm) and Lys$^{12}$(Fmoc) with piperidine, and cyclization using BOP reagent (25) as outlined in Figure 1. Molecular modeling of the resultant cyclo (Asp$^8$-Lys$^{12}$)-GRF analogs revealed compatibility of the side-chain to side-chain i-(i+4) lactam with backbone $\alpha$-helical conformation. These predictions were in good agreement with conformational analysis using circular dichroism and 2-dimensional n.m.r. studies. Furthermore, molecular dynamics calculations of the parent cyclic analog, cyclo-(Asp$^8$-Lys$^{12}$) [Asp$^8$,Ala$^{15}$]-GRF(1-29)-$NH_2$, in 75% methanol revealed extensive $\alpha$-helicity between residues 2-29. In water the cyclic peptide also had more helical character than the linear analog since residues 7-17 and 21-25 retain $\alpha$-helicity. As observed for the corresponding linear peptide, [Ala$^{15}$]-GRF(1-29)-$NH_2$, multiple replacements of residues on cyclo (Asp$^8$-Lys$^{12}$)-GRF with D-Ala$^2$,desNH$_2$-Tyr$^1$ or N-MeTyr$^1$ resulted in more potent analogs (Table I).

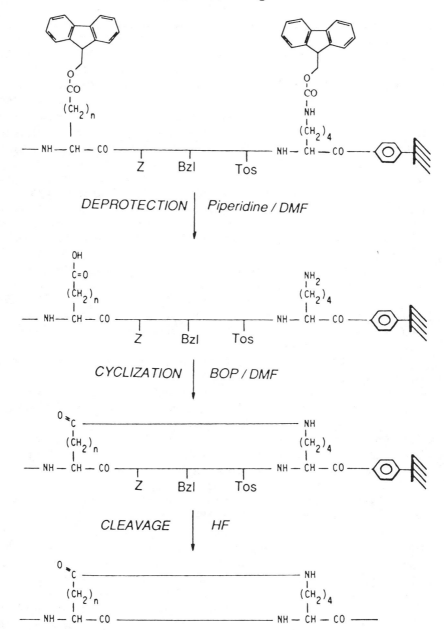

Figure 1. Schematic representation for the Boc/benzyl solid phase synthesis with selective deprotection for the general synthesis of cyclic (lactam) peptides.

Molecular modeling of the stereoisomeric replacement analogs of $Asp^8$ or $Lys^{12}$ with $D-Asp^8$ or $D-Lys^{12}$ in $cyclo^{8,12}[D-Ala^2,Ala^{15}]-GRF(1-29)-NH_2$ predicts only minimal disruption of the $\alpha$-helical conformation since the i-(i+4) lactam ring overcomes any potential loss of helicity imposed by the $D-Asp^8$ or $D-Lys^{12}$ residues. These predictions were confirmed by circular dichroism spectra of the stereoisomers which were nearly superimposable with that of the parent cyclic GRF analog (Figure 2). The $cyclo-(Asp^8-Lys^{12})-[D-Ala^2,Ala^{15}]-GRF(1-29)-NH_2$ stereoisomeric replacement analogs demonstrated relative potencies that were nearly comparable to the parent peptide (Table I). These observations support our conclusion that the biological activity of GRF may be explained in part on the basis of retention of a preferred bioactive conformation which includes $\alpha$-helicity in the region between residues 7-17. The lactamization of $Asp^8$ to $Lys^{12}$ is expected to decrease the hydrophilic character of both residues and enhance the overall amphiphilic character which may play an important role in receptor-ligand interaction. At least two distinct regions of hydrophobic character are predicted as shown in Figure 3.

Molecular modeling of the isosteric analog, $cyclo-(Lys^8-Asp^{12})-[D-Ala^2,Ala^{15}]-GRF(1-29)-NH_2$, also predicts minimal disruption of the $\alpha$-helical conformation. The circular dichroism spectrum of $cyclo-(Lys^8-Asp^{12})-[D-Ala^2,Ala^{15}]-GRF(1-29)-NH_2$ was also nearly superimposable with $cyclo-(Asp^8-Lys^{12})-[D-Ala^2,Ala^{15}]-GRF(1-29)-NH_2$. Our observation of nearly identical biological activity of cyclo $(Lys^8-Asp^{12})-[D-Ala^2,Ala^{15}]-GRF(1-29)-NH_2$ and $cyclo-(Asp^8-Lys^{12})-[D-Ala^2,Ala^{15}]-GRF(1-29)-NH_2$ (Table I) also supports our conclusion that a preferred bioactive conformation retains $\alpha$-helicity by virtue of the i-(i+4) lactam ring.

GRF has been reported to undergo initial enzymatic degradation by dipeptidylpeptidase which results in the conversion of $GRF(1-44)-NH_2$ to the biologically inactive form, $GRF(3-44)-NH_2$ (26). Replacement analogs of GRF have been evaluated for their *in vitro* plasma stability. $[Ala^{15}]-GRF(1-29)-NH_2$ demonstrates increased plasma stability (Table II). Further enhancement of plasma stability was observed for the N-terminal replacement analogs since they are poor substrates for dipeptidylpeptidase cleaves for the $Tyr^1-Ala^2$-sequence (25). Cyclic peptides have been reported to have enhanced stability and resistance to enzymatic degradation. Cyclo $(Asp^8-Lys^{12})-[Ala^{15}]-GRF(1-29)-NH_2$ was more resistant to *in vitro* plasma degradation than any of the linear GRF analogs. It has been reported that enzymatic degradation in plasma occurs primarily by dipeptidyl-peptidase (Type IV) and to a lesser extent by trypsin-like enzymes (27). Maximum stabilization to enzymatic degradation was observed when both cyclization and N-terminal substitution were incorporated in GRF. Therefore $cyclo-(Asp^8-Lys^{12})-[D-Ala^2,Ala^{15}]-GRF(1-29)-NH_2$ underwent only 7% degradation in human plasma after 60 min at 37°C, compared to $GRF(1-44)-NH_2$ which undergoes 92% degradation under these conditions.

The novel cyclic GRF analogs have been used to gain insights into the bioactive conformation of GRF and stabilization to enzymatic degradation. This information

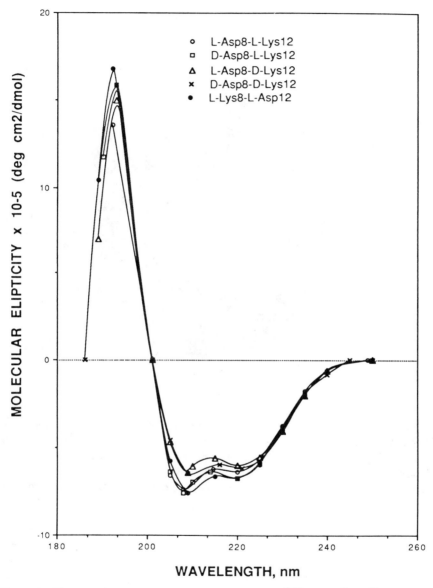

Figure 2. Circular dichroism of stereoisomers of cyclo$^{8,12}$-[D-Ala$^2$,Ala$^{15}$]-GRF(1-29)-NH$_2$ in 75% methanol/phosphate buffer (pH 6). Concentration 2 x 10$^{-5}$M.

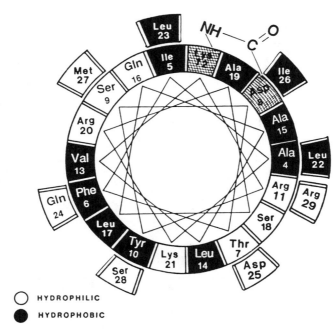

Figure 3. Schematic representation of the amphiphilic character of cyclo-(Asp[8]-Lys[12])-[Ala[15]]-GRF(1-29)-NH₂. The region between residues 4-29 which was determined to be essentially α-helical (75% methanol) is shown. Hydrophobic residues are shown in black, hydrophilic residues are shown in white. Asp[8] and Lys[12] are shown as intermediate since the side-chains are altered (lactamized).

is expected to be useful in designing additional analogs of GRF with increased potency and prolonged duration of biological activity.

Table II. *In vitro* Plasma Stability of GRF and Analogs

| Peptide | % Degraded[a] |
|---|---|
| GRF(1-44)-NH$_2$ | 92% |
| GRF(1-29)-NH$_2$ | 94% |
| [Ala$^{15}$]-GRF(1-29)-NH$_2$ | 65% |
| [D-Ala$^2$,Ala$^{15}$]-GRF(1-29)-NH$_2$ | 46% |
| [desNH$_2$Tyr$^1$]-GRF(1-44)-NH$_2$ | 44% |
| [desNH$_2$Tyr$^1$]-GRF(1-29)-NH$_2$ | 39% |
| [desNH$_2$Tyr$^1$,D-Ala$^2$,Ala$^{15}$]-GRF(1-29)-NH$_2$ | 28% |
| Cyclo-(Asp$^8$-Lys$^{12}$)-[Ala$^{15}$]-GRF(1-29)-NH$_2$ | 12% |
| Cyclo-(Asp$^8$-Lys$^{12}$)-[D-Ala$^2$,Ala$^{15}$]-GRF(1-29)-NH$_2$ | 7% |

[a]Incubation in human plasma for 60 min at 37°C

## Acknowledgments

The authors acknowledge Professor Lawrence Frohman (University of Cincinnati Medical Center) for carrying out the dipeptidylpeptidase stability studies on the GRF analogs. We also thank Dr. V. Toome for the circular dichroism studies, Dr. F. Scheidl and his staff for the amino acid analyses, Dr. W. Benz for the mass spectrometry and Dr. Y.-C. Pan for sequence analyses. We also thank Dr. R. Campbell for the *in vitro* bioassays.

## Literature Cited

1. Rivier, J.; Spiess, J.; Thorner, M.; Vale, W. Nature 1982, 300, 276-8.
2. Guillemin, R.; Brazeau, P.; Böhlen, P.; Esch, F.; Ling, N.; Wehrenberg, W. B. Science 1982, 218, 585-7.
3. Gubler, U.; Monahan, J. J.; Lomedico, P. T.; Bhatt, R. S.; Collier, K. J.; Hoffman, P. J.; Böhlen, P.; Esch, F.; Ling, N.; Zeytin, F.; Brazeau, P.; Poonian, M. S.; Gage, L. P. Proc. Natl. Acad. Sci. USA, 1983, 80, 4311-4.
4. Mayo, K. E.; Vale, W.; Rivier, J.; Rosenfeld, M. G.; Evans, R. M. Nature 1983, 306, 86-8.
5. Gelato, M.; Ross, J.; Pescovitz, O.; Cassorla, F.; Skeeda, M.; Merriam, G. R. Pediat. Res. 1984, 18, 167A.
6. Thorner, M. O.; Reschke, J.; Chitwood, J.; Rogol, A. D.; Furlanetto, R.; Rivier, J.; Vale, W.; Blizzard, R. M. New Engl. J. Med. 1985, 312, 4-9.

7. Peticlerc, D.; Pelletier, G.; Lapierre, H.; Gaudreau, P.; Couture, Y.; Dubreuil, P.; Morisset, J.; Brazeau, P. J. Anim. Sci. 1987, 65, 996-1005.

8. Lance, V. A.; Murphy, W. A.; Sueiras-Diaz, J.; Coy, D. H. Biochem. Biophys. Res. Commun. 1984, 119, 265 72.

9. Ling, N.; Baird, A.; Wehrenberg, W. B.; Ueno, N.; Munegumi, T.; Chiang, T.-C.; Regno, M.; Brazeau, P. Biochem. Biophys. Res. Commun. 1984, 122, 304-10.

10. Ling, N.; Baird, A.; Wehrenberg, W. B.; Ueno, N.; Munegumi, T.; Brazeau, P. Biochem. Biophys. Res. Commun. 1984, 123, 854-61.

11. Grossman, A.; Savage, M. O.; Lytras, N.; Preec, M. A.; Suerias-Diaz, J.; Coy, D. H.; Rees, L. H.; Besser, G. M. Clin. Endocrinol. 1984, 21, 321-30.

12. Felix, A. M.; Heimer, E. P.; Mowles, T. F.; Eisenbeis, H.; Leung, P.; Lambros, T.; Ahmad, M.; Wang, C.-T.; Brazeau, P. In Proceedings of the 19th European Peptide Symposium; Theodoropoulos, D., Ed.; Walter de Gruyter: Berlin, 1987; pp 481-4.

13. Heimer, E. P.; Ahmad, M.; Lambros, T.; McGarty, T.; Wang, C.-T.; Mowles, T.; Maines, S.; Felix, A. M. In UCLA Symposium on Molecular and Cellular Biology; Tam, J.; Kaiser, E. T., Eds.; Alan R. Liss, Inc.: New York, 1988; New Series, Vol. 86, pp 309-19.

14. Felix, A. M.; Heimer, E. P.; Wang, C.-T.; Lambros, T. J.; Fournier, A.; Mowles, T. F.; Maines, S.; Campbell, R. M.; Wegrzynski, B. B.; Toome, V.; Fry, D.; Madison, V. S. Int. J. Pept. Prot. Res. 1988, 32, 441-54.

15. Barany, G.; Merrifield, R. B. In The Peptides: Analysis, Synthesis, Biology; Gross, E.; Meienhofer, J., Eds.; Academic Press: New York, 1980; Vol. 2, pp 1-284.

16. Kaiser, E.; Colescott, R. L.; Bossinger, C. D.; Cook, P. I. Anal. Biochem. 1970, 34, 595-8.

17. König, W.; Geiger, R. Chem. Ber. 1970, 103, 788-98.

18. Bolin, D.; Wang, C.-T.; Felix, A. M. Org. Prep. Proc. Int. 1989, 21, 67-74.

19. Felix, A. M.; Wang, C.-T.; Heimer, E.; Fournier, A.; Bolin, D. R.; Ahmad, M.; Lambros, T.; Mowles, T.; Miller, L. In Proceedings of the 10th American Peptide Symposium; Marshall, G., Ed.; ESCOM: Leiden, 1988; pp 465-7.

20. Brooks, B. R.; Bruccoleri, R. E.; Olafson, B. D.; States, D. J.; Swaminathan, S.; Karplus, M. J. Comput. Chem. 1983, 4, 187-217.

21. Brazeau, P.; Ling, N.; Böhlen, P.; Esch, F.; Ying, S.-Y.; Guillemin, R. Proc. Natl. Acad. Sci. USA 1982, 79, 7909-13.

22. Campbell, R.; Su, C.-M.; Stricker, P.; Jensen, L.; Heimer, E.; Felix, A.M.; Mowles, T. Amer. Dairy Assoc. & Amer. Soc. Anim. Sci. Joint Meeting, Lexington, KY; Aug., 1989.

23. Madison, V.; Berkovitch-Yellin, Z.; Fry, D.; Greeley, D.; Toome, V. In UCLA Symposium on Molecular and Cellular Biology; Tam, J.; Kaiser, E. T., Eds.; Alan R. Liss, Inc.: New York, 1988; New Series, Vol. 86, pp 109-23.
24. Chou, P. Y.; Fasman, G. D. Ann. Rev. Biochem. 1978, 47, 251-76.
25. Castro, B.; Dormay, J. R.; Evin, G.; Selve, C. Tetrahedron Lett. 1975, 1219-22.
26. Frohman, L. A.; Downs, T. R.; Williams, T. C.; Heimer, E. P.; Pan, Y.-C.; Felix, A. M. J. Clin. Inv. 1986, 78, 906-13.
27. Frohman, L.A.; Downs, T.R.; Heimer, E.P.; Felix, A.M. J. Clin. Inv. 1988, 83, 1533-40.

RECEIVED August 27, 1990

# Chapter 16

# Tumor-Secreted Protein Associated with Human Hypercalcemia of Malignancy

## Biology and Molecular Biology

**Michael P. Caulfield[1], Roberta L. McKee[1], John E. Fisher[1],
Noboru Horiuchi[2], David D. Thompson[1], Sevgi B. Rodan[1],
Mark A. Thiede[1], Jay J. Levy[1], Ruth F. Nutt[1], Thomas L. Clemens[2],
and Gideon A. Rodan[1]**

[1]Parathyroid Hormone Laboratory and Department of Bone Biology and
Osteoporosis, Merck Sharp and Dohme Research Laboratories,
West Point, PA 19486
[2]Bone Center, Cedars–Sinai Medical Center, Los Angeles, CA 90048

The tumor secreted factor associated with humoral hypercalcemia of malignancy (HHM) has recently been purified and cloned. This factor has limited homology at its amino terminus to parathyroid hormone (PTH) and has been termed PTH-related protein (PTHrP). By analogy to PTH it was reasoned that the biological activity of PTHrP would reside within the 1-34 region. We have chemically synthesized this peptide and evaluated it biologically *in vitro* and *in vivo* and found it to be comparable to PTH in its biological responses. *In vivo* it can produce all the symptoms of HHM. The finding that two peptides with substantially different amino acid sequences work through the same receptor permits a new approach in the design of more potent antagonists of PTH and PTHrP which may be of therapeutic utility in the treatment of HHM and PTH-related disorders.

The paraneoplastic syndrome of humoral hypercalcemia of malignancy (HHM) is associated with certain types of non-metastatic tumors. Recently a tumor-secreted factor thought to be responsible for this syndrome was purified and cloned. From the deduced amino acid sequence it was found to have limited homology to parathyroid hormone (PTH) in its amino terminus where eight of the first thirteen amino acids are identical. Because the 1-34 fragment of PTH retains all the biological activity of the parent hormone, and with consideration of the similarities between the syndrome of HHM and hyperparathyroidism, the amino terminal 1-34 peptide of the hypercalcemia factor (PTH-related protein, PTHrP) was synthesized and the biological properties compared to bovine and human PTH. *In vitro*, at the major targets for PTH, bone and kidney, PTH and PTHrP were found to have similar properties and both peptides appear to work through the PTH receptor. *In vivo*, PTHrP(1-34)NH$_2$ exhibited all of the properties associated with hypercalcemia of malignancy: increased serum calcium, increased serum vitamin D levels, decreased serum phosphate, and increased urinary excretion of phosphate and cAMP. With respect to all of these parameters PTHrP(1-34)NH$_2$ was shown to be equivalent to or more potent than bPTH(1-84). PTHrP(1-34)NH$_2$ was shown to be equipotent with bPTH(1-34) in its ability to resorb bone. Actions of PTHrP(1-34)NH$_2$ could be inhibited by the PTH antagonist, [Tyr$^{34}$]bPTH(7-34)NH$_2$, indicating that PTHrP exerts its effects through the PTH receptor. The finding that two peptides with different amino acid sequences in the receptor binding domain interact with the same receptor and with equal affinities suggests a new approach to the design of more potent PTH/PTHrP antagonists which may be of utility in the treatment of HHM.

Parathyroid hormone (PTH) is the principal mammalian hormone involved with regulation of blood calcium levels. It is secreted from the parathyroid gland in response to decreases in circulating calcium levels and exerts its effects on bone and kidney.

In bone, PTH interacts with its receptors on bone-forming osteoblasts and through some presently unknown cellular mechanism(s) brings about the recruitment of multinucleated bone resorbing osteoclasts. Among suggested mechanisms are the release of cytokines or collagenase from the osteoblasts (1,2) and osteoblast retraction after stimulation with PTH to allow access of osteoclasts to bone (3). Regardless of how it occurs, the net effect of PTH on bone cells is to initiate bone resorption which, in turn, results in release of calcium to the extracellular fluid.

In kidney, PTH exerts its effects in two ways; firstly, it causes an increase in the reabsorption of calcium by the proximal tubules concomitant with an increase in the excretion of phosphate and cAMP; secondly, it stimulates the vitamin D 1, alpha hydroxylase which increases the amount of the active form of vitamin D (1,25 dihydroxy vitamin D$_3$), the role of which is to stimulate uptake of dietary calcium by the gut.

### Identification of the Hypercalcemia Factor

In hyperparathyroidism the secretion of excess PTH from the parathyroid gland, due to adenoma or hyperplasia, results in increased serum calcium levels and various symptoms associated with this condition. In humoral hypercalcemia of malignancy (HHM) similar high serum calcium levels are seen but with normal parathyroid gland function. The similarity of these disorders led to speculation that the latter disease state might arise from the ectopic production of PTH by these tumors (4). It has subsequently been shown from studies employing molecular biological techniques examining expression of PTH mRNA by these tumors (5) and by the use of antibodies raised against PTH (6,7) that this is not the case. However, these tumors have been shown to secrete a factor(s) possessing activity that can be inhibited *in vitro* by PTH antagonists (7-9); this has suggested that the factor, although not PTH, does interact with PTH receptors on target tissues (bone and kidney) to mimic the response of PTH.

The use of PTH receptor responses in assays for the hypercalcemic factor has enabled a number of groups to purify and subsequently clone this factor (6,10-12). The deduced amino acid sequence includes an N-terminal signal sequence and pro-region, similar to PTH, and a mature protein of 141 amino acids, compared with 84 amino acids for PTH. The N-terminus has limited homology to PTH, eight of the first thirteen amino acids are identical, however, beyond residue 21, there is virtually no sequence similarity (10).

Subsequent cloning from other tumors showed that the C-terminal region could vary in length resulting in three different mature peptides of 139, 141 and 173 amino acids (12). The factor is encoded by a single gene in man and collectively it has been shown that the multiplicity of PTHrP mRNAs results from alternative RNA splicing mechanisms (12,13). In addition, the 3' untranslated regions contain AU rich motifs which are common to messenger RNAs encoding a variety of cytokines and protooncogenes such as c-myc (12).

The phenomenon of two different peptides being recognized by the same receptor is well established, as in the case of insulin and insulin-like growth factors I and II (14). In these examples, however, affinities of the receptors for their respective ligands is higher than their affinity for the ligands with which they cross-react. From initial studies, using semi-purified material, it has appeared that PTHrP has an affinity for the PTH receptor which is similar to that of PTH itself (7-9).

### In vitro Properties of Synthetic PTHrP(1-34)NH$_2$

The biological activity of the intact 1-84 PTH molecule has been shown to reside within the amino terminal 1-34 region (15). To determine whether PTHrP had a structure-function profile similar to that of PTH we synthesized the amino terminal 1-34 region of PTHrP and compared it directly with PTH(1-34). At issue were three questions: firstly, would this peptide be biologically active?; secondly, would it have a similar biological profile to PTH?; thirdly, could the activity of this peptide account for all of the biological characteristics of HHM without the necessity of invoking the involvement of alternative factors, as had been proposed (16).

Direct comparisons of the PTH and PTHrP peptides were initially performed using bovine renal cortical membranes and rat osteosarcoma (ROS 17/2.8) cells.

The two 1-34 peptides of bovine or human PTH and PTHrP were very similar in their ability to inhibit PTH binding having low nanomolar potencies in the bone ($K_b$ 9±1 and 6±1 nM for hPTH(1-34) and PTHrP(1-34)$NH_2$, respectively) and kidney assays ($K_b$ 13±3, 0.8±0.2 and 1.2±0.2 nM for PTHrP(1- 34)$NH_2$ and bovine and human PTH(1-34), respectively) (17). In stimulating adenylate cyclase activity the two peptides were again similar in rat bone cells ($K_m$ 1.0±0.1, 1.1±0.1 and 2.0±0.5 nM for PTHrP(1- 34)$NH_2$ and bovine and human PTH(1-34), respectively) (17). As with binding, PTHrP was slightly less potent than PTH in adenylate cyclase stimulation with bovine renal membranes ($K_m$ 37±4, 1.8±0.1 and 5.4±0.2 nM for PTHrP(1-34) and bovine and human PTH(1-34), respectively) (17). Similar results to these have been reported by others (18, additional references can be found in reference 19).

As in the binding to the PTH receptor and stimulation of adneylate cyclase, the post receptor effects of the two peptides in ROS 17/2.8 cells were virtually identical (20). Both peptides, at similar doses, caused increased cAMP production in dexamethasone treated cells and had equal inhibitory effects on cell growth in the presence of dexamethasone. Both peptides also decreased alkaline phosphatase activity in the absence and in the presence of dexamethasone and, at higher concentrations of peptide, reduced the levels of alkaline phosphatase mRNA.

## In vivo Properties of Synthetic PTHrP(1-34)$NH_2$

*In vivo* the peptides were compared in a phosphaturic assay and a bone-dependent assay (17). In the phosphaturic assay renal parameters are measured in rats that have been thyroparathyroidectomized (TPTX'd) and infused with a nutrient solution. Administration of the nutrient solution and peptides is accomplished via femoral vein cannulae. In this assay PTHrP(1-34)$NH_2$ caused a dose-dependent (infusion of between 10 and 500 pmoles/hr) increase in the urinary excretion of phosphate, elevating levels from 0.5 to 8.0 mg $P_i$/mg creatinine and increasing cAMP excretion from 5 to 100 nmol/mg creatinine. In this assay PTHrP(1-34)$NH_2$ was approximately 6-10 times more potent than bPTH(1-84). In effects on serum parameters PTHrP(1-34)$NH_2$, over the same dose range, was again more potent than bPTH(1-84) in increasing ionized calcium levels (from 1.0 to 2.4 mM), decreasing phosphate levels (from 11 to 7.5 mg%) and elevating vitamin $D_3$ levels (from approximately 40 to over 200 pg/ml). In this last aspect PTHrP(1-34)$NH_2$ was significantly more potent than bPTH(1-84). The vitamin D result is of interest because it has been reported that in many patients with HHM the levels of vitamin $D_3$ are reduced (21), yet this peptide is a potent stimulator of the renal vitamin D 1,alpha hydroxylase activity. Interestingly, not all patients with HHM show decreased vitamin $D_3$ levels (22,23) and, quite possibly, the variability in levels detected is the result of compounding factors associated with the disease, such as renal damage. The discrepant potencies of PTHrP(1-34)$NH_2$

and bPTH(1-84) may be explained by different rates of clearance of the two peptides; perhaps the intact PTH molecule is preferentially cleared.

In a bone-dependent assay in which rats were TPTX'd and nephrectomized ensuring that only calcium released from bone could cause a rise in serum calcium, PTHrP(1-34)NH₂ caused a dose-dependent increase in serum calcium (17), raising levels by 15% over basal in a dose range of 0.1 to 1.0 nmole/hr; bPTH(1-34) gave an equivalent increase at the 1.0 nmole/hr infusion rate. Further comparison of the actions of bPTH(1-34) and PTHrP(1-34)NH₂ on bone have been performed (24). In these studies TPTX'd rats were maintained on low calcium diets and the peptides were infused over extended periods of time (up to 48 hr). Due to the TPTX and low calcium diet the serum calcium levels were decreased before infusion of agonist to about 5.5 mg%. Under these conditions bPTH(1-34) and PTHrP(1-34)NH₂ were found to be equipotent, producing a rise in serum calcium of over 5 mg% within 24 hr with doses of 0.06 nmol/hr and above. Increased calcium levels were delayed by 24 h at 0.03 nmol/h and not detected by 48 h at 0.01 nmol/h. A parallel dose-response effect was seen in the decrease of serum phosphate levels. Quantitative histomorphometric examination of the bones indicated that the two peptides had identical activities, causing concentration-dependent increases in the number of trabecular bone-associated osteoclasts (24).

Similar studies have been performed by other groups using the same or longer fragments of PTHrP. In all cases the findings are in good agreement, i.e., there is little or no difference between the actions and potencies of PTH and PTHrP (18, for additional references see 19).

From these data it would appear that PTHrP(1-34)NH₂ is capable of mimicking all effects of PTH and producing all of the symptoms associated with HHM. This is also supported by the experiments of Kukreja and co- workers (25) who showed that the hypercalcemia observed in nude mice bearing xenographs of tumors from patients with HHM could have the high serum calcium levels reduced transiently, to near normal levels, by the infusion of neutralizing antibodies raised against PTHrP.

### Inhibition of PTHrP(1-34)NH₂ by PTH and PTHrP Antagonists

The similar affinities of the PTH receptor for PTH and PTHrP, despite their limited sequence homology, suggests that the binding domain recognized by the receptor may not be the primary amino acid sequence but, rather, the tertiary structure of the peptides. This idea is supported by the observation that the receptor binding domain of PTH lies within the 25-34 region (26), a region with little homology to PTHrP (17).

The binding region of these peptides is of interest in the design of antagonists of PTH/PTHrP which may be of clinical utility in the treatment of hyperparathyroidism and HHM. Previous structure-function studies have elucidated the minimum requirements for a competitive antagonist of PTH(1-34). Initial studies indicated that truncation of the 1-34 peptide to 3-34 resulted in loss of activation of adenylate cyclase with retention of receptor binding (27). *In vivo*, however, this peptide was subsequently found to display weak agonism (28,29). Further truncation to the 7-34 peptide resulted in an antagonist which competitively occupies the PTH receptor and is active both *in vitro* and *in vivo* (30,31).

$[Tyr^{34}]$bPTH(7-34)NH$_2$ blocks the PTH- and PTHrP- stimulated cAMP production in ROS 17/2.8 cells and in bovine renal cortical membranes (BRCM): versus PTH agonist $K_i$(ROS) = 6300±1800 nM and $K_i$(BRCM) = 840±65 nM versus PTHrP agonist $K_i$(ROS) = 5530±130 nM and $K_i$(BRCM) = 430±170 nM (32).

The 7-34 sequence of PTHrP was also synthesized and analyzed for its ability to antagonize PTH and PTHrP effects. As with $[Tyr^{34}]$bPTH(7-34)NH$_2$, this peptide blocked cAMP production but in a similar dose range in kidney and in bone; versus PTH agonist $K_i$(ROS) = 790±220 nM and $K_i$(BRCM) = 609±133 nM and versus PTHrP agonist $K_i$(ROS) = 470±90 nM and $K_i$(BRCM) = 410±50 nM (32). PTHrP(7-34)NH$_2$, however, unlike its PTH counterpart was found to be a partial agonist in the rat bone cell assay, although only stimulating cAMP production to a level 5% of that seen with a full agonist (32). No partial agonism for this peptide was seen with bovine renal membranes. These observations lend credence to the hypothesis that PTH and PTHrP interact with the same receptor so that the different sequences present in the 7-34 regions have different effects on the activation of cAMP production via this receptor.

To further demonstrate that the PTHrP(1-34)NH$_2$ works through the PTH receptor, $[Tyr^{34}]$bPTH(7-34)NH$_2$ was used to antagonize the hypercalcemic response of PTHrP(1-34)NH$_2$ seen in TPTX'd rats.

*In vivo* responses to bPTH(1-84) have previously been shown to be blocked by a 200-fold molar excess of the PTH antagonist (31). Using the same molar excess $[Tyr^{34}]$bPTH(7-34)NH$_2$ was shown to block the phosphaturia and urinary cAMP excretion induced by PTHrP(1-34)NH$_2$ (33). In addition, this dose of antagonist inhibited increases in serum calcium seen in longer-term agonist infusions (16 hr) using a similar model (33). Although PTHrP(7- 34)NH$_2$ infused into TPTX'd rats was found to be a weak agonist, confirming *in vitro* results for cAMP production in ROS cells, this compound blocked the effects of PTHrP(1-34)NH$_2$ at doses below those required for the PTH antagonist (34). PTHrP(7-34)NH$_2$ has also been shown to block a serum calcium increase associated with infusion of PTHrP(1-34) in TPTX'd rats (35).

From the results presented above it is clear that PTHrP by itself can cause hypercalcemia of malignancy, apparently by exerting its effects through the PTH receptor present in bone and in kidney. Similarity in potencies and the use of a common receptor might well have evolved to allow for PTHrP to exert its physiological effects, which presumably involve the movement of calcium, through a pre-existing hormone receptor system. Conceivably the converse could also be true, that is, PTH has evolved to work through the PTHrP receptor system.

## Physiological Role of PTHrP

Recent work to elucidate the physiological role of PTHrP has revealed a number of possible functions. The roles that have been proposed for PTHrP are summarized in Table I.

Table I. Possible Physiological Roles of PTHrP

| Observation | Proposed Function | Reference |
|---|---|---|
| PTHrP causes differentiation | Differentiation factor | 36,38,39 |
| PTHrP but not PTH restores fetal serum calcium levels | Placental calcium gradient maintenance during fetal growth | 41 |
| PTHrP expressed in lactating mammary tissue | Calcium mobilization during lactation | 42 |

PTHrP has been shown to have transforming growth factor-like properties in normal rat kidney cells, however, the transforming properties were dependent on epidermal growth factor (EGF) (36). In a human osteosarcoma cell line (SaOS-2/B-10), which was found to express PTHrP in early passage, the expression of PTHrP mRNA was found to be up regulated by EGF and phorbol esters (37). Similarly, EGF-dependent differentiation of primary keratinocytes resulted in the expression of PTHrP (38). In retinoic acid stimulated differentiation of embryonal carcinoma cells into a parietal endoderm phenotype PTHrP was also found to be expressed (39). These observations raise the possibility of PTHrP involvement in differentiation, working in an autocrine or paracrine fashion. PTHrP has also been shown to be expressed in a number of different tissues and in various cancers, including adenomas of the parathyroid gland (40). The significance of this is still unclear.

PTHrP has also been proposed to act as a regulator of the calcium gradient maintained between the mother and fetus across the placenta. This hypothesis stems from the observation that sheep fetuses that have been TPTX'd and maintained in utero develop hypocalcemia. This situation is not reversed by the infusion of PTH into the fetus but is rectified by the infusion of an extract from the fetal parathyroid gland and conditioned media obtained from the PTHrP secreting lung carcinoma (BEN) cells (41). This suggests that fetal PTHrP in this system might act to increase calcium uptake across the placenta thus restoring normal skeletal growth.

In an evaluation of rat tissues expressing PTHrP it was observed that relatively high levels of PTHrP mRNA were present in lactating mammary tissue but were absent in non-lactating tissue (42). Expression of this message was found to be directly related to the suckling stimulus since the removal of pups from their mothers for four hours resulted in the disappearance of PTHrP mRNA. Levels of PTHrP mRNA rebound within four hours following the return of the litters (42). Further studies have shown that the expression of PTHrP mRNA in lactating mammary tissue is produced in response to the elevations in serum prolactin induced by suckling (43).

The expression of PTHrP in response to the suckling stimuli and the large demands for calcium during lactation suggest a correlate with PTHrP physiology, namely that mammary tissue secretes PTHrP to enable large amounts of calcium to be mobilized from the skeleton. Alternatively PTHrP in mammary tissue may be acting locally to increase the translocation of calcium into milk. With the development of radioimmunoassays for PTHrP (44 and Kao, P.C.; Klee, G. G.; Taylor, R.; Heath, H. III., unpublished data) it has been shown that levels of PTHrP are not raised in serum from lactating women (44), while large amounts of PTHrP have been detected in milk (45). The purpose that PTHrP serves in milk, whether it is taken up by the neonate and/or what role if any it serves there remains to be determined.

In conclusion, the structure of the tumor-secreted peptide associated with HHM has recently been identified and found to share limited sequence homology to PTH. Synthetic amino terminal fragments of PTHrP show very similar biological profiles to PTH both *in vivo* and *in vitro*. PTHrP is capable of eliciting all of the symptoms associated with the syndrome of HHM. Both PTH and PTHrP appear to have a common receptor in bone and in kidney. The presence of two peptides interacting with the same receptor with equal affinity suggests new directions for the design of more potent PTH/PTHrP antagonists.

## Literature Cited

1. McSheehy, P. M. J.; Chambers, T. J. Endocrinology 1986, 119, 1654-9.
2. Heath, J. K.; Atkinson, S. J.; Meikle, M. C.; Reynolds, J. J. Biochim. Biophys. Acta 1984, 802, 151-4.
3. Rodan, G. A.; Martin, T. J. Calcif. Tissue Int. 1981, 33, 349-51.
4. Case records of the Massachusetts General Hospital, Case 27461. New Engl. J. Med. 1941, 225, 789-91.
5. Simpson, E. L.; Mundy, G. R; D'Souza, S. M.; Ibbotson, K. J.; Bockman, R.; Jacobs, J. W. New Engl. J. Med. 1983, 309, 325-30.
6. Mosely, J. M.; Kubota, M.; Diefenbach-Jagger, H.; Wettenhall, R. E. H.;Kemp, B. E.; Suva, L. J.; Rodda, C. P.; Ebeling, P. R.; Hudson, P. J.; Zajac, J. D.; Martin, T. J. Proc. Natl. Acad. Sci. U.S.A. 1987, 84, 5048-52.
7. Strewler, G. J.; Williams, R. D.; Nissenson, R. A.; J. Clin. Invest. 1983, 71, 769-74.
8. Rodan, S. B.; Insogna, K. L.; Vignery, A. M.-C.; Stewart, A. F.; Broadus, A. E.; D'Souza, S. M.; Bertolini, D. R.; Mundy, G. R.; and Rodan, G. A. J. Clin. Invest. 1983, 72, 1511-15.

9. Stewart, A. F.; Insogna, K. L.; Goltzman, D.; Broadus, A. E. Proc. Natl. Acad. Sci. U.S.A. 1983, 80, 1454-58.

10. Suva, L. J.; Winslow, G. A.; Wettenhall, R. E. H.; Hammonds, R. G.; Moseley, J. M.; Diefenbach-Jagger, H.; Rodda, C. P.; Kemp, B. E.; Rodriguez, H.; Chen, E. Y.; Hudson, P. J.; Martin, T. J. Science 1987, 237, 893-6.

11. Mangin, M.; Webb, A. C.; Dreyer, B. E.; Posillico, J. T.; Ikeda, K.; Weir, E. C.; Stewart, A. F.; Bander, N. H.; Milstone, L.; Barton, D. E.; Francke, U.; Broadus, A. E. Proc. Natl. Acad. Sci. U.S.A. 1988, 85, 597-601.

12. Thiede, M. A.; Strewler, G. J.; Nissenson, R. A.; Rosenblatt, M.; Rodan, G. A.; Proc. Natl. Acad. Sci. U.S.A. 1988, 85, 4605-9.

13. Mangin, M.; Ikeda, K.; Dreyer, B. E.; Milstone, L.; Broadus, A. E. Molec. Endocrinol. 1988, 2, 1049-55.

14. Czech, M.P. Cell, 1989, 59, 235-8.

15. Potts, J. T. Jr.; Tregear, G. W.; Keutmann, H. T.; Niall, H. D.; Sauer, R.; Deftos, L. J.; Dawson, B. F.; Hogan, M. L.; Aurbach, G. D. Proc. Natl. Acad. Sci. U.S.A. 1971, 68, 63-7.

16. Mundy, G. R. J. Clin. Invest. 1988, 82, 1-6.

17. Horiuchi, N.; Caulfield, M. P.; Fisher, J. E.; Goldman, M. E.; McKee, R. L.; Reagan, J. E.; Levy, J. J.; Nutt, R. F.; Rodan, S. B.; Schofield, T. L.; Clemens, T. L.; Rosenblatt, M. Science 1987, 238, 1566-8.

18. Kemp, B. E.; Moseley, J. M.; Rodda, C. P.; Ebeling, P. R.; Wettenhall, R. E. H.; Stapleton, D.; Diefenbach-Jagger, H.; Ure, F.; Michelangeli, V. P.; Simmons, H. A.; Raisz, L. G.; Martin, T. J. Science 1987, 238, 1568-70.

19. McKee, R. L.; Caulfield, M. P. Peptide Res. 1989, 2, 161-6.

20. Rodan, S. B.; Noda, M.; Wesolowski, G.; Rosenblatt, M.; Rodan, G. A. J. Clin. Invest. 1988, 81, 924-7.

21. Stewart, A. F.; Horst, R.; Deftos, L. J.; Cadman, E. C.; Lang, R.; Broadus, A. E. New Engl. J. Med. 1980, 303, 1377-83.

22. Ralston, S. H.; Cowan, R. A.; Gardner, M. D.; Fraser, W. D.; Marshall, E.; Boyle, I. T. Clin. Endocrinol. 1987, 26, 281-91.

23. Yamamoto, I.; Kitamura, N.; Aoki, J.; Kawamura, J.; Dokoh, S.; Morita, R.; Torizuka, K. J. Clin. Endocrin. Metab. 1987, 64, 175-9.

24. Thompson, D. D.; Seedor, J. G.; Fisher, J. E.; Rosenblatt, M.; Rodan, G. A. Proc. Natl. Acad. Sci. 1988, 85, 5673-7.

25. Kukreja, S. C.; Shevrin, D. H.; Wimbiscus, S. A.; Ebeling, P. R.; Danks, J. A.; Rodda, C. P.; Wood, W. I.; Martin, T. J. J. Clin. Invest. 1988, 82, 1798-802.

26. Rosenblatt, M.; Segre, G. V.; Tyler, G. A.; Shepard, G. L.; Nussbaum, S. T.; Potts, J. T. Jr. Endocrinol. 1980, 107, 545-50.

27. Rosenblatt, M.; Callahan, E. N.; Mahaffey, J. E.; Pont, A.; Potts, J. T. Jr. J. Biol. Chem. 1977, 252, 5847-51.

28. Segre, G. V.; Rosenblatt, M.; Tully, G. L. III; Lahgharn, J.; Reit, B.; Potts, J. T. Jr. Endocrinol. 1985, 116, 1024-9.

29. Horiuchi, N.; Rosenblatt, M.; Keutmann, H. T.; Potts, J. T. Jr.; Holick, M. F. Am. J. Physiol. 1983, 244, E589-95.

30. Mahaffey, J. E.; Rosenblatt, M.; Shepard, G. L.; Potts, J. T. Jr. J. Biol. Chem. 1979, 254, 6496-8.

31. Horiuchi, N.; Holick, M. F.; Potts, J. T. Jr.; Rosenblatt, M. Science 1983, 220, 1053-5.

32. McKee, R. L.; Goldman, M. E.; Caulfield, M. P.; Dehaven, P. A.; Levy, J. J.; Nutt, R. F.; Rosenblatt, M. Endocrinol. 1988, 122, 3008-10.

33. Horiuchi, N.; Hongo, T.; Clemens, T. L. J. Bone Mineral Res. 1990, 5, 541-5.

34. Horiuchi, N.; Caulfield, M. P.; Nutt, R. F.; Levy, J. J.; Rosenblatt, M.; Clemens, T. L. 71st Annual Meeting of the Endocrine Society, 1989, Abstract #618.

35. Nagasaki, K.; Yamaguchi, K.; Miyake, Y.; Hayashi, C.; Honda, S.; Urakami, K.-I.; Miki, K.; Kimura, S.; Watanabe, T.; Abe, K. Biochem. Biophys. Res. Comm. 1989, 158, 1036-42.

36. Insogna, K. L.; Stewart, A. F.; Morris, C. A.; Hough, L. M.; Milstone, L. M.; Centrella, M. J. Clin. Invest. 1988, 83, 1057-60.

37. Rodan, S. B.; Wesolowski, G.; Ianacone, J.; Thiede, M. A.; Rodan, G. A. J. Endocrinol. 1989, 122, 219-27.

38. Ernst, M.; Rodan, G. A.; Thiede, M. A. J. Bone Mineral Res. 1989, 4, S194 (abstract).

39. Chan, S. D. H.; Nissenson, R. A.; King, K.; Strewler, G. J. J. Bone Mineral Res. 1989, 4, S135 (abstract).

40. Ikeda, K.; Weir, E. C.; Mangin, M.; Dannies, P. S.; Kinder, B.; Deftos, L. J.; Brown, E. M.; Broadus, A. E. Molec. Endocrinol. 1988, 2, 1230-6.

41. Rodda, C. P.; Kubota, M.; Heath, J. A.; Ebeling, P. R.; Moseley, J. M.; Care, A. D.; Caple, I. W.; Martin, T. J. J. Endocrinol. 1988, 117, 261-71.

42. Theide, M. A.; Rodan, G. A. Science 1988, 242, 278-80.

43. Thiede, M. A. Molec. Endocrinol. 1989, 3, 1443-7.

44. Budyar, A. A.; Nissenson, R. A.; Klein, R. F.; Pun, K. K.; Clark, O. H.; Diep, D.; Arnaud, C. D.; Strewler, G. J. Ann. Intern. Med., 1989, 111, 807-12.

45. Budyar, A. A.; Halloran, B. P.; King, J. C.; Diep, D.; Nissenson, R. A.; Strewler, G. J. Proc. Natl. Acad. Sci. USA 1989, 86, 7183-5.

RECEIVED August 27, 1990

# Chapter 17

# Lytic Peptides from the Skin Secretions of *Xenopus laevis*: A Personal Perspective

**Bradford W. Gibson**

**Department of Pharmaceutical Chemistry, School of Pharmacy, University of California, San Francisco, CA 94143–0446**

Peptides with lytic and antimicrobial activities comprise a large portion of the peptide content of the skin secretions in *Xenopus laevis*, or African clawed toad. When these positively charged peptides are presented as Edmundson wheel projections, they all have in common a predicted amphipathic $\alpha$-helical structure. It is likely that the ability to form such an $\alpha$-helical conformation while spanning cellular membranes is the molecular basis for their lytic activity, although the precise organization of these peptides in membranes is not known. In this paper, a short personal perspective of the events that led up to the discovery of these peptides and their associated activity is discussed.

In late July of 1987, a report from researchers at NIH announcing the discovery of a novel class of amphibian peptides (the magainins) with potent antimicrobial activity aroused considerable media and scientific interest (1-4). Indeed, a front page headline in The New York Times (1) and a major feature article in The Washington Post (2) hailed this as a major discovery or "medical miracle," and comparisons were made to the discovery of penicillin. Even NIH director Dr. James Wyngaarden had this to say about Dr. Michael Zasloff, the NIH researcher who had discovered these peptides, "It is the kind of experience that a scientist has once in a lifetime, and then only if he is lucky" (1). Dr. Zasloff himself made the claim at a press conference at NIH on July 30, 1987 that, "It is the first time a chemical defense system separate from the immune system has been discovered in vertebrate animals" (1).

At the time, I found the reports of the discovery of a new class of peptides from the skin of *Xenopus laevis* (or African clawed toad) somewhat surprising. In the laboratory of Dr. Dudley Williams at Cambridge University, my colleagues and I had just spent over two years (1984-1985) attempting to identify all peptides secreted from the skin of *Xenopus laevis* (see Figure 1). We had undertaken this project to identify novel peptide hormones and other bioactive peptides and to learn something about the biosynthesis or processing of these peptides from

0097–6156/91/0444–0222$06.00/0

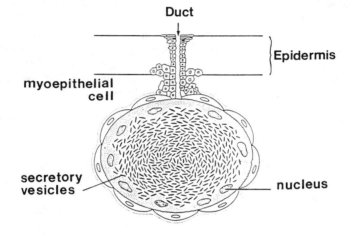

Figure 1. A schematic diagram of the granular gland in the skin of *X. laevis.* As with many other amphibians, these contents are secreted (or exuded) when the amphibian is under stress or by direct adrenergic stimulation.

their precursor polypeptides. The rationale for this approach came largely from the work of the Italian pharmacologist, Vittorio Erspamer, who was the first to show that a wealth of bioactive peptides exists in amphibian skin, many of which share considerable structural and functional similarities to mammalian neuropeptides (5). We had, in fact, managed to isolate and sequence (using a combination of HPLC and mass spectrometric techniques) (6) a number of novel peptides, some of which had no known counterparts in mammalian or amphibian systems (7-10). However, when Dr. Zasloff published the actual sequences of the two magainins in the August 1987 issue of the Proc. Natl. Acad. Sci. USA (11), it was apparent (at least to me and a few others in the scientific community) that these peptides were neither novel in their structure nor in their associated biological activity (12,13). In fact, the two magainins had been previously isolated and sequenced by our group in Cambridge in 1985 and are members of a much broader class of lytic peptides with predicted amphipathic structures (8,10). Needless to say, this point was not brought up during the flurry of press activity surrounding the announcement of this "discovery" at NIH. However, to understand fully the efforts that led up to the discovery of the magainins (or PGS peptides), one must go back over twenty years in the scientific literature.

### The Discovery of Bombinin, The First Vertebrate Antimicrobial Peptide

In the 1960's, Michl and colleagues at the University of Vienna published a series of papers beginning in 1962 (14) that detailed the isolation and sequencing of the active "bacteriostatic" agent(s) from the skin secretions of the orange-bellied European toad, *Bombina variegata* L. A summary publication appeared in Monatshefte fur Chemie in 1970 (15) which presented the sequence of bombinin, a 24 amino acid peptide (see Table I). This positively charged peptide contains several lysine residues that punctuate the structure at somewhat regular intervals and, as pointed out by the authors, "resembles somewhat that of melittin," a lytic $\alpha$-helical peptide from honey bee venom. Michl and Csordas went on to add that this peptide also possesses hemolytic and bacteriostatic action and is quickly hydrolyzed to smaller peptides by proteolytic enzymes that are also present in the slimy skin secretions.

At an international conference several years earlier on animal toxins in Los Angeles (First International Symposium on Animal Toxins, 1967), Michl and his associates presented data that showed relative bacteriostatic activities of several amphibian skin toxins, using agar plates containing several types of bacteria, yeasts and molds (16). By measuring the radius of growth inhibition, these researchers showed that *Bombina* contained components in its skin secretion that profoundly inhibit the growth of *Staphylococcus aureus* and some other microorganisms, but they found no such activity in the skin secretions of *Xenopus laevis*. In an accompanying article by Kaiser and Kramar entitled "The Biochemistry of the Cytotoxic Action of Amphibian Poisons" (17), a more thorough evaluation was made of the activity of these secreted toxins on cells and organelles, such as fibroblasts, mast cells, tumor cells, and liver mitochondria.

Table I. Lytic Peptides from the Skin of *X. laevis* and *B. variegata*

| | |
|---|---|
| Bombinin | GIGALSAKGALKGLAKGLAEHFAN-NH$_2$ |
| PGLa | GMASKAGAIAGKIAKVALKAN-NH$_2$ |
| XPF | GWASKIGQTLGKIAKVGLKELIQPK |
| PGK | GWASKIGQTLGKIAKVGLQGLMQPK |
| PGS | GIGKFLHSAKKFGKAFVGEIMNS |
| $G^{10}K^{22}$-PGS | GIGKFLHSAGKFGKAFVGEKMNS |
| CPF(1+5)* | GFGSFLGKALKAALKIGANALGGSPQQ |
| CPF(4+9)* | GLASLLGKALKAALKIGTHFLGGSPQQ |

* In the case of CPF, only two of the numerous homologs are shown

In their summary, Michl and Csordas concluded that "The skin secretions of *Bombina* and *Titurus* species protect the animals not only from desiccation and predators, but they also control the growth of microorganisms on their skin. Direct hemolyzing proteins and antibiotic- active peptides are responsible for these activities" (16). Considering what we now know about components in the skin secretions of *X. laevis*, it may seem a bit surprising that this frog's secretions were excluded from their summary statement. But, as will be discussed later, this contradiction can be readily explained, and is related to the earlier observation that bombinin (and presumably other secreted peptides) can undergo rapid proteolysis after secretion.

## PGLa, The Second Member

After the discovery of bombinin, it appears that the group at Vienna headed by Michl apparently did not continue this work with the discovery of any new members of this antimicrobial peptide family. Indeed, thirteen years passed before, in 1983, Gunther Kreil and his colleagues at the Institut fur Molekularbiologie in Salzburg published a paper (18) showing a novel cDNA sequence encoding a small polypeptide precursor whose processing was predicted to yield a 24 amino acid peptide, PYLa.

Kreil and co-workers had come across this novel sequence somewhat by accident while in the process of screening clones for the precursor to caerulein, an amphibian counterpart to the gastrin/CCK peptide family (19). Conventional doctrine at the time (20) suggested that processing of this unknown precursor should occur at the two sets of double basic amino acid pairs, Lys34-Arg35 and Arg61-Arg62, which flanked a 25 amino acid peptide containing residues Tyr36-Gly60. However, further processing at the C-terminal Gly was also predicted to occur, converting Gly60 to an amide for the neighboring Leu59. The resulting peptide, Tyr36-Leu59-amide (PYLa, or peptide tyrosine leucine-amide), was therefore theorized to be the major product of this small precursor. Kreil and his colleagues went on to suggest that "If one arranges PYLa into an $\alpha$-helical configuration, it is striking that this peptide yields a highly amphipathic structure. Several examples are known where peptides capable of forming amphipathic helices have been shown to have a high affinity for cell membranes and to have cytotoxic, bactericidal or lytic effects." Kreil further states that they expect PYLa, if it indeed exists, to possess one or more of these activities.

When this paper appeared in 1983, I had just arrived at Cambridge University to begin my postdoctoral studies. In the first four or five months of 1984, Linda Poulter (a graduate student) and I began to catalogue peptide molecular weights and their partial sequences from the skin secretions of *Xenopus laevis* using fast atom bombardment (FAB) mass spectrometry and other techniques (6), and the search for PYLa was at the top of our agenda. By analyzing these complex peptide mixtures after size separation and HPLC purification by FAB mass spectrometry, we were able to quickly determine the accurate molecular weights of these peptide components by measuring their protonated molecular ions ($MH^+$) with an accuracy of better than $\pm 0.2$ Da. While molecular ion data allowed us to determine the number of peptide components present in these secretions and their corresponding masses, this information in and of itself had limited value. However, when combined with other information, such as a precursor structure or a partial peptide sequence, this data became exceedingly useful. For example, we could search for potential processing and proteolytic events within a precursor structure by matching these accurate molecular weights to precursor subfragments, taking into account reasonable processing sites. In fact, as a graduate student at M.I.T., I had developed such a scheme to confirm and correct preliminary DNA sequences of several large aminoacyl-tRNA synthetases (21). Surprisingly, when this scheme was applied to the hypothetical PYLa precursor, we could find no evidence for the existence of PYLa ($MH^+$ 2385, predicted nominal mass), but instead found several smaller subfragments of this peptide. The largest of these fragments stretched from Gly39 to the same C-terminal residue expected for PYLa, i.e., Leu-amide (residue 59 according to the precursor numbering scheme) (see Figure 2).

Through the grapevine at the neighboring MRC laboratories in Cambridge, we had learned that Kreil and his colleagues had since synthesized PYLa and found no associated bioactivity, but had also discovered that this peptide did not exist in *X. laevis* skin secretions. Instead, a truncated form of PYLa, resulting from additional cleavage at the single Arg38 residue near the N-terminus, was producing a smaller 21 amino acid peptide termed PGLa (peptide Gly39-Leu59-amide) (22). While we had also identified PGLa in our secretions using FAB mass spectrometric data, a number of smaller fragments were additionally isolated and sequenced and appeared to result from further cleavages of PGLa at the N-terminal side of several internal lysine residues. The significance of this observation was unclear at the time, but this rather novel cleavage site was to become a common and recurring theme in the processing or, as we were to find out later, degradation of many of these skin peptides.

## The Xenopsin and Caerulein Precursor Fragments

As we continued our analysis of the peptide content of *X. laevis* in the early part of 1984, two important papers appeared in the literature that proved to lead the way for a breakthrough in our research efforts. Sures and Crippa, from the Department of Molecular Embryology at the University of Geneva in Switzerland, had published the complete cDNA sequence encoding the precursor structure of xenopsin (23), a peptide hormone secreted by *X. laevis* with structural and pharmacological properties similar to the mammalian peptide neurotensin (24). From the cDNA deduced precursor structure, we could immediately identify the origin of several peptides for which we had exact mass and partial (and sometimes complete) sequence data. These peptides originated from a spacer region of the xenopsin precursor that was located between the N-terminal signal sequence peptide and the C-terminal region encoding xenopsin itself. These peptides were apparently arising from a combination of processing events that included cleavages at a single arginine site, a double basic site that flanked the N-terminal portion of xenopsin, and several lysine residues which were undergoing cleavages on their N-terminal side. Of course, this was very similar to the events we had just seen for PGLa and its fragments.

A seminar given at the MRC laboratories later that spring by Gunther Kreil turned out to be even more interesting. In his talk, Prof. Kreil gave some of his recent data on the sequence and processing of the caerulein precursor, and suggested that it, like the xenopsin and PGLa precursors, was undergoing additional processing at single arginine residues. After the seminar was over, Linda Poulter and I introduced ourselves and suggested that if these processing events were indeed occurring, we probably already had the peptide data in hand that would confirm it. Since at the time we had not yet set up our DNA sequencing laboratory at the University Chemistry Laboratory and Kreil did not have a mass spectrometry facility in Salzburg, this encounter began a very fruitful collaboration between our two laboratories. By sharing our data and expertise, we were both able to explore the processing events in these precursors with much greater insight.

What we discovered over the next year was a complicated picture. From the xenopsin precursor, a large 25 amino acid peptide was present in the skin secretions of *X. laevis* ($MH^+$ 2662), which was found to be identical with residues Gly38-Lys62 (7) of the xenopsin precursor published by Sures and Crippa (23). This peptide was being formed via a single base cleavage at Arg37 to yield the N-terminal Gly38 of this peptide, followed by cleavage between the C-terminal flanking double basic pair, Lys62-Arg63 (see Table I, Figure 3). Presumably, the presence of Pro61 next to the Lys62 protects the new C-terminal lysine residue from being removed via carboxypeptidase B-type activity, a common processing event. However, there were many more abundant subfragments of this 25 amino acid peptide (also called xenopsin precursor fragment, or XPF) that were being formed by subsequent cleavage at the N-terminal side of two of the three internal lysine residues (7). This was very similar to what was observed for PGLa, and suggested that, like bombinin, XPF and PGLa undergo rapid proteolysis after secretion. Experiments carried out shortly in our laboratory in Cambridge confirmed this hypothesis and, when the peptides were isolated from intact vesicles, only these larger intact species (i.e., PGLa and XPF) were found (8). Furthermore, when care was taken to minimize the time interval between the secretion and the freezing step ($^{<}$5 min), much larger amounts of the intact peptides (and smaller amounts of the subfragments) were isolated from these secretions.

A partial sequence for the cDNA encoding the precursor for caerulein (or preprocaerulein) had also been published in 1984 by G. Kreil's group in Salzburg (25). From this partial structure, we identified a single large 27 residue peptide encompassing amino acids Gly41-Gln67 (see Table I), and several smaller fragments that originated from cleavage primarily at one of the three internal lysines (7) (see Figure 4). As in the case of PGLa and XPF, the N-terminus was formed by cleavage at a single arginine (Arg40), but in this case the C-terminus was formed from a second single basic site, Arg68. Unlike the first N-terminal site, cleavage at Arg68 was occurring on the N-terminal side or, more likely, by a two-step process consisting of a C-terminal cleavage between Arg68-Glu69 followed by a carboxypeptidase B-type removal of the now terminal Arg68. Besides this caerulein precursor fragment (or CPF), a series of very homologous subfragments was also identified which differed from one another by one or more amino acid substitutions. Once again, Kreil's group was able to supply the key information for the origin of these additional peptides by providing us with unpublished cDNA-deduced sequences for the complete caerulein precursor. It turned out that there were at least four different caerulein precursors, each carrying multiple copies of caerulein as well as one to four copies of the 27 amino acid spacer peptides or CPF's (26,27). These 27 amino acid peptides, which generally differed from one another by three or four amino acid substitutions, all contain three lysine residues that punctuate the sequence in a manner suggesting an amphipathic α-helical structure. When these peptides are drawn according to an Edmundson wheel projection (28), it is quite striking that they all have a distinct hydrophobic and hydrophilic face (Figure 5).

In addition to the CPF peptides and the single peptide from the xenopsin precursor (or XPF), we also isolated and sequenced a peptide that differed by only three amino acid substitutions from XPF. At first we assumed that this peptide represented a peptide from a second xenopsin precursor, reminiscent of the multiple members of the caerulein precursor family. This time we decided to isolate and sequence the cDNA encoding this precursor ourselves in our newly set-up molecular biology facility. Starting from a cDNA skin library supplied by Kreil's group, we isolated and sequenced several cDNA clones encoding the 25 amino acid peptide that was homologous to XPF. However, instead of encoding xenopsin, this precursor was found to encode a novel 14 amino acid hormone-

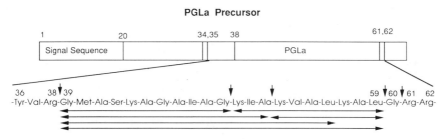

Figure 2. The precursor encoding PGLa showing the processing of this lytic peptide.

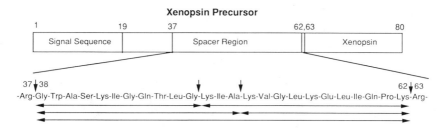

Figure 3. The precursor to xenopsin showing the additional lytic peptide, XPF, that is also formed during the processing steps.

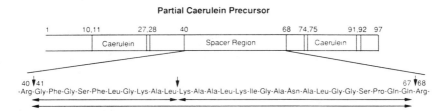

Figure 4. The partial precursor to caerulein showing one of the several additional lytic peptides, CPF's that are formed along with caerulein during the processing of this precursor polypeptide.

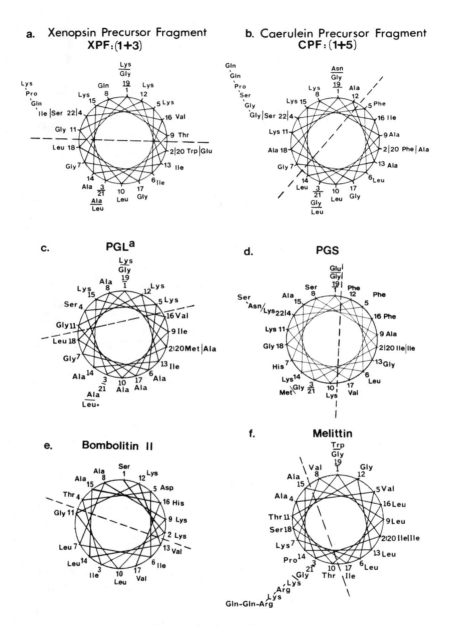

Figure 5. Edmundson wheel projections of several lytic peptides secreted by *X. laevis*. For comparison, the structures of bombolitin II (from bumble bee) and melittin (from the European honey bee) are also shown.

like peptide, levitide, that we had previously isolated and sequenced over a year earlier: PyroGlu-Gly-Met-Ile-Gly-Thr-Leu-Thr-Ser-Lys-Arg-Ile-Lys-Gln-NH$_2$(9). The associated lytic peptide was renamed PGK (for peptide glycine lysine) to be consistent with peptide nomenclature proposed by Tatemoto and Mutt (29).

## PGS and Gly$^{10}$-Lys$^{22}$-PGS (the Magainins)

In our fairly exhaustive search, we found one last class of lytic (and amphipathic) peptides in *X. laevis* skin secretions. These two peptides, PGS and Gly$^{10}$Lys$^{22}$-PGS, were both 23 amino acids in length and did not result from the processing of any precursor or precursor family that was known at the time. To determine their origin, oligonucleotide probes were made and a *X. laevis* cDNA library was screened as had been carried out for the XPF-like (renamed PGK) peptide described previously. In this case, several clones were isolated and their cDNA revealed the existence of a single precursor that contained five copies of PGS and one copy of Gly$^{10}$Lys$^{22}$-PGS (10). In each case, these peptides were thought to arise in an analogous manner to the other lytic peptides from *X. laevis*, i.e., through single Arg cleavages at the N-terminus and double basic excisions (Lys-Arg) at the C-terminus (see Figure 6). In addition to encoding the PGS peptides (or magainins), a second class of much smaller peptides is also processed and observed in the skin secretions, i.e., two acidic peptides, Asp-Glu-Asp-Leu-Asp-Glu and Asp-Glu-Asp-Met-Asp-Glu. The significance of these small acidic peptides is not known, although it is possible they could be highly antigenic or possess some other type of activity of some value in predator defense.

## A Common Origin and Function?

Thus, it appears that there are at least two types of precursors that encode these amphipathic helical peptides: peptide hormone-like precursors, such as those for caerulein, xenopsin and levitide that encode both the lytic and peptide hormone type peptides, and ones that encode only (or primarily) the lytic peptides, such as PGLa and the PGS peptides (or magainins). Whether or not these classes have two distinct origins is not known, but a strong homology (94-76%) found among their 5′ DNA upstream flanking regions strongly suggests a common origin for all these precursors (10).

As to the actual function or functions of these lytic peptides, a bacteriostatic or antimicrobial function as first suggested by Michl (14-16), and more recently by Kreil (18), Zasloff (11) and ourselves, certainly seems attractive (see Table II). A possible mechanism for this action might be the formation of an anion channel, as suggested by Zasloff and co-workers (30), that would lead to the disruption of membrane-linked energy transduction (31). Zasloff has also stated that magainins are specific for prokaryotic membranes, and are not hemolytic to human red blood cells. Our own work using rat erthyrocytes suggests otherwise (unpublished data), but since hemolytic susceptibilities can vary greatly in different

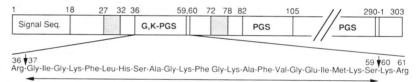

**PGS Precursor (prepromagainin)**

Figure 6. The precursor to PGS (or prepromagainin) showing the processing of one of the six copies of these lytic peptides, $Gly^{10}Lys^{22}$– PGS (or magainin 1). The shaded areas indicate the positions of the acidic peptides (see text).

mammals, it is quite possible that we are both correct in our observations. The underlying mechanism by which this specificity is based has not yet been elucidated, but studies on another class of insect antimicrobial peptides, the cecropins (32), suggests that the cholesterol content of eukaryotic cell membranes is likely to be a crucial factor. If such a channel were to form, at least three or four of these peptides would have to aggregate so as to present their hydrophilic (polar) faces inward and their hydrophobic (apolar) faces outward. Such a possible scheme is illustrated in Figure 7.

Table II. Growth Inhibition of Selected Pathogens by PGS ($\mu$g/ml)

| Strain | 0 | 10 | 25 | 50 | 100 |
|---|---|---|---|---|---|
| *Escherichia coli D31* | + | − | − | − | − |
| *Enterobacter amnagenesis* | + | + | + | − | − |
| *Pseudomonas florescens* | + | + | + | + | + |
| *Bacillus brevis* | + | + | + | + | − |
| *Saccharomyces sp.* | + | + | + | + | − |

My own bias is towards a dual or multiple function theory, where these peptides may also play a role in the rapid adsorption of other skin-secreted toxins by a predator, possibly by lysing or disrupting oral or gastrointestinal membranes. Kreil has also postulated such a dual role in his paper on the caerulein precursor family by suggesting "they could act as lytic agents or permeability factors to facilitate the uptake of potent hormone-like peptides" or "as antimicrobial agents to combat infections in an aqueous environment" (23). Such a dual role is certainly consistent with these peptides being released during periods of stress (predator attack) or locally, as might be expected if the amphibian were to suffer a cut or other wound.

The role of the co-secreted proteolytic degrading enzymes is also a mystery. We have previously postulated that they may serve to inactivate the lytic peptides once the secreted vesicles have lysed on the surface of the frog (7). Unpublished studies conducted at Cambridge have shown that subfragments of these larger intact lytic peptides no longer possess any lytic or hemolytic action. Thus, rapid proteolysis might serve to de-toxify these secretions once they have served their (brief) purpose. There do, however, appear to be different rates by which the co-secreted proteases degrade these peptides, being very fast in *Xenopus* and *Rana*, but somewhat slower in *Bombina* (personal observation). This would tend to explain why Michl found no lytic or bacteriostatic activity in *Xenopus laevis* skin secretions.

## Coming Full Circle in the Helix, the Bombinins

As might seem appropriate, I shall end this discussion by returning to bombinin,

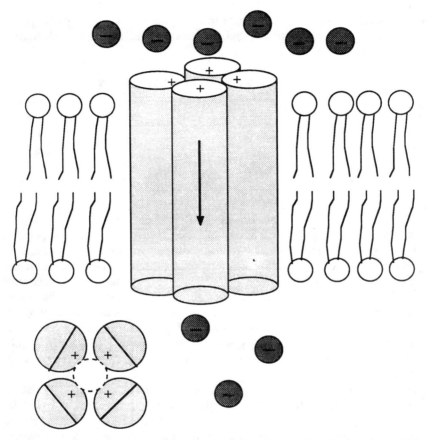

Figure 7. A hypothetical model for the formation of an anion channel in a cellular membrane from an assembly of four amphipathic helices. In this orientation, all the lysine (and other polar) residues face inward and the hydrophobic residues face outward.

the first of the amphibian antimicrobial peptides. In the last year, Dr. Eliot Spindel (currently at the Oregon Primate Center, Beaverton, Oregon) and I have examined the skin secretion of *Bombina orientalis*, the Asian relative of the European *Bombina variegata* L. While we undertook this analysis primarily to study the processing of bombesin, an important regulatory peptide and tissue-specific growth factor (33) that is also contained in the skin secretions of *Bombina orientalis*, we took this opportunity to look for other members of the lytic family of amphibian peptides. Although our results are somewhat preliminary, we have isolated five new members of this lytic peptide class, all of which share significant, but not complete, homology with bombinin. They vary in size (25-27 amino acids) and have some limited internal sequence differences with bombinin. Preliminary circular dichroism studies also indicate that they have random structures in water, but significant $\alpha$-helical structures in the helix-inducer solvent, trifluoroethanol. We will soon attempt to determine the origin of this family by screening an already constructed *Bombina orientalis* cDNA skin library for their precursors. It will certainly be interesting to see if these new (and original) bombinin-like peptides have some similarity to the other precursor classes and, possibly, a common origin as well.

## Acknowledgments

I would like to thank Dr. Dudley Williams for giving me the opportunity to carry out this research under his tutorship as a postdoctoral fellow at Cambridge University and the Science and Engineering Research Council for funding this research. Additional thanks must also go to my colleagues at Cambridge University, Drs. Linda Poulter, Maria Giovanni, John E. Maggio, Adrian Terry and Chris Moore. I would also like to thank Dr. A. F. Findeis, who was on sabbatical from the National Science Foundation, for many helpful discussions on the nature of amphipathic helices and Dr. Daniel Witt (Repligen Corporation, Cambridge, MA) for the antimicrobial data on PGS. Lastly, I would like to thank Prof. Gunther Kreil and his former students, Drs. W. Hoffman and K. Richter, for sharing their data and intellectual insights in a completely honest and selfless manner.

## Literature Cited

1. Altman, L. K. The New York Times July 31, 1987.
2. Okie, S. The Washington Post July 30, 1987.
3. Cannon, M. Nature 1987, 328, 478.
4. Dagani, R. Chem. Eng. News 1987, 65, 27.
5. Erspamer, V.; Melchiorri, P. Trends Pharmac. Sci. 1980, 1, 391.
6. Gibson, B. W.; Poulter, L.; Williams, D. W. Peptides 1985, 6, Suppl. 3, 23.

7. Gibson, B. W.; Poulter, L.; Williams, D. H.; Maggio, J. J. Biol. Chem. 1986, 261, 5341.
8. Giovannini, M. G.; Poulter, L.; Gibson, B. W.; Williams, D. H. Biochem J. 1987, 243, 113.
9. Poulter, L.; Terry, A.; Williams, D. H.; Giovannini, M. G.; Moore, C. H.; Gibson, B. W. J. Biol. Chem. 1988, 263, 3279.
10. Terry, S. A.; Poulter, L.; Williams, D. H.; Nutkins, J. C.; Giovannini, M. G.; Moore, C. H.; Gibson, B. W. J. Biol. Chem. 1988, 263, 5745.
11. Zasloff, M. Proc. Natl. Acad. Sci. USA 1987, 84, 5449.
12. Cannon, M. Nature 1987, 329, 494.
13. Williams, D. H. Chem. Eng. News 1987, 65, 4.
14. Kiss, G.; Michl, H. Toxicon 1962, 1, 33.
15. Csordas, A.; Michl, A. Monatshefte fur Chemie 1970, 101, 182.
16. Bachmayer, H.; Michl, H.; Roos, B. In Animal Toxins; Russell, F. E.; Saunders, P. R., Eds., Pergamon Press: Oxford, 1967, p 395-399.
17. Kramer, R.; Kaiser, E. In Animal Toxins; Russell, F. E.; Saunders, P. R., Eds., Pergamon Press: Oxford, 1967, p 389-394.
18. Hoffman, W.; Richter, K.; Kreil, G. EMBO J. 1983, 1, 711.
19. Erspamer, V.; Falconieri Erspamer, G.; Mazzati, G.; Endean, R. Comp. Biochem. Physiol. 1984, 77C, 99.
20. Turner, A. J. Trends Neuro. Sci. 1984, 258.
21. Gibson, B. W.; Biemann, K. Proc. Natl. Acad. Sci. USA 1984, 81, 1956.
22. Andreu, D.; Aschauer, H.; Kreil, G.; Merrifield, R. B. Eur. J. Biochem. 1985, 149, 531.
23. Sures, I.; Crippa, M. Proc. Natl. Acad. Sci. USA 1984, 81, 380.
24. Arakai, K.; Tachibana, S.; Uchiyama, M.; Nakajima, T.; Yasuhara, T. Chem. Pharm. Bull. 1973, 21, 2801.
25. Richter, K.; Kawashima, E.; Egger, R.; Kreil, G. EMBO J. 1984, 3, 617.
26. Richter, K.; Egger, R.; Kreil, G. J. Biol. Chem. 1986, 261, 3676.
27. Wakabayashi, T.; Kato, H.; Tachibana, S. Gene 1984, 31, 295.
28. Schiffer, M.; Edmundson, A. B. Biophys. J. 1967, 7, 121.
29. Tatemoto, K.; Mutt, V. Nature 1980, 285, 417.
30. Cruciani, R. A.; Stanley, E. F.; Zasloff, M.; Lewis, D. L.; Barker, J. L. Biophys. J. 1988, 532, 9.
31. Westerhoff, H. V.; Juretic, D.; Hendler, R. W.; Zasloff, M. Proc. Natl. Acad. Sci. USA 1989, 86, 6597.
32. Christensen, B.; Fink, J.; Merrifield, R. B.; Mauzerall, D. Proc. Natl. Acad. Sci. USA 1988, 85, 5072.
33. Sunday, M. E.; Kaplan, L. M.; Motoyama, E.; Chin, W. W.; Spindel, E. R. Lab Invest. 1988, 59, 5.

RECEIVED August 27, 1990

# Chapter 18

# Peptide Antibiotics from the Animal Kingdom

## Cecropins and Synthetic Analogs

D. Wade[1], R. B. Merrifield[1], and H. G. Boman[2]

[1]Rockefeller University, New York, NY 10021
[2]Department of Microbiology, University of Stockholm, Stockholm
S–10691, Sweden

The cecropins are a family of small, basic peptides which have been isolated from insects and mammals, and shown to possess broad-spectrum antibacterial activities. This article reviews recent research into the structures and functions of cecropins, and concentrates on work involving the solid-phase synthesis and study of cecropin analogs, model peptides, and hybrids of cecropin and melittin, the toxic peptide of bee venom. Cecropins contain an N-terminal amphipathic $\alpha$-helix linked to a C-terminal hydrophobic $\alpha$-helix by a flexible segment of amino acids. All of these secondary structure elements were shown to be required for antibacterial activity. Their disruption of liposomes, and effects on conductivities of lipid bilayers suggested that cecropins act by forming pores in the bacterial cell membrane. Melittin resembles the cecropins in size, components of secondary structure, and antibacterial activity. However, unlike the cecropins, it lyses red blood cells. A synthetic hybrid of cecropin and melittin was developed which exhibited enhanced antibacterial activity, greatly reduced red cell lytic activity, and potent antimalarial activity.

The resistance of bacteria to antimicrobial agents has been recognized as an increasingly important health problem and an economic burden to society (1). This fact necessitates the development of new antibacterials and strategies to counter the problem of resistance. Some recently discovered microbicidal compounds and their sources include: the cecropins, a family of peptides isolated from insects and mammals (2,3), the magainins from frog skin (4), and the defensins from various mammalian species and insects (5-7). In addition to these antibacterial

0097–6156/91/0444–0237$06.00/0
© 1991 American Chemical Society

agents, the bee venom toxin, melittin, was recently shown to have antibacterial activity (8).

Interest in the cecropins arose from the study of cell-free immunity in insects, and since the early 1970's there has been much progress in elucidating the molecular mechanisms of the immune response of insects to bacterial infection (9).

Boman and colleagues have purified and extensively characterized several of the inducible, bactericidal proteins from hemolymph of pupae of the giant silkmoth, *Hyalophora cecropia* (2,10). One of the proteins was shown to be a lysozyme, with a molecular weight of about 15,000 and activity against some Gram-positive organisms (2). A second group of proteins, the attacins, were of higher molecular weight (20-23,000), and were active against some Gram-negative bacteria (10). The third group of bactericidal proteins, the cecropins, were shown to be low molecular weight (about 4,000), heat-stable, basic proteins which were highly potent against several species of Gram-positive and Gram-negative bacteria (8).

## Structural Properties of Cecropins

Seven low molecular weight antibacterial proteins were isolated from *H. cecropia* pupae, and designated alphabetically, A through G (2,8,11). Cecropins A, B, and D were found to comprise most of the antibacterial activity of these proteins in hemolymph whereas cecropins C, and E-G were minor components. The primary structures of cecropins A-D were determined, and partial amino acid sequences were obtained for cecropins E and F (8,11). Homology between the primary structures of the major cecropins was found to be 65% for A and B, 64% for A and D, and 39% for B and D (Figure 1) (11). More than half of the amino acid replacements could be explained by single-base changes, and it was therefore suggested that the three principal forms of cecropin evolved through gene duplication (8).

Primary structure information showed that the cecropins were very basic, and that the amino acids in the carboxyl terminal half of the molecules were strongly hydrophobic (8,11). Theoretical predictions showed that the amino terminal portions of cecropins had strong potentials to form amphipathic $\alpha$-helices, and $\alpha$-helices were also predicted for the C-terminal segments (Figures 1 and 2) (11,12). Circular dichroism spectra of cecropins A and B demonstrated that these peptides existed as random coils in dilute aqueous solutions, but adopted more helical conformations in hydrophobic environments (13). Recently, the three-dimensional structure of cecropin A in a hydrophobic solvent was determined by two-dimensional nuclear magnetic resonance and dynamical simulated annealing techniques (14). Analyses agreed well with previous predictions (12,13) and indicated that the structure contained two helical regions, residues 5-21 and residues 24-37, separated by a bend (Figure 1).

The nucleotide sequences of cDNA clones for cecropins A, B and D have also been determined, and the peptides were shown to be synthesized *in vivo* as precursor proteins of 62-64 residues (15,16). With the aid of chemically

```
                         1      5        10        15       20        25       30        35
        Cec. A:  K W K L F K K I E K V G Q N I R D G I I K A G P A V A V V G Q A T Q I A K *
                 + o + o o + + o - + o     o + -   o o + o   o o o o o o       o       o o +
                 Amphipathic      Flexible              Hydrophobic

        Cec. B:  - - - V - - - - - - M - R - - - N - - V - - - - - I - - L - E - K A L *

        Cec. D:  - N P - - E L - - - - - R V - - A V - S - - - - - - T - A - - - A L - - *
     A(1-11)-
       D(12-37): - - - - - - - - - - - - - R V - - A V - S - - - - - - T - A - - - A L - - *

          MP I:  - - - - - - - - - - - A K K - K E A - E - - L E - - - K L L K E A K E I A K *
                             Amphipathic            Amphipathic
         MP II:  - - - - - - - - - - - A K K - K E A - E - - L E - I - - L A L - L A L *
                             Amphipathic            Hydrophobic
        MP III:  - - - - - - - - - - - R - G - N - - V - - - - - I - - L A L - L A L *
                             Flexible               Hydrophobic
         MP IV:  - - - - - - - - - - - *
   CA(1-13)-
      M(1-13):   - - - - - - - - - - - - - G I G A V L K V L T T G L *

       Melittin: G I G A V L K V L T T G L P A L I S W I K R K R Q Q *
                 o   o o o + o o       o o o o o   o o + + + +
                 Hydrophobic             Basic
```

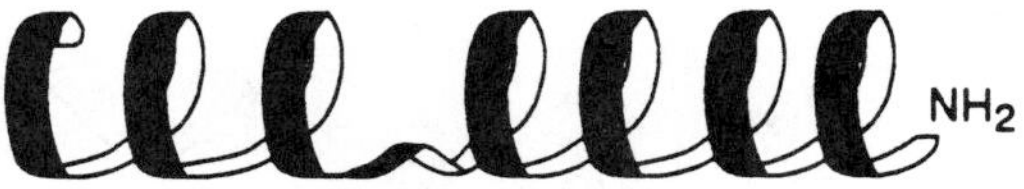

Figure 1.  Amino acid sequences of cecropins A, B, and D, of the hybrid cecropin
AD, of model peptides MP I-IV, of the cecropin A-melittin hybrid,
CA(1-13)M(1-13), and of melittin. Asterisks indicate a C-terminal amide,
dashes indicate that the amino acid is identical to the amino acid in
that position in cecropin A, and charged or hydrophobic amino acids
are indicated by +/- or o. Ribbon diagrams indicate regions of primary
structure corresponding to helices or bends in cecropin A or melittin.
Analyses by two-dimensional NMR and x-ray crystallography indicate
that the helices of cecropin A include residues 5-21 and 24-37, with
a bend located at residues 22-23 (14), and the helices of melittin include
residues 1-10 and 13-26, with a bend at residues 11-12 (28-29). Adapted
from references 9, 14, and 22-29.

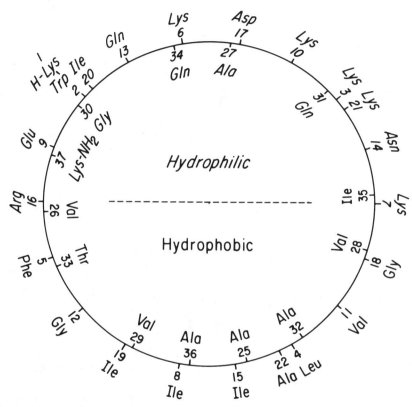

Figure 2. Amphipathic α-helix model for cecropin A. The structure represented is a planar projection from the N-terminus down the axis of the entire molecule, when constructed as two ideal α-helical segments. In this model, the N- and C-terminal helices are separated by Gly[23] and Pro[24], and are depicted as being coaxial. The helices have been adjusted arbitrarily to maintain the hydrophobic and hydrophilic surfaces of the two helices in the same orientation. Reprinted from ref. 12. Copyright 1982 American Chemical Society.

synthesized, radiolabelled pre-procecropins, evidence was obtained to show that the precursor proteins undergo post-translational processing, first by a microsomal enzyme which removes the 1-22 signal peptide and then by a dipeptidyl aminopeptidase to yield, in two steps, the mature peptides (17). This was the first example of a synthetic pre-proprotein and its *in vitro* processing to biologically active products. In addition, it was shown that the amide group located on the C-terminal residues of cecropins probably derives from a glycine residue that corresponds to termination of the coding part of the DNA sequence.

## Antibacterial Assays and Activities of Cecropins

The antibacterial activities of cecropins have been measured by a variety of techniques; however, the inhibition-zone assay is the most convenient and accurate method (9). In this assay (11), test bacteria are suspended in a thin layer of agar which contains a rich culture medium. A dilution series of a test peptide is then applied to small wells in the agar. During incubation, the bacteria multiply to form a turbid layer of small colonies, but diffusion of the peptide produces zones without bacterial growth around the wells. A plot is constructed of the squares of the zone diameters versus the logarithms of the concentrations of test peptide, and the slope and intercept of the regression line obtained from this plot are then used to calculate a lethal concentration (LC). LC-values are used to compare the antibacterial potencies of test peptides, and smaller values indicate higher potencies.

The major cecropins A and B have been shown to be highly active, with lethal concentrations in the $\mu$molar range against several Gram-positive and -negative bacteria, whereas the other major form, cecropin D, is active against only a few Gram-negative organisms (Table I).

## Structure and Function Studies of Cecropins

Based upon the ability of cecropins to lyse bacteria (8), their predicted potentials for forming amphipathic $\alpha$-helices, and their propensities for adopting helical conformations in hydrophobic environments (12,13), Steiner suggested that the mechanism of attachment of cecropins to membranes might be similar to that proposed for the bee venom toxin, melittin (13). According to this hypothesis, the amphipathic $\alpha$-helix of the molecule anchors to the surface of the bacterial membrane while the central part of the molecule penetrates deeper into the lipid.

The importance of the N-terminal amphipathic $\alpha$-helix (residues 1-11) to the bactericidal activity of cecropins was investigated in a series of experiments by Merrifield, Boman, and colleagues (12,18,19). Solid-phase peptide synthesis techniques (20,21) were used to prepare cecropin A, and various N-terminal and truncated analogs of cecropin A. The synthetic peptides were then compared by assaying their antibacterial activities against representative Gram-positive and -negative bacteria (Table I). Natural and synthetic cecropin A had identical

Table I Antibacterial activities of the major cecropins, and of several synthetic analogs, hybrids, and model peptides against representative Gram-positive and Gram-negative organisms.

| Peptide: | Lethal Concentrations ($\mu$M): | | | |
| | Gram-negative: | | Gram-positive: | |
| | D21: | OT97: | Bm11: | Ml11: |
|---|---|---|---|---|
| Cecropin A (N) | 0.3 ± 0.1(3) | 3.6 ± 1.1(3) | 0.6 ± 0.1(3) | 2.6 ± 1.7(3) |
| Cecropin A (S) | 0.4 ± 0.2(3) | 2.0 ± 0.7(3) | 0.8 ± 0.4(3) | 1.6 ± 0.6(3) |
| CA (1-37)COOH | 0.5 | 8.2 | 10.8 | 17.8 |
| Cecropin B (N) | 0.4 ± 0.1(2) | 1.8 ± 0.2(2) | 0.7 ± 0.3(2) | 1.4 ± 0.1(2) |
| Cecropin B (S) | 0.6 ± 0.1(2) | 1.4 | 0.8 | 1.3 |
| Cecropin D (N) | 2.5 ± 2.1(2) | >95 ±5 (2) | 34.5 ± 6.5(2) | 28 ±7 (2) |
| Cecropin D (S) | 2.6 | 66 | 31 | 12 |
| CA(1-33)COOH | 0.43 | 13 | 24 | 27 |
| CA(2-37) | 0.37 | 8.6 | 1.5 | 7.3 |
| CA(3-37) | 2.6 | 90 | 13 | >110 |
| CA(Phe$^2$) | 0.34 | 3.5 | 0.78 | 7.4 |
| CA(Glu$^2$) | 3.2 | 170 | 39 | >170 |
| CA(Pro$^4$) | 0.36 | 8.1 | 11 | 87 |
| CA(Leu$^6$) | 0.56 | 120 | 0.78 | 7.3 |
| CA(Glu$^6$) | 0.58 | 34 | 2.2 | 4.7 |
| CA(Pro$^8$) | 0.50 | 15 | 31 | 80 |
| CD(Lys$^1$) | 1.1 | 37 | 16 | 7 |
| CD(Gln$^3$,Leu$^4$) | 3.2 | 86 | 24 | 11 |
| CD(9-37) | >390 | >390 | >390 | >390 |
| CA(1-11)-CD(12-37) | 0.5 | 1.2 | 0.83 | 0.71 |
| MP I | 15 | 34 | 28 | 23 |
| MP II | >206 | >206 | >206 | >206 |
| MP III | 0.7 | 4 | 4 | 2 |
| MP IV | 39 | 71 | 13 | 71 |

Abbreviations for organisms: D21, Escherichia coli D21; OT97, Pseudomonas aeruginosa OT 97; Bm11, Bacillus megaterium 11; Ml11, Micrococcus luteus 11. Other abbreviations: (N), natural; (S), synthetic; CA, CB, CD cecropins A, B, and D; COOH, peptide with C-terminal carboxyl rather than amide; (Phe$^2$), CA with Phe substituted at postition 2. Number in parentheses represents number of values used to calculate the mean ± std. dev. (3) or ± range (2). From references 9, 11, 13, 15, 17-19, 23, and 24.

antibacterial activities, and assays with the truncated analogs indicated that the complete molecule was needed for full activity against all test bacteria, although good activity against *Escherichia coli* was retained when Lys[1] and residues 34-37 of cecropin A were deleted. Tryptophan at position 2 is conserved in all cecropins, and an aromatic amino acid at that position seemed to be essential for antibacterial activity. Activity did not require an *extended* N-terminal α-helix, but broad spectrum activity was lost when the helix was disrupted by substitution with proline. Good activity against all test bacteria also required an amidated C-terminus; without it, activity was reduced 2-13 fold.

Steiner et al. studied the mechanism of action of cecropins by using unilamellar liposomes as a model system (22). These artificial membranes were disrupted by cecropins, and there was a strong correlation between the abilities of synthetic cecropin analogs to lyse liposomes and kill bacteria. The latter result suggested that the phospholipid bilayer of the bacterial membrane is the site of cecropin action. Cecropin analogs with impaired N-terminal amphipathic α-helices were not as efficient as cecropin itself in membrane disruption, and this result was attributed to a decrease in the affinities of these peptides for binding to membranes.

Fink et al. (23,24) examined the structure of cecropins A, B, and D in an attempt to determine the basis for differences in their antibacterial potencies. Analyses of charge distribution, hydrophobicities, and predicted secondary structures revealed several similarities between cecropins A and B. However, cecropin D was shown to be more hydrophobic, and it had a greater potential for α-helix formation in its central and C-terminal segments. To determine whether or not these factors were related to the biological activity of cecropins, Fink et al. (23) synthesized and assayed the antibacterial activities of cecropin D and analogs of D, and several cecropin-like model peptides. Synthetic cecropin D was identical to its natural counterpart in physical and chemical properties and antibacterial assays (Table I). Comparisons of the activities of cecropin D analogs with those of cecropins A and B indicated that a strongly basic N-terminal segment was required for antibacterial activity. A synthetic hybrid of cecropins A and D produced the first analog to exhibit greater activity than either of the parental peptides. Its enhanced activity was attributed to a combination of the basic, amphipathic, α-helical, N-terminal segment of cecropin A (residues 1-11) with the helical and highly hydrophobic C-terminal segment of cecropin D (residues 12-37).

The design of model peptides was based on the sequences of natural cecropins, and included modifications of whole segments of the peptides in order to mimic special structural features of cecropins (24). Circular dichroism measurements of the model peptides in a hydrophobic solvent showed good agreement between predicted and actual conformations. The results of antibacterial assays indicated that high antibacterial activity in a cecropin-like peptide required a helical, amphipathic, N-terminal segment connected to a helical, hydrophobic C-terminal segment by a flexible, non-helical hinge region (Table I).

Further support for this hypothesis was obtained in studies of the effect of several of these peptides on the electrical conductivity of lipid bilayers (25). The cecropins were shown to form large, time-variant and voltage-dependent ion channels in planar lipid bilayer membranes, and a flexible segment between the N-terminal amphipathic and C-terminal hydrophobic regions was required for conductance. As in antibacterial assays, the cecropin AD hybrid was the most effective pore-forming peptide. The presence of cholesterol in the lipid bilayer reduced conductances due to the peptides, a behavior consistent with the known insensitivity of eucaryotic cells to cecropins. Overall, the results suggested that antibacterial activity of cecropins was due to formation of large pores in bacterial cell membranes (Figure 3).

## Comparisons of Cecropins and Melittin

Steiner et al. first observed similarities in the structures and activities of cecropins and the bee venom toxin, melittin (8). They noted that both types of molecules were short polypeptides containing distinct basic and hydrophobic segments of amino acid sequences (Figure 1). The N-terminal portion of cecropins is strongly basic and the C-terminal segment is predominantly hydrophobic whereas, in melittin, this sequence arrangement is reversed. The C-terminus of melittin had been shown to be amidated (26), and the C-terminal blocking group of cecropins was later shown to also be a primary amide (19). Secondary structure predictions for these molecules indicated that they each contained two helical segments, which were joined by a flexible region, or bend, and that the helices were of an amphipathic nature (12,13,27). Results of recent analyses by X-ray crystallography (28) and two-dimensional NMR techniques (14,29) support these structure predictions (Figure 1). Although the cecropins and melittin both exhibited antibacterial activity (8,30), only melittin lysed eucaryotic cells (8). Similarities in the structures of cecropins and melittin, but differences in their specificities, led Boman et al. to suggest that a synthetic program exploring the properties of cecropin-melittin hybrids would yield information concerning the structural bases for the specificities of these molecules (30,31). This investigation is currently in progress, and some preliminary findings are discussed below.

Cecropin A, melittin, and analogs and hybrids of these peptides were synthesized as peptide amides by solid-phase methods, and purified by reverse-phase HPLC. The peptides were assayed for antibacterial activity against Gram-positive and -negative bacteria as described earlier, except that agarose was used instead of agar (17), and for their ability to lyse red cells (32,33). The results (Table II) indicated that melittin was a better broad spectrum antibiotic than cecropin A; however, melittin was hemolytic and cecropin A was not. Lysis of eucaryotic cells is not a desirable property of an antibacterial agent.

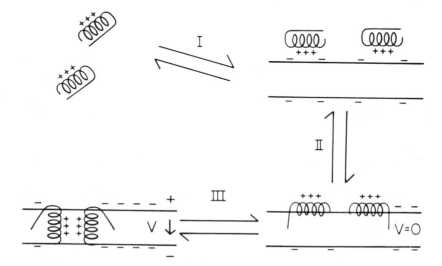

Figure 3. Model for the interaction of cecropins with a lipid bilayer membrane. Peptides adsorb to the bilayer-water interface by electrostatic forces (I), and the hydrophobic segment of each molecule then inserts into the membrane core (II). Upon application of a voltage (positive on the side of peptide addition), a major conformational rearrangement takes place resulting in channel formation (III).

Reprinted with permission from ref. 25. Copyright 1988 National Academy of Sciences.

Table II.  Lethal and Lytic Concentrations ($\mu$M) of Cecropin A and Melittin, of Analogs of These Peptides, and of the Cecropin A - Melittin Hybrid Against Bacteria and Red Blood Cells[a]

| Peptide Amides: | Gram-negative: | | Gram-positive: | | | SRC: |
|---|---|---|---|---|---|---|
| | D21: | OT97: | Bs11: | SaC1: | Sp1: | |
| CA(1-37) | 0.2 | 2 | 4 | >200 | 4 | >200 |
| M(1-26) | 0.8 | 3 | 0.2 | 0.2 | 0.5 | 4-8 |
| CA(25-37)(1-13) | 200 | 300 | 20 | >300 | 20 | >300 |
| M(16-26)(1-13) | 0.7 | 8 | 0.7 | 10 | 1 | >200 |
| CA(1-13)M(1-13) | 0.5 | 1 | 0.7 | 2 | 1 | >200 |

[a] Abbreviations: D21, *Escherichia coli* D21; OT97, *Pseudomonas aeruginosa* OT97; Bs11, *Bacillus subtilis* 11; SaC1, *Staphylococcus aureus* Cowan 1; Sp1, *Streptococcus pyogenes* 1; SRC, sheep red cells; CA, cecropin A; M, melittin. Adapted from references 32 and 33.

Analog CA(25-37)(1-13) is a shortened version of cecropin A with transposed amino acid segments, and with its central flexible region removed. This analog has a hydrophobic N-terminal segment and a basic C- terminal segment; an arrangement that mimics that of melittin. The antibacterial activity of this analog was greatly reduced with respect to cecropin A, and it was non-hemolytic for sheep red cells.

The M(16-26)(1-13) analog is a shortened version of melittin with inverted segments. The omitted amino acids were those predicted to comprise the flexible hinge region of melittin (27). This analog mimicked the amino acid sequence arrangement of cecropin A, with a basic N-terminal part and a hydrophobic C-terminal part. It retained substantial antibacterial activity against four species, and it had the desirable property of being non-hemolytic.

Hybrid CA(1-13)M(1-13) was designed to contain the basic N-terminal segment of cecropin A linked to the hydrophobic N-terminal portion of melittin, a sequence arrangement that mimics cecropin A. This hybrid had approximately the same antibacterial activity as the parent peptides against Gram-negative bacteria, but was up to 100 fold more active than cecropin A against certain Gram-positive species, a very significant improvement. In addition, it was non-hemolytic.

Two analogs of cecropin B were recently reported to inhibit the blood stream forms of the malaria parasite, *Plasmodium falciparum* (34), and magainin 2 and cecropin B were also found to inhibit the mosquito forms of the parasite (35). We therefore tested the CA(1-13)M(1-13) hybrid for its ability to inhibit the growth of the blood stream forms of this parasite, and found that it was ten-fold more potent than cecropin B or magainin 2 in its ability to inhibit the reinvasion of human red cells by *P. falciparum* and was non-hemolytic (32).

In summary, these results demonstrate that the general approach of synthesis and study of analogs and hybrids of these peptides can lead to an understanding of the structural features that are important for their biological activities and may lead to improved antibiotics.

## Literature Cited

1. Holmberg, S. D.; Solomon, S. L.; Blake, P. A. Revs. Infectious Dis. 1987, 9, 1065-78.
2. Hultmark, D.; Steiner, H.; Rasmuson, T.; Boman, H. G. Eur. J. Biochem. 1980, 106, 7-16.
3. Lee, J.-Y.; Boman, A.; Sun, C.; Andersson, M.; Jörnvall, H.; Mutt, V.; Boman, H. G. Proc. Natl. Acad. Sci. USA 1989, 86, 9159-62.
4. Zasloff, M. Proc. Natl. Acad. Sci. USA 1987, 84, 5449-53.
5. Selsted, M. E.; Harwig, S. S. L. Infect. Immun. 1987, 55, 2281-6.
6. Matsuyama, K.; Natori, S. J. Biol. Chem. 1988, 263, 17112-6.
7. Lambert, J.; Keppi, E.; Dimarcq, J.-L.; Wicker, C.; Reichhart, J.-M.; Dunbar, B.; Lepage, P.; Dorsselaer, A. V.; Hoffmann, J.; Fothergill, J.; Hoffman, D. Proc. Natl. Acad. Sci. USA 1989, 86, 262-6.
8. Steiner, H.; Hultmark, D.; Engström, Å.; Bennich, H.; Boman, H. G. Nature 1981, 292, 246-8.
9. Boman, H. G.; Hultmark, D. Ann. Rev. Microbiol. 1987, 41, 103-26.
10. Boman, H. G. In Chemical Recognition in Biology, Chapeville, F.; Haenni, A.-L. Eds.; Springer-Verlag: New York, 1980; pp 217-28.
11. Hultmark, D.; Engström, Å.; Bennich, H.; Kapur, R.; Boman, H. G. Eur. J. Biochem. 1982, 127, 207-17.
12. Merrifield, R. B.; Vizioli, L. D.; Boman, H. G. Biochemistry 1982, 21, 5020-31.
13. Steiner, H. FEBS Lett. 1982, 137, 283-7.
14. Holak, T. A.; Engström, Å.; Kraulis, P. J.; Lindeberg, G.; Bennich, H.; Jones, T. A.; Gronenborn, A. M.; Clore, G. M. Biochemistry 1988, 27, 7620-9.
15. von Hofsten, P.; Faye, I.; Kockum, K.; Lee, J.-Y.; Xanthopoulos, K. G.; Boman, I. A.; Boman, H. G.; Engström, Å.; Andreu, D.; Merrifield, R. B. Proc. Natl. Acad. Sci. USA 1985, 82, 2240-3.
16. Lidholm, D.-A.;Gudmundsson, G. H.; Xanthopoulos, K. G.; Boman, H. G. FEBS Lett. 1987, 226, 8-12.
17. Boman, H. G.; Boman, I. A.; Andreu, D.; Li, Z.-Q.; Merrifield, R. B.; Schlenstedt, G.; Zimmermann, R. J. Biol. Chem. 1989, 264, 5852-60.
18. Andreu, D.; Merrifield, R. B.; Steiner, H.; Boman, H. G. Biochemistry 1985, 24, 1683-8.
19. Andreu, D.; Merrifield, R. B.; Steiner, H.; Boman, H. G. Proc. Natl. Acad. Sci. USA 1983, 80, 6475-9.
20. Merrifield, R. B. J. Am. Chem. Soc. 1963, 85, 2149-54.
21. Barany, G.; Merrifield, R. B. In The Peptides Vol. 2, Part A., Gross, E.; Meienhofer, J., Eds.; Academic Press: New York, 1980, pp 1-284.

22. Steiner, H.; Andreu, D.; Merrifield, R. B. Biochim. Biophys. Acta 1988, 939, 260-6.
23. Fink, J.; Merrifield, R. B.; Boman, A.; Boman, H. G. J. Biol. Chem. 1989, 264, 6260-7.
24. Fink, J.; Boman, A.; Boman, H. G.; Merrifield, R. B. Int. J. Peptide Protein Res. 1989, 33, 412-21.
25. Christensen, B.; Fink, J.; Merrifield, R. B.; Mauzerall, D. Proc. Natl. Acad. Sci. USA 1988, 85 5072-6.
26. Habermann, E.; Jentsch, J. Hoppe-Seyler's Z. Physiol. Chem. 1967, 348, 37-50.
27. Dawson, C. R.; Drake, A. F.; Helliwell, J.; Hider, R. C. Biochim. Biophys. Acta 1978, 510, 75-86.
28. Terwilliger, T. C.; Eisenberg, D. J. Biol. Chem. 1982, 257, 6016-22.
29. Bazzo, R.; Tappin, M. J.; Pastore, A.; Harvey, T. S.; Carver, J. A.; Campbell, I. D. Eur. J. Biochem. 1988, 173, 139-46.
30. Boman, H. G. Fortschr. Zool. 1982, 27, 211-22.
31. Boman, H. G.; Faye, I.; v. Hofsten, P.; Kockum, K.; Lee, J.-Y.; Xanthopoulos, K. G.; Bennich, H.; Engström, Å.; Merrifield, R. B.; Andreu, D. In Immunity in Invertebrates, Brehelin, M. Ed.; Springer-Verlag: Berlin, 1986; pp 63-73.
32. Boman, H. G.; Wade, D.; Boman, I.A.; Wåhlin, B.; Merrifield, R. B. FEBS Lett. 1989, 259, 103-6.
33. Wade, D.; Merrifield, R. B.; Boman, H. G. In Peptides: Chemistry, Structure and Biology. Rivier, J.E.; Marshall, G. R., Eds.; ESCOM Science Publishers: Leiden, The Netherlands, 1990; pp 120-1.
34. Jaynes, J. M.; Burton, C. A.; Barr, S. B., Jeffers, G. W.; Julian, G. R.; White, K. L.; Enright, F. M.; Klei, T. R.; Laine, R. A. FASEB J. 1988, 2, 2878-83.
35. Gwadz, R. W.; Kaslow, D.; Lee, J.-Y.; Maloy, W. L.; Zasloff, M.; Miller, L. H. Infect. Immun. 1989, 57, 2628-33.

RECEIVED August 27, 1990

# Chapter 19

# Inhibition of Ice Crystal Growth by Fish Antifreezes

**James A. Raymond[1] and Arthur L. DeVries[2]**
[1]**Alaska Department of Fish and Game, Fairbanks, AK 99701**
[2]**Department of Physiology and Biophysics, University of Illinois, Urbana, IL 61801**

Polar and subpolar marine fishes have peptide and glycopeptide antifreezes in their body fluids that protect them from freezing. The antifreezes have a high affinity for ice and prevent it from growing by a non-colligative process. Despite wide variations in composition and structure, the antifreezes cause ice single crystals to develop unusual and strikingly similar habits. This indicates that the antifreezes have affinities for similar crystal faces of ice.

In many parts of the world's oceans, water temperatures fall to as much as 1° below the equilibrium freezing point of fishes' body fluids. If a temperate water fish is placed in these waters, it will die immediately from freezing. One can actually see the ice propagating through its tissues. However, many fishes inhabit the polar and subpolar regions where freezing occurs. Avoidance of freezing in these fishes has been linked to the presence of a class of serum peptides and glycopeptides (1). These antifreezes are not found in temperate water fishes and they disappear in summer in fishes that experience warmer summer temperatures (1-3). Some polar fishes do not have an antifreeze, and avoid freezing by existing in a supercooled state in ice-free deeper waters (4). Fishes that have an antifreeze are usually found in shallow waters where ice particles are abundant.

## Composition and Structure

The first antifreezes to be identified were a series of glycopeptides in the antarctic fishes that consisted of a repeating tripeptide-disaccharide unit (1):

-Ala-Ala-Thr-
galactose N-acetylgalactosamine

The glycopeptides are found in eight sizes ranging in molecular mass from 2300 to 34,000. In the smaller glycopeptides, glycopeptides 6-8, the first alanine is occasionally replaced with proline.

The freezing point depressing activity of the glycopeptide antifreezes is shown as a function of concentration (Figure 1). There are a number of features in this figure that show the antifreezes are unusual freezing point depressants. First, the curve is non-linear. Most solutes, such as NaCl, produce linear freezing point depressing curves. Second, the freezing point depression is, for a given molar concentration, about two orders of magnitude greater than one would expect. Another unusual feature is a difference in the freezing and melting curves; if a 10mg/ml solution freezes at -0.6°C, the temperature would have to be raised to close to 0°C before it would melt. This is called a melting point-freezing point hysteresis. These features indicate that the antifreezes are quite different from normal solutes in their effect on ice.

Do the antifreezes merely slow down the growth of ice, or do they completely halt it? We have placed ice crystals in glycopeptide antifreeze solutions at subzero temperatures for up to 13 days and have not detected any growth at 40X magnification (5). The glycopeptide antifreeze thus appears to completely stop growth of ice.

One approach to understanding the antifreeze mechanism was to compare antifreezes from different species of fish. If different antifreezes were found, an analysis of their common features would help to focus on the active part of the molecules. Two of the first non-antarctic fish to be examined were the saffron cod, *Eleginus gracilis* and the sculpin *Myoxocephalus verrucosus*.

The antifreeze of the saffron cod was found to be almost identical to glycopeptides 6-8 (6,7). One difference was the occasional substitution of arginine for threonine. This was quite surprising because the cod family and the antarctic nototheniid fishes, from which glycopeptides 1-8 were obtained, are widely separated, both taxonomically and geographically.

Circular dichroism spectra of the glycopeptide antifreezes indicated either an extended coil configuration or a $3_1$ helix configuration similar to that found in collagen. In order to distinguish between the two spectra, a solution of glycopeptides 1-5 was gradually heated from 0 to 95°. An abrupt loss of CD signal at around 40°C would have indicated the melting of a helix. However, only a gradual loss in signal was observed, indicating that the extended coil was a more likely configuration (8).

The antifreezes of two other northern fishes, the sculpin *Myoxocephalus scorpius* and the winter flounder, *Pseudopleuronectes americanus*, were found to be very similar to one another (1,9). Like the glycopeptide antifreezes, these antifreezes consisted of more than 60% alanine, repeating sequences of amino acids, and were of relatively small molecular mass. On the other hand, the repeats consisted of 11 amino acids and the molecules did not contain a sugar moiety. Although not sequenced, the antifreeze of *M. verrucosus* was similar to these peptides in amino acid composition.

Circular dichroism spectra of the antifreeze of *M. verrucosus* and *P. americanus* were very different from the glycopeptide CD spectra. The CD spectra

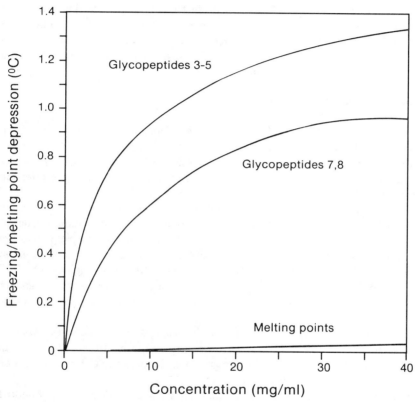

Figure 1. Freezing and melting point depressions of aqueous solutions of glycopeptide antifreezes.

indicated that approximately 85% of these molecules were in the alpha-helical configuration (8,10).

More recently, other antifreezes from the sculpin and zoarcid families have been characterized and have been found to differ from the above glycopeptide and peptide antifreezes. Antifreezes from three different zoarcids from the arctic and the antarctic have nonrepeating sequences, with between 61 and 64 amino acid residues, and have homologies ranging from 56 to 69% (11-13). CD analysis of one of these antifreezes indicated a distinct tertiary structure that lacks both alpha-helical and beta components (12). The antifreeze of the sea raven, a northern sculpin, has a molecular mass of 17,000 Da and a nonrepeating sequence that differs from the zoarcid sequences (14). Its secondary structure includes both beta and alpha-helical components.

## Physical Chemical Properties

How do the antifreezes affect the growth of ice? Normally, when ice grows from solution, it grows in a direction parallel to the a-axes and forms large plates. In the presence of most of the antifreezes, however, ice grows in long spicules that are parallel to the ice c-axis (15). In other words, growth in the direction of the a-axes is inhibited.

Another unusual property of the antifreezes is that they are not completely excluded from ice as it grows, as are most solutes. The distribution coefficient quantifies the exclusion, with a value of zero corresponding to complete exclusion and a value of 1.0 corresponding to no exclusion, i.e., equal concentrations in ice and liquid. The antifreezes have distribution coefficients ranging from 0.17 to 0.89 (15). In general, the larger the molecular mass of the antifreeze, the greater is its incorporation in ice. For comparison, the distribution coefficient for bovine serum albumin was 0.10.

Because of the affinity of the antifreezes for ice, it is reasonable to ask whether the structure of the ice is in any way changed by the antifreeze. However, an x-ray diffraction pattern of spicular ice grown in the presence of the glycopeptide antifreeze showed normal hexagonal ice (5).

Ideas underlying the antifreeze mechanism were developed over a hundred years ago when Lord Kelvin found that the equilibrium of a system depended not only on the bulk phases but also on the surface that separated them. Working from Kelvin's equation, Kuhn (16) derived an expression for the lowering of the equilibrium freezing temperature of water due to surface effects. We applied Kuhn's equation to the case of adsorbed antifreeze molecules on an ice surface (15). Because the growth of crystals occurs mainly through the advancement of steps, adsorbed molecules will force the steps to grow between them, thus increasing the curvature and surface area of the steps. The result is a freezing point depression that is proportional to the square root of the product of the antifreeze concentration and the antifreeze's distribution coefficient. The predicted freezing point depressions for various concentrations agree well with the observed freezing point depressions.

Various mechanisms have been proposed to account for the binding of the antifreezes to ice. Sequences of the alpha-helical antifreezes show that they are amphiphilic, i.e., polar and nonpolar residues are on opposite sides of the molecule (1,17). If the polar groups are spaced a distance that corresponds to a fundamental repeat distance in the ice lattice, as has been shown for the antifreeze of the winter flounder (1), the polar side of the antifreeze would have a higher affinity for water molecules in the lattice than water molecules in solution. This type of affinity is called lattice matching. In addition, the nonpolar groups on the antifreeze would project out into the water and tend to keep other water molecules from joining the lattice. The existence of amphiphilicity and lattice matching in other, non-alpha helical antifreezes has not yet been confirmed or ruled out. Another binding mechanism recently proposed for the alpha helical antifreezes is a dipole- dipole interaction (18). In this model, an electric dipole associated with the alpha helix induces an opposite dipole in the surface of the ice and is then attracted to it.

In an attempt to see more clearly how the antifreezes bind to ice, recently we have used large single crystals of ice with well-defined basal (0001) and prism (10$\bar{1}$0) faces (19). We found a number of unusual growth features in the ice. The antifreezes completely inhibit growth on the prism faces, but allow limited growth on the basal plane. As new layers are deposited on the basal plane, pyramidal planes develop on the outside of the crystal, and large, hexagonal pits form within the basal plane (Figure 2). This growth eventually leads to the disappearance of the basal plane, at which point all growth ceases. An unusual feature of the pits was that their orientation is rotated 30° with respect to the normal orientation of hexagonal ice crystals. These features were found to occur with 6 antifreezes having widely varying compositions and structures. The unusual faces that develop in ice single crystals indicate that they are sites of binding by the antifreezes. As each new layer is deposited on an adjacent faster growing face, the retarded face increases slightly in area. The result is a crystal that is dominated by its slowest growing faces. That the crystal habits were similar for different antifreezes suggests that they share a common binding mechanism.

Some features of the pits suggested that they were associated with dislocations: spiral steps occasionally occurred in the pits and rows of pits often were aligned with low-index crystal planes. The number density of the pits was also inversely proportional to the temperature at which the basal plane growth occurred. These observations suggest the possibility that dislocations may play a role in the binding of the antifreezes to ice. However, further study is needed to confirm this.

Figure 2. A single crystal of ice grown in the presence of the glycopeptide antifreeze of the antarctic fish *Dissostichus mawsoni*. In this view, the basal plane, shown as small flat areas, is deeply pitted with hexagonal pits. The pit faces are likely sites of antifreeze adsorption. The antifreeze inhibits growth on the pit faces but not on the basal plane. Growth ceases when the pit faces completely cover the basal plane. Width of pits is approximately 500 $\mu$.

## Literature Cited

1. DeVries, A. L. Ann. Rev. Physiol. 1982, 45, 245-60.
2. Duman, J. G.; DeVries, A. L. J. Exp. Zool. 1974, 190, 89-97.
3. Fletcher, G. L.; Haya, K.; King, M. J.; Reisman, H. M. Mar. Ecol. Prog. Ser. 1985, 21, 205-12.
4. Scholander, P. F.; van Dam, L.; Kanwisher, J. W.; Hammel, H. T.; Gordon, M. S. J. Cell Comp. Physiol. 1957, 49, 5-24.
5. Raymond, J. A. Ph.D. Thesis, University of California, San Diego 1976.
6. Raymond, J. A.; Lin, Y.; DeVries, A. L. J. Exp. Zool. 1975, 193, 125-30.
7. O'Grady, S. M.; Schrag, J. D.; Raymond, J. A.; DeVries, A. L. J. Exp. Zool. 1982, 224, 177-85.
8. Raymond, J. A.; Radding, W.; DeVries, A. L. Biopolymers 1977, 16, 2575-8.
9. Hew, C. L.; Fletcher, G. L.; Ananthanarayanan, V. S. Can. J. Biochem. 1980, 58, 377-83.
10. Ananthanarayanan, V. S.; Hew, C. L. Biochem. Biophys. Res. Commun. 1977, 74, 685-9.
11. Schrag, J. D.; Cheng, C.-H. C.; Panico, M.; Morris, H. R.; DeVries, A. L. Biochim. Biophys. Acta 1987, 915, 357-70.
12. Li, X. M.; Trinh, K. Y.; Hew, C. L.; Buettner, B.; Baezinger, J.; Davies, P. L. J. Biol. Chem. 1985, 260, 12904-9.
13. Cheng, C.-H. C.; DeVries, A. L. Biochim. Biophys. Acta 1989, 997, 55-64.
14. Hew, C. L.; Joshi, S.; Wang, N. C.; Kao, M. H.; Ananthanarayanan, V. S. Eur. J. Biochem. 1985, 151, 167-172.
15. Raymond, J. A.; DeVries, A. L. Proc. Natl. Acad. Sci. USA 1977, 74, 2589-93.
16. Kuhn, W. Helv. Chim. Acta 1956, 39, 1071-86.
17. Hew, C. L.; Fletcher, G. L. In Proceedings in Life Science; Gilles, R., Ed.; Springer-Verlag, Heidelberg, 1985; pp 553-563.
18. Yang, D. S.; Sax, M.; Chakrabartty, A.; Hew, C. L. Nature 1988, 333, 232-7.
19. Raymond, J. A.; Wilson, P.; DeVries, A. L. Proc. Natl. Acad. Sci. USA 1989, 86, 881-5.

RECEIVED August 27, 1990

# Chapter 20

# Marine Metal Chelating Proteins

**Douglas C. Hansen and J. Herbert Waite**

**College of Marine Studies, University of Delaware, Lewes, DE 19958–1298**

The attachment strategy of the common sea mussel *Mytilus edulis* relies on the formation of byssal threads which originate in the animal at one end and attach to a surface at the other. The point of attachment for these threads are small plaques which contain an adhesive protein. This protein contains the amino acid 3,4-dihydroxyphenylalanine (DOPA). DOPA-containing proteins have been isolated from other mytilid bivalves, namely *Mytilus californianus* and *Geukensia demissa*. Being an o-diphenol, this amino acid enables the adhesive plaque to displace water molecules through hydrogen bonding and to bind to metal ions that are present in the substrata. This chemisorption effectively anchors the mussel to the surface and allows it to resist the physical forces to which it is exposed. The byssal threads of these bivalves have demonstrated preferential accumulation and sequestration of trace metals. The adhesive qualities and the metal binding properties of this protein in aqueous environments present today's researcher with numerous potential uses and applications.

The common sea mussel *Mytilus edulis* has an attachment strategy that relies on the formation of byssal threads which originate in the animal at one end and attach to a surface at the other. The ability of mytilid bivalves to attach themselves to objects underwater has been an object of curious fascination for man for centuries. Indeed, Aristotle mused that byssal threads were some sort of parasite or symbiotic plant that grew into the animal (1). These byssal threads are actually extracorporeal tendons that originate in the byssal retractor muscles and attach to any object that the animal can find (2). Mussels, like barnacles, are adhesive opportunists, and these animals will attach themselves to anything that provides them with a secure, solid holdfast. The points of attachment for these threads are small, flattened ovoid discs, or plaques, which contain the adhesive protein. The threads fuse with the plaque in such a way as to increase the surface area with the plaque material (2).

0097–6156/91/0444–0256$06.00/0

The formation of the thread and plaque is the result of a very complicated secretory process that is orchestrated by the foot of the mussel. This organ is very muscular and is covered with cilia (3). On its ventral surface, the foot has a groove extending from the base to the tip where it ends in a distal depression. Byssal threads are formed in the following sequence of events. The valves of the animal gape and the foot emerges. After a brief exploratory interval, the foot pushes the distal depression against a surface until the bulk of the water is displaced (4). Then, using a series of contracting muscles, the ceiling of the depression is lifted creating a vacuum within. Adhesive protein is discharged into the evacuated cavity through a series of 'injector ports' (4). After a few minutes the foot retracts and, as it does, the plaque emerges from the distal depression and the thread from the ventral groove to form a contiguous attachment structure. When the thread has completely disengaged from the groove, the sequence is ready to be repeated. At the outset the thread and plaque are a milky-white color, but over time they take on a golden-brown luster. This is evidence of the sclerotization that is taking place (5).

The foot has a number of exocrine glands, which produce the various chemical components of the byssal thread. These are the phenol gland, the collagen gland and the accessory gland. The phenol gland is responsible for secreting the adhesive protein at the interface; the collagen gland secretes the tensile element present in the core of the threads, and the accessory gland secretes the coating that forms the cortex surrounding the thread (2). The protein that is present in the plaques and also in the cortex of the thread is the cornerstone of the adhesive strategy of these organisms.

### Isolation and Characterization of Adhesive Protein

Simple hydrolysis of the byssus of *M. edulis* yields an amino acid composition that is high in glycine and lower in hydroxyproline and proline for the whole thread, as expected for a structural, fibrous protein complex (6). A similar composition is found in the plaques except that there is a higher amount of L-dopa present. This amino acid is named dopa from the German nomenclature "dioxyphenylalanin" and is a post-translational modification of tyrosine. Intuitively, the existence of a precursor protein containing this amino acid was suggested, and an assay to facilitate its location sought. In 1937, L.E. Arnow (7) described an assay for the colorimetric determination of the components of 3,4-dihydroxyphenylalanine and tyrosine mixtures. This assay, despite its age, is very specific and yields a dinitrocatechol that absorbs strongly at 500 nanometers. With this assay then, it was quite easy to demonstrate that the adhesive, or dopa-containing, protein was indeed present in extracts of the foot, thread and plaque (8).

Extraction of the dopa protein from the tip of the foot yields a pure protein as determined by acidic polyacrylamide gel electrophoresis (8). The amino acid composition of this purified dopa protein contains hydroxyproline, although it is not collagen as previously suspected due to its low glycine content and high levels of hydroxylated amino acids, such as dopa and lysine. The molecular weight of this protein as determined by polyacrylamide gel electrophoresis in the presence of cetylpyridinium bromide is 130,000 ($\pm$ 10,000) and it has a pI above 10 (9). Tryptic digestion of the protein yields a consensus decapeptide that is

1    2    3    4    5    6    7    8    9    10

NH2 Ala Lys Pro Ser Tyr Hyp Hyp Thr Dopa Lys  COOH
(Dopa)

and is repeated approximately 80 times (10). Researchers at Genex Corp. (11) and at Boston University (Laursen and Connors, unpublished studies) have sequenced about 80% of a cDNA prepared from dopa-protein mRNA and have found 76 repeats of either hexa- or deca-peptides with many of the positions being highly or completely conserved therefore corroborating the repeat structure predicted earlier for this protein.

The dopa protein isolated from other mytilid bivalves, namely *Mytilus californianus* and *Geukensia demissa,* have similar amino acid sequences in their peptides. Consensus decapeptides of *M. californianus* and *G. demissa* are, respectively:

1    2    3    4    5    6    7    8    9    10

NH2 Ile Thr Tyr Hyp Hyp Thr Dopa Lys Hyp Lys  COOH
(Dopa)

and

1    2    3    4    5    6    7    8    9    10

NH2  Gln Thr Gly Tyr Ser* Ala** Gly Dopa Lys   COOH
(Dopa)
*(Val/Asp) **(Pro/Hyp)

(12,13). The similarity between *M. edulis* protein and *M. californianus* proteins is quite striking. Positions 2 to 8 of the *M. californianus* decapeptide are essentially indistinguishable from *M. edulis* decapeptide positions 4 to 10 (12). Comparing the *M. edulis* consensus peptide to that of *G. demissa*, it is clear that there is a superimposibility of the second dopa and lysine residues. In all three cases, dopa/tyrosine positions are found three residues apart and the second dopa/tyrosine residue, located at the carboxy end, is followed by a lysine. This illustrates the continuity that seems to be present in the adhesive proteins of mytilid bivalves.

The secondary structure of this polypeptide in solution has yet to be elucidated. It has a CD spectrum that is characteristic of a random coil, either in the presence of sea water-like conditions or in 6M guanidine hydrochloride. To date, laboratory results indicate that when a peptide consisting of 20-30 consensus repeats is exposed to ascorbate, oxygen and mushroom tyrosinase at a pH of 7.2, only the second tyrosine of each consensus repeat was hydroxylated 100%. This clearly suggests that the conformation of the polypeptide is not a random coil, but that tyrosines particularly in position 5 are sheltered possibly by a $\beta$-turn (14).

## Adhesion and Complexation of Metals

o-Diphenols form hydrogen bonds with hydrogen acceptors that are more powerful than water so displacement of water molecules or dehydration of adsorbent surfaces can occur (15). This phenomenon is well known in the technology of tanning of animal skins where the use of polyphenolic vegetable tannins is applied to cure the collagen present in the skin to produce leather (16). One of the actions of the mussel adhesive in sticking to a film that is made out of a protein or carbohydrate (as in the case of organic films that form on surfaces in seawater) would be to locally dehydrate the surface thereby displacing the water (2). Another property is the ability to chelate metals (17). The bonds between the oxygens present in the quinone at ring positions 3 and 4 and a metal (iron in this case) are very short and energetically favorable (18).

There are three stability constants in the chelation of iron (III) by o-diphenols, with log $K_s$ values of 20, 15, 10, and these are associated with mono-, bis- and tris-complexes, respectively (19). These complexes result in three distinct colors - the mono-complex being a green color in solution, the bis- resulting in a blue and the tris- in a red color. The mono-complex occurs at an optimum pH of 5.0, the bis at 7.0 and 9.5-10.0 for the tris-complex (20). Green and blue iron complexes are formed *in vitro* by the mussel protein, but the stability of these is hard to assess due to the insolubility and multifunctionality of each molecule (Waite, unpublished results). The question remains as to whether the complex that is being formed is between phenolic residues on two different molecules or whether there is a bend in the protein and an ion is shared by two different residues of the same chain.

o-Diphenols also have a high cumulative stability constants for other metals such as aluminum, zinc, titanium oxide, silicon dioxide and manganese and iron oxides (19,21,22). McBride (22) has demonstrated that catechol and hydroquinone are indeed adsorbed and oxidized by iron and manganese oxides. This chemisorption has also been demonstrated on aluminum oxide containing minerals such as gibbsite and boehmite and is dependent on their crystal structure (23). Environmental toxicologists have been using mussel byssus as a bioindicator of chemical contamination since the 1970's. Contaminants accumulated in the byssus of *M. edulis* include halogens and radionuclides such as U, Pu, Po and Am (24,25,26), arsenic (27) and transition metals including Cu, Co, Zn Ti, Mn and V (28,29,30,31). That these metals are strongly bound in byssus is suggested by the fact that they remain bound even after extensive washes with 1N HCl or

1N NaOH (29). There is controversy with respect to how these metals are deposited in the byssus. The presence of metals in *M. edulis* byssus greatly exceeds the concentration of metals in seawater by a factor of $10^4$ to $10^6$ (29). Tateda and Kayanagi (31) have shown that when the bifurcate mussel, *Septifer virgatus* and *M. edulis* share the same seawater, *S. virgatus* seems to preferentially accumulate Co and Mn whereas *M. edulis* accumulates Fe and Zn. If the adhesive protein of these two mussels is largely responsible for the metal binding, then protein sequence and secondary structure may play a role in determining the metal ion-affinity of various dopa-containing proteins.

Why would an animal make a polymer that has such high affinities for so many different metals? Although there is, of course, no pat answer to this, mussels are unquestionably adhesive opportunists. It is therefore no accident that three of the most abundant metals in the earth's crust, Si, Al and Fe (32) are precisely those for which the o-diphenolic moieties show their highest affinity.

## Potential Uses and Applications

The mussel protein adhesive qualities and its metal binding properties in aqueous environments present today's researcher with numerous potential uses and applications. It's not surprising that mans' attention should turn to agents in the natural environment that routinely adhere to surfaces in the presence of water. Adhesive technologists go to great lengths and expense to prepare surfaces and seal joints to protect them from moisture. An organic adhesive that adheres and cures in environments such as the mouth, orthopaedic implants and underwater construction is highly sought. This niche may be filled in the future by the dopa-containing bioadhesives.

A good bioadhesive spreads spontaneously on its substrate, especially when adhesion is derived largely from van der Waals dispersion forces between the two materials. Good spreading means that the surface tension of the adhesive is lower than that of the adherend. In the marine mussels, the adhesive is always in direct competition with seawater for a particular surface. On non-polar surfaces, water interacts through weak dispersive forces. Since the adhesive protein is larger than a molecule of water, it offers the surface more dispersive interactions per molecule than does water, so the water is easily displaced (33). When the mussel is exposed to polar vs. non-polar substrates such as slate and teflon, the mussel will "select" the slate over the teflon to attach itself. If the mussel is only offered teflon, it will attach itself to that as well (33). On slate, however, due to its high surface energy with abundant charged polar groups, water tends to interact strongly through hydration spheres, hydrogen bonding, dipoles and induced dipoles as well as dispersive forces. Due to the strength of these interactions, the mussel cannot hope to spontaneously push aside the water with its adhesive protein. It can, however, establish a foothold that is fortified with interactions more energetic than the best that water has to offer, thus the higher bonding strength (4). The high catechol-Fe stability constants suggest that the proteins may have potential as organic anticorrosive compounds in aqueous environments. This idea is supported by a number of commercial rust-proofing procedures that currently involve addition of catechols to coating resins (34-38). The catechols may work as inhibitors by forming an adsorbed layer on the metal surface by means of the constituent hydroxyl groups. The advantage of using large natural proteins with catecholic groups is compelling. First, proteins are less soluble than the low molecular weight catechols presently in industrial use, and therefore are more likely to stay where they are deposited. Secondly, the mussel glue proteins have multiple (up to 150) catecholic groups per molecule. The stability of the protein-metal interface is much higher than the stability of the single catechol:metal complex. This is reinforced by the finding that the degree of corrosion inhibition for iron in 1N HCl varied systematically with chain length of the inhibiting molecule (39). Thirdly, the mussel glue protein can be easily imitated by enzymatic modification of genetically engineered analogues (40). It is well known that the mussel glue penetrates to the bare metal in making its adhesive bond to the metal surface (41). Combining this fact with the high stability constants of the catechol:Fe system, some anticorrosive properties of the mussel glue may be involved in reducing corrosion rates of steel surfaces exposed to seawater.

Another potential use of the metal-binding properties of the mussel protein is the semi-specific sequestration of soluble metals from seawater. A microfibrous material containing the protein may be used to remove metals from aqueous systems. This area of potential research centers on the metal binding properties of the byssus and the possibility of imparting this property to other materials by grafting or coupling the mussel glue protein to other polymers such as cellulose, chitin, dextran, polyacrylamide or sulfonated polystyrene.

## Literature Cited

1. Van der Feen, P. J. Basteria 1949, 4, 66-71.
2. Waite, J. H. Biol. Rev. 1983, 58, 209-31.
3. Tamarin, A.; Lewis, P.; Askey, J. J. Morph. 1976, 149, 199-222.
4. Waite, J. H. Int. J. Adhesion and Adhesives 1987, 7, 9-14.
5. Waite, J. H. In The Mollusca Vol. I. Hochachka, P. W.; Wilbur, K. M. Eds.; Academic Press: New York, 1983; pp 467-504.
6. Benedict, C. V.; Waite, J. H. J. of Morph. 1986, 189, 261-70.
7. Arnow, L. E. J. Biol. Chem. 1937, 118, 531-7.
8. Waite, J. H.; Tanzer, M. L. Science 1981, 212, 1038-40.
9. Waite, J. H. J. Biol. Chem. 1983, 258, 2911-5.
10. Waite, J. H.; Housley, T. J.; Tanzer, M. L. Biochemistry 1985, 24, 5010-14.
11. Strausberg, R. L.; Anderson, D. M.; Filipula, D.; Finkelman, M.; Link, R.; McCandliss, R.; Orndorff, S. A.; Strausberg, S. L.; Wei, T. ACS Symp. Ser. 1989, 385, 453-64.
12. Waite, J. H. J. Comp. Physiol. 1986, 156, 491-6.

13. Waite, J. H.; Hansen, D. C.; Little, K. T. J. Comp. Physiol. 1989, 159, 517-25.
14. Williams, T.; Marumo, K.; Waite, J. H.; Henkens, R. Arch. Biochem. Biophys. 1989, 269, 415-22.
15. Hagerman, A. E.; Butler, L. G. J. Biol. Chem. 1981, 256, 4494-7.
16. Bienkiewicz, K. In Physical Chemistry of Leathermaking, Krieger Pub. Comp.: Malabar, FL, 1983, pp 385-406.
17. Pierpont, C. G.; Buchanan, R. B. Coordination Chem. Rev. 1981, 38, 45-87.
18. Raymond, K. N.; Isied, S. S.; Brown, L. D.; Fronczek, F. R.; Nibert, J. H. J. Am. Chem. Soc. 1976, 98, 1767.
19. Sillen, L. G.; Martell, A. E. Chem. Soc. London 1971, Suppl. No. 1, Spec. Publ., 25.
20. Hider, R. C.; Howlin, B.; Miller, J. R.; Mohd-Nor, A. R.; Silver, J. Inorg. Chim. Acta 1983, 80, 51-56.
21. Bartels, H. Helv. Chim. Acta. 1964, 47, 1605-9.
22. McBride, M. B. Soil Sci. Soc. Am. J. 1987, 51, 1466-72.
23. McBride, M. B.; Wesselink, L. J. Env. Sci. Tech. 1988, 22, 703-8.
24. Koide, M.; Lee, D. S.; Goldberg, E. D. Est. Coast. Shelf. Sci. 1982, 15, 679-95.
25. Hamilton, E. I. Mar. Ecol. Prog. Ser. 1980, 2, 61-73.
26. Hodge, V. F.; Koide, M.; Goldberg, E. D. Nature 1979, 277, 206-9.
27. Unlu, M. Y.; Fowler, S. W. Mar. Biol. 1979, 57, 209-19.
28. Pentreath, R. J. J. Mar. Biol. Assoc. U.K. 1973, 53, 127-43.
29. Coombs, T. L.; Keller, P. J. Aquat. Toxicol. 1981, 1, 291-300.
30. Coulon, J.; Trutchet, M.; Martoja, R. Ann. Inst. Ocean. Paris 1987, 63, 89-100.
31. Tateda, Y.; Kayanagi, T. Bull. Jap. Soc. Fish. 1986, 52, 2019-26.
32. Rochow, E. G. Silicon and Silicones. 1987, Springer-Verlag: Berlin.
33. Crisp, D. J.; Walker, G.; Young, G. A.; Yule, A. B. J. Colloid and Interface Sci. 1985, 104, 40-50.
34. Faulkner, R. N.; O'Neill, L. A. S.C.I. Monograph 1964, 26, 177-88.
35. Pizzi, A. J. Appl. Polym. Sci. 1979, 24, 1247-55.
36. Toyota Motor Company. Rustproofing coating material, Japan Patent 57 139155. 1982.
37. Vacchini, D. Anti-Corros. Methods Mater. 1985, 32, 9-11.
38. Tury, B.; John, G. R.; Scovell, E. G. Corrosion Inhibition. Eur. Patent 239288. 1987.
39. Carrol, M. J. B.; Travis, K. E.; Noggle, J. H. Corrosion 1975, 31, 123-7.
40. Marumo, K.; Waite, J. H. Biochem. et Biophys. Acta 1986, 872, 98-103.
41. Selin, H. H. Biologiya Morya 1980, 3, 97-9.

RECEIVED August 27, 1990

# Chapter 21

# Corrosion Inhibition by Thermal Polyaspartate

Brenda J. Little[1] and C. Steven Sikes[2]

[1]Naval Oceanographic and Atmospheric Research Laboratory, Stennis Space Center, MS 39529–5004

[2]Mineralization Center, University of South Alabama, Mobile, AL 36688

Polyanhydroaspartic acid, synthesized by thermal condensation of L-aspartic acid at 190 °C for 24 to 96 hours and converted to polyaspartate by mild alkaline hydrolysis, was evaluated as an inhibitor of corrosion. Inhibitor efficiency was judged on mild steel coupons exposed to salt water by visual inspection and by electrochemical testing. Electrochemical potential, polarization resistance and impedance spectra were evaluated. The mild steel surfaces were also examined using scanning electron microscopy/energy-dispersive x-ray spectrometry (SEM/EDAX) before and after exposure. The results indicated that thermal polyaspartate binds to surfaces of mild steel and moderately suppresses both anodic and cathodic corrosion reactions.

The use of polymers as additives to water to control corrosion and mineral scaling has become widespread (1-3). Total organic and total polymer formulations are replacing traditional corrosion-control additives like zinc and chromate, principally due to concern about toxic effects (4, 5). To date, polyacrylate, polyacrylate-acrylamide copolymers, and their derivatives have been most widely studied and used (6,7).

Toxicity and biodegradability are still areas of concern, however, in that polyacrylate-acrylamides are not biodegradable and have toxic effects at doses used in process water (8,9). The use of biodegradable and non-toxic polypeptides as water treatment additives presents a promising alternative to hydrocarbon-based polymers (10-14). These polypeptides, like polyacrylate polymers, are polyanionic. They have a variety of interesting surface-reactive properties, including inhibition of crystallization (15), dispersant activity (16, 17), and antifouling activity (10).

Based on studies of the mechanism of polyanionic peptides as inhibitors of crystallization (18, 19), it is likely that the peptides will bind to metal surfaces. The purpose of the present study is to demonstrate that the simplest of the

0097–6156/91/0444–0263$06.00/0

polyanionic peptides, polyaspartate, has corrosion inhibiting activity when bound to surfaces of mild steel. In addition, a simple thermal polymerization to produce polyaspartic acid is decribed. The method is based on the early work by Fox et al. (20, 21) on the abiotic origin of polyamino acids and proteins in the primitive atmosphere and ocean.

## Preparation of Thermal Polyaspartate

L-aspartic acid (500 g, Sigma Chemical Co.) was placed in a two-liter, round-bottom reaction vessel, originally designed as the evaporator vessel in a rotary evaporator apparatus. The reaction vessel was partially immersed in cooking oil in a deep-fryer set at $190 °C \pm 2 °C$. The reaction vessel was coupled by a ground glass fitting to a condenser vessel attached to a rotator shaft driven by a rheostated electric motor. The fittings were sealed with tape and fastened with hose clamps. A stream of nitrogen was continuously purged into the condenser vessel to eliminate $O_2$ and lessen the possibility of charring. The water of dehydration during peptide bond formation collects in the condenser vessel and is a visible indicator of the progress of the reaction. The product converts to a tawny color during polymerization. The reaction was allowed to proceed for up to 96 hours.

At the end, the bulk product was weighed, then dialyzed against 2 l distilled $H_2O$ for two hours at least 4 times to remove residual unreacted aspartic acid and small reaction products. Dialysis tubing (Spectrapor) with a pore size to retain molecules larger than 1000 daltons was used. The dialysate was lyophilized and the purified product weighed. The yield after 24 hours of polymerization was about 30% peptide and about 50% by 72 hours with little change with further reaction. The sample used in this study was polymerized for 48 hours with a yield of about 40% peptide.

The purified thermal product was in the form of the polyimide, polyanhydroaspartic acid (22) in which a 5 membered imide ring forms by a condensation reaction between the free carboxyl group of the aspartic acid R-group and the secondary amine of the peptide backbone. Prior to experimentation, the polyanhydroaspartic acid was converted to polyaspartate by opening the imide rings by mild aqueous alkaline hydrolysis: pH 10, 60 °C, 1 h. This treatment yields a mixed L,D polymer of alpha and beta residues (23).

Asp $\xrightarrow{\Delta}$   anhydro - Asp   $\xrightarrow[H_2O]{OH^-}$   $\alpha$ - linkage   $\beta$ - linkage

Polyaspartate was characterized by gel permeation chromatography (Toya Soda PW 3000 XL), amino acid analysis (Waters, Picotag), and automated amino acid sequencing (Applied Biosystems model 477A) with on-line identification of phenylthiohydantoin derivatives (ABI, model 120). The purified thermal polyaspartate used in the corrosion measurements had an apparent molecular weight of 5000 daltons based on standard of polyaspartate molecules made by solid-phase methods (12). The amino acid analysis and sequence analysis confirmed that the product consisted of a pure polyaspartate.

**Corrosion Testing**

**Inhibitor Solution.** Mild steel coupons were exposed to a 0.01% solution of polyaspartate (PASP) in a synthetic salt solution (Instant Ocean) made up to a salinity of 35 ppt, pH 8.0. Results were compared to control samples which were exposed under identical conditions to blank synthetic salt solutions and solutions containing 10 mM potassium dichromate ($K_2Cr_2O_7$), a known anodic corrosion inhibitor.

**Corrosion Samples.** Mild steel corrosion coupons were prepared by cleaning in 20% HCl, polishing with 400 emery paper and degreasing in acetone and hexane. They were then either totally immersed in the test solution or placed into an electrochemical cell. One $cm^2$ coupons of mild steel were used for polarization resistance measurements, while 20 $cm^2$ coupons were used for impedance measurements.

**Electrochemical Testing.** Corrosion potentials ($E_{corr}$) were measured using a saturated calomel reference electrode (SCE), a high impedance volt meter and a recording device. The impact of inhibitor on $E_{corr}$ was evaluated by allowing the mild steel electrode to equilibrate 16 to 20 hours before polymer was added.

Polarization resistance ($R_p$) was measured as the slope of a curve of potential (E) vs. current density (i) at $E_{corr}$, where i = 0 using an EG&G Parc, Model 173, potentiostat/galvanostat with a three electrode system - a saturated calomel reference electrode, graphite rod counter electrodes and a mild steel working electrode (24). The corrosion current density ($i_{corr}$) was calculated from $R_p$ by:

$$i_{corr} = B/R_p \qquad (1)$$
$$\text{with } B = b_a\, b_c/2.303(b_a + b_c)$$

where $b_a$, $b_c$ are the Tafel slopes.

Impedance spectra were obtained at $E_{corr}$ with a Solartron model 1250 frequency response analyzer (FRA) and a model 1286 potentiostat. The data were collected and analyzed with software developed at the Corrosion and Environmental Effects Laboratory (CEEL) at the University of Southern California.

**Surface Analysis.** Surface topography and chemistry were documented using a Kevex 7000 energy-dispersive x-ray spectrometer (EDAX) coupled to an AMRay 1000A scanning electron microscope (SEM). Electrodes were removed from the electrolytes (salt water or salt water + polymer), air dried and examined.

## Results

Visual examinations showed severe corrosion in the blank salt solutions, mild corrosion in the salt solutions containing 0.01% polyaspartate (PASP) and corrosion inhibition in the dichromate-containing solutions (Figure 1). In the absence of polymer, the carbon steel typically had a corrosion potential of − 684 mv ± 20 mV. Addition of 0.01% PASP caused the corrosion potential to move 20 to 50 mV in the positive direction (Figure 2). The anodic and cathodic polarization behavior of the electrode was essentially unchanged by the addition of the polymer (Figure 3). The cathodic reaction became diffusion limited at a c.d. of 50 $\mu$Acm$^{-2}$ in the absence of polymer and 20 $\mu$Acm$^{-2}$ in the presence of the polymer. Anodic and cathodic Tafel slopes remained essentially unchanged in the presence of the polymer. Addition of the polymer reduced the corrosion rate from 0.17-0.5 mm/y to 0.05-.13 mm/y (Table I).

Table I. Typical Corrosion Rate Data for Mild Steel

|  | 35 ppt artificial seawater | 35 ppt artificial seawater +0.01% PASP |
|---|---|---|
| E (i = 0) (mV) | − 686 | − 667 |
| i-corr ($\mu$A/cm$^2$) | 15.5 | 4.9 |
| Corr Rate (mm/y) | 0.18 | 0.06 |

Table II shows the values of E$_{corr}$ (vs. SCE), R$_p$ and corrosion rates from the analyses of the impedance data obtained after two hours. As before, an estimate of the corrosion rate is given by the polarization resistance, R$_p$, which is inversely proportional to the corrosion rate. A value of R$_p$ = 1000 ohm·cm$^2$ corresponds to a corrosion rate of about 0.2 mm/y, assuming B = 20 mV (eq 1). The corrosion rates for the blank, 0.01% PASP, and 10 mM dichromate solutions were 0.17, 0.16 and 0.08 mm/y, respectively. The impedance data are presented in Figure 4 as complex plane plots where Z$_{real}$ is the real part of the impedance (Z) and Z$_{imag}$ is the imaginary part (Z = Z$^2$real + Z$^2$imag). In a first approximation, R$_p$ is given by the diameter of the semicircles. The impedance data in Figure 4 reflect mild corrosion inhibition in the presence of 0.01% PASP.

Figures 5, 6 and 7 are SEM micrographs and EDAX spectra of the mild steel surfaces after exposure to salt water, salt water plus PASP and salt water

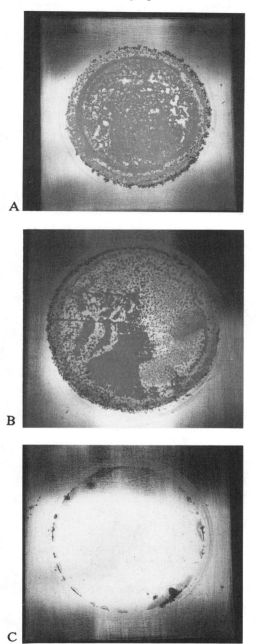

Figure 1. Photographs of mild steel exposed to (A) blank synthetic salt solution, (B) 0.01% PASP, (C) 10 mM $K_2Cr_2O_7$.

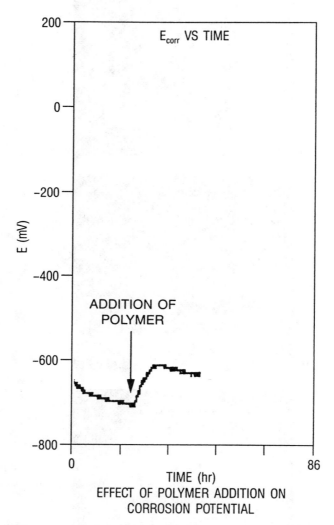

Figure 2. $E_{corr}$ vs. time for mild steel electrode before and after PASP addition.

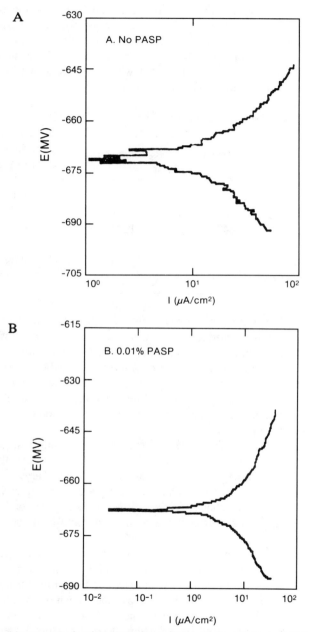

Figure 3. Linear polarization data for mild steel electrodes in (A) blank synthetic
salt solution, (B) 0.01% PASP.

plus dichromate, respectively. In the absence of polymer the corroded mild steel surface was crystalline and chlorine dominated the spectrum. In the presence of the polymer, some areas of the surface were covered with an organic film, and sulfur and calcium have accumulated on the surface. Surface bound chlorine decreased in the presence of the polymer. In the presence of the dichromate inhibitor, the polished surface had not reacted with the constituents in the salt water. Only peaks for iron were found on the surface.

Table II. Typical EIS Data for Mild Steel (2 hr/Instant Ocean)

| Electrolyte | $E_{corr}$ (mV) | $R_p$ (ohm·cm$^2$) | Corrosion Rate (mm/y) |
|---|---|---|---|
| Blank synthetic salt solution | – 666 | 1172 | 0.17 |
| 0.01% PASP in synthetic salt solution | – 648 | 1268 | 0.16 |
| 10 mM dichromate in synthetic salt solution | – 676 | 2660 | 0.08 |

The corrosion of mild steel in aerated seawater involves an anodic process whereby metal ions from the surface are oxidized and passed into solution and a cathodic process in which oxygen is reduced. Complete inhibition of corrosion of mild steel in seawater is difficult, if not economically impossible. Most systems make use of inhibitors to give marked reductions in corrosion rates. Furthermore, the majority of inhibitors are specific in their actions in metal/electrolyte environments. An inhibitor may decrease the rate of the anodic process, the cathodic process or both processes.

Adsorbed inhibitors can retard corrosion reactions by forming a surface film which acts as a physical barrier to restrict diffusion of ions or molecules to or from the metal surface. These types of inhibitors are usually large molecules, e.g. proteins, such as gelatine; polysaccharides, such as dextrin; or compounds containing long hydrocarbon chains. Surface films of these types of inhibitors give rise to resistance polarization affecting both anodic and cathodic reactions (25).

Inhibition in neutral solutions like aerated seawater can be due to compounds which form or stabilize protective surface films. An inhibitor may form a surface film of an insoluble salt by precipitation or reaction. Inhibitors forming films of this type include: (a) salts of metals such as zinc, magnesium, manganese and nickel form insoluble hydroxides at cathodic areas which are more alkaline due to the hydroxyl ions produced by reduction of oxygen; (b) soluble calcium salts can precipitate as calcium carbonate at cathodic areas; (c) polyphosphates in the presence of zinc or calcium produce a thin amorphous salt film. Salt

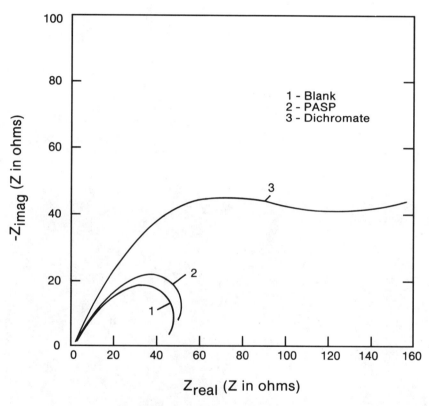

Figure 4. Complex plane plots for mild steel.

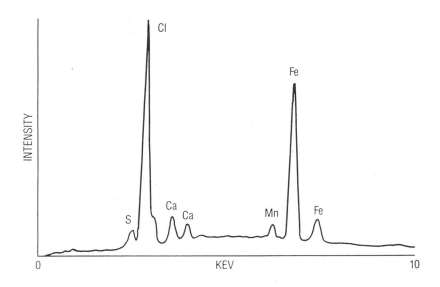

Figure 5. Micrograph and EDAX spectrum of mild steel after exposure to salt solution.

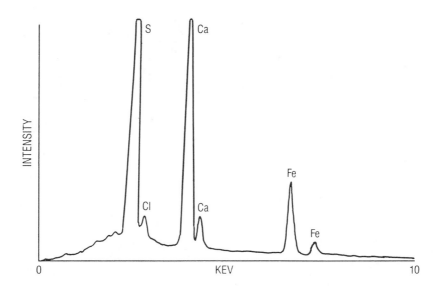

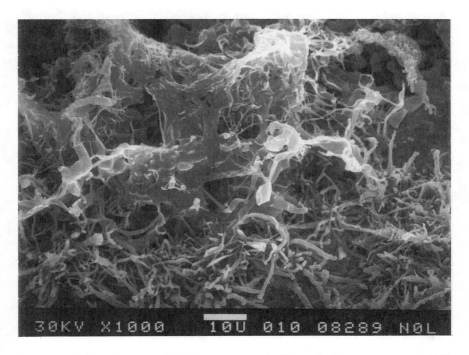

Figure 6.  Micrograph and EDAX spectrum of mild steel after exposure to 0.01%
         PASP.

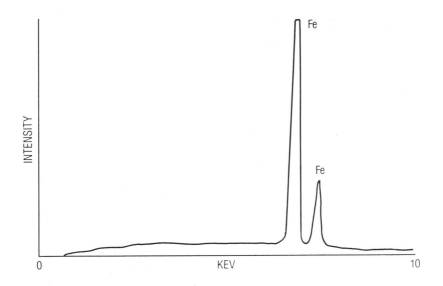

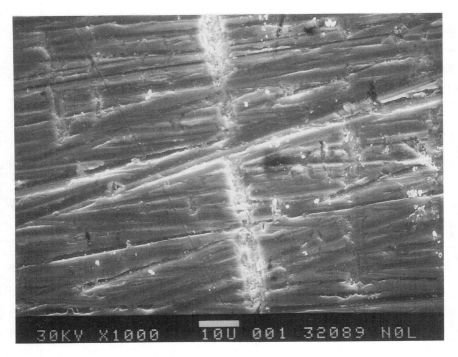

Figure 7. Micrograph and EDAX spectrum of mild steel after exposure to 10 mM $K_2Cr_2O_7$.

films are usually thick and visible, restricting diffusion, particularly of dissolved oxygen to the metal surface (26). Since salt films are poor electronic conductors restricting oxygen reduction at the film surface, they are referred to as cathodic inhibitors.

Another type of inhibitor that is effective in near-neutral solutions stabilizes oxide films on metals to form thin protective passivating films. These inhibitors are the anions of weak acids such as chromate, nitrite, benzoate, silicate, phosphate, and borate. Passivating oxide films on metals inhibit metal dissolution by preventing diffusion of metal ions. Because the anodic reaction is inhibited, these anions are referred to as anodic inhibitors. Anodic inhibitors are more frequently used than cathodic inhibitors to inhibit the corrosion of iron, zinc, aluminum, copper and their alloys in seawater (26).

It is obvious from the visual inspection of the mild steel electrodes that the 0.01% solution of PASP reduces the corrosion of mild steel in a synthetic salt solution. The SEM/EDAX data confirm that the polymer is surface active, adsorbing to the surface and decreasing the amount of surface bound chlorine.

The effects of adsorbed inhibitors on individual electrode processes of corrosion may be determined from electrochemical data. The measured potential of a corroding metal is a mixed potential resulting from the superposition of the anodic and cathodic reactions and is known as the corrosion potential or $E_{corr}$. The change in the corrosion potential due to addition of the inhibitor is an indication of which process is retarded. Displacement of the corrosion potential in the positive direction indicates retardation of the anodic process (anodic control), whereas displacement in the negative direction indicates mainly retardation of the cathodic process (cathodic control). Little change in the corrosion potential suggests mixed control whereby both anodic and cathodic processes are retarded (26). The observed shift of $E_{corr}$ in the positive direction upon addition of PASP to the synthetic salt solution suggests that the polymer mainly retards the anodic dissolution reaction.

While measurements of $E_{corr}$ provide clues as to electrochemical behavior, the corrosion potential alone provides no information on the corrosion rate. Skold and Larson (27) found in studies of the corrosion of steel and cast iron in natural seawater that a linear relationship existed between potential and the applied anodic and cathodic current densities in the region of $E_{corr}$ ($\pm$ 20 mV). Stern and his co-workers (28, 29) used the term "linear polarization" to describe the linearity. The slope of this linear curve, $\Delta E$-$\Delta i$, is termed the polarization resistance, $R_p$, and has units of ohms·cm$^2$.

On the basis of a detailed analysis of the polarization curves of the anodic and cathodic reactions involved in the corrosion of a metal, the following expression was derived:

$$\frac{1}{R_p} = \left(\frac{\Delta i}{\Delta E}\right)_{E_{corr}} = \left[\frac{2.3(b_a + b_c)}{b_a\,b_c}\right]i_{corr}$$

where $R_p$ is the polarization resistance determined at $E_{corr}$, and $b_a$, $b_c$ are the Tafel slopes. This equation shows that the corrosion rate is inversely proportional to $R_p$. It is evident from equations 1 and 2 that the evaluation of $i_{corr}$ from $R_p$ requires a knowledge of Tafel slopes $b_a$ and $b_c$. Figure 8 is a graphic representation of potential vs. log current density curves showing the relationship between $E_{corr}$, $i_{corr}$, $b_a$, $b_c$ and the limiting diffusion current density ($i_1$). The Tafel slopes must be derived from complete E-i curves ( $\geq 60$ mV in each direction) for each system studied. In the experiments described Tafel slopes were determined independently for the anodic and cathodic reactions. As stated previously, the Tafel constants were not altered by the addition of polymer. Using polarization resistance data to calculate corrosion rates for 0.01% PASP in synthetic salt solution, a decrease of the corrosion rate from 0.17-0.50 mm/y to 0.05-0.13 mm/ y was observed.

Thorough reviews of polarization resistance ($R_p$) techniques for the measurement of corrosion currents have been reported elsewhere (30). The use of polarization resistance for determining corrosion rates has a number of significant advantages. It provides a method for rapidly monitoring instantaneous corrosion rates, requiring small changes in potential that do not significantly disturb the system.

Electrochemical impedance spectroscopy (EIS) is a relatively new technique in corrosion research which has found many successful applications. A recent summary of the background of EIS and applications in many areas of electrochemistry can be found in the proceedings of the First International Symposium on EIS which was held in 1989 (31). Mansfeld (32, 33) has discussed some basic concepts concerning the recording and analysis of impedance data, and with Lorenz (34) has compared the results obtained with ac and dc techniques for iron in the presence of corrosion inhibitors.

In the EIS technique the impedance data are recorded as a function of the frequency of the applied signal at a fixed working point (E,i) of a polarization curve. This working point often is the corrosion potential (E = $E_{corr}$, i= 0). Usually a very large frequency range must be investigated to obtain the complete impedance spectrum, ranging from 65 kHz to 10 or lower mHz. The impedance data are usually determined with a three-electrode system. A potentiostat is used to apply the potential at which the data are to be collected. A frequency response analyzer (FRA) is programmed to apply a series of sine waves of varying frequency and constant amplitude small enough to remain in the linear potential range. Impedance data are determined by the FRA at each frequency and stored in memory. Properties of the corrosion system at a fixed potential (or current) can be determined through the analysis of the frequency dependence of the impedance. One of the advantages of EIS is that only very small signals are applied.

The corrosion rates resulting from the EIS technique were similar to those obtained using linear polarization. Mild corrosion inhibition was observed with addition of the polymer to the synthetic salt solution.

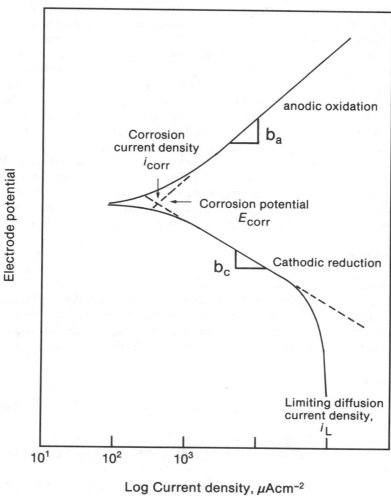

Figure 8. E vs. log $i_{corr}$ curve.

## Conclusion

PASP is a surface active polymer that adsorbs to mild steel after brief exposure. The polymer appears to increase the concentration of calcium and sulfur on the surface of the metal and to form a diffusion barrier that inhibits chloride ions from reacting with the surface. The shift of the corrosion potential in the positive direction upon addition of polymer indicates decreased anodic dissolution processes, suggesting that PASP is an anodic inhibitor. Corrosion rate measurements by linear polarization and EIS confirm that PASP is a mild inhibitor of corrosion on mild steel exposed to aerated seawater. It did not inhibit corrosion to the extent that the anodic inhibitor, potassium dichromate did. In this regard, it will be interesting to evaluate derivatives of polyaspartate for improved performance. That is, it is clear that the nature and distribution of both anionic and nonionic residues are critical to the activity of polyanionic polymers (35-38). Phosphorylated, phosphonated, and sulfonated residues, as well as neutral and hydrophobic residues may impart increased activity to the molecules as regulators of crystallization (12,36). Therefore, it is likely that polyamino acids with improved performance as inhibitors of corrosion will be forthcoming.

## Acknowledgments

This work was supported by the Office of Naval Research and the Naval Oceanographic and Atmospheric Research Laboratory (NOARL), Program Element 61153N, Ms. H. E. Morris, program director, NOARL Contribution SP 005:333:90. Additional funding was provided by the Mississippi-Alabama Sea Grant Consortium. Approved for public release; distribution is unlimited.

## Literature Cited

1. Smyk, E.B.; Hoots, J.E.; Fivizzani, K.P.; Fulks, K.E. Corrosion 88, 1988, paper 14, 20 p.
2. Lipinski, R.J.; Chang, K.Y. Corrosion 88, 1988, paper 12, 7p.
3. Fivizzani, K.P.; Dubin, L.; Fair, B.E.; Hoots, J.E. Corrosion 89, 1989, paper 433, 15p.
4. Environmental Protection Agency. Federal Register 1985, 15, 111, Notice ADL-FRL 2784-4.
5. Environmental Protection Agency. Federal Register 1985, 50, 134, Notice OW-FRL-2815-5.
6. Rohm and Haas Company. Polymer additions for aqueous systems; 1985, 21 p.
7. Nalco Chemical Company. Prism AAA-LH06; advance technical data sheet.
8. Cekolin, C.S. Echinoderm Conference abstract, Dauphin Island Sea Lab, 1989.
9. Cekolin, C.S. M.S. thesis, University of South Alabama, Mobile, 1990.

10. Sikes, C.S.; Wheeler, A.P. U.S. Patent 4 534 881, 1985.
11. Sikes, C.S.; Wheeler, A.P. In Chemical Aspects of Regulation of Mineralization: Sikes, C.S.; Wheeler, A.P., Eds.; Univ. S. Alabama Publ. Services: Mobile, 1988; pp 15-20.
12. Sikes, C.S.; Yeung, M.L.; Wheeler, A.P. In Surface Reactive Peptides and Polymers: Discovery and Commercialization; Sikes, C.S.; Wheeler, A.P., Eds; American Chemical Society; Washington, D.C., 1990.
13. Fujimoto, Y.; Teranishi, M. West German Patent 2 253 190, 1973.
14. Boehmke, G. West German Patent 3 626 672, 1988.
15. Wheeler, A.P.; Sikes, C.S. In Biomineralization: Chemical and Biochemical Perspectives; Mann, S., Webb, J.; Williams, R.J.P., Eds.; VCH Publishers; Weinheim, FRG, 1989; 95-132.
16. Barbucci, R.; Marabini, A.M.; Barbaro, M.; Nocentini, M.; Corezzi, S. Reagents Miner. Ind. 1984, 245-250.
17. Garris, J.; Sikes, C.S. Ocean's '89 abstract, 1989.
18. Low, K.C. M.S. Thesis, Univ. S. Ala., Mobile, 1990.
19. Wheeler, A.P.; Low, K.C.; Sikes, C.S. In Surface Reactive Peptides and Polymers: Discovery and Commercialization; Sikes, C.S.; Wheeler, A.P., Eds.; American Chemical Society, Washington, DC, 1990.
20. Fox, S.W.; Harada, K. J. Am. Chem. Soc. 1960, 82, 3745-81.
21. Fox, S.W.; Harada, K.; Rohlfing, D. In Polyamino Acids, Polypeptides, and Proteins; Stahman, M.A., Ed.; Univ. Wisconsin Press: Madison, 1962; pp 47-53.
22. Kokufuta, E.; Suzuki, S.; Harada, K. Biosystems 1977, 9, 211-214.
23. Saudek, V.; Pivkova, H.; Drobnik, J. Biopolymers 1981, 20, 1615-1623.
24. Kendig, M.; Mansfeld, F.; Tsai, S. Corr. Sci. 1983, 23, 317.
25. Manchu, W. First European Symposium on Corrosion Inhibitors, 1961, p 111.
26. Shrier, L. L. Corrosion; Newnes-Butterworths: Boston, MA, 1979; Chapter 16.3, 18.3.
27. Skold, R. V.; Larson, T. E. Corrosion 1957, 13, 1392.
28. Stern, M.; Geary, A. L. J. Electrochem. Soc. 1957, 104, 56.
29. Stern, M. Corrosion 1958, 14, 440.
30. Mansfeld, F. In Advances of Corrosion Science and Technology; Plenum Press: New York, 1976, Vol. 6, p 163.
31. Proc. 1st Intern. Symp. on EIS, 1989.
32. Mansfeld, F. Corrosion 1981, 37, 301.
33. Mansfeld, F.; Kendig, M. W.; Tsai, S. Corrosion 1982, 38, 570.
34. Lorenz, W. J.; Mansfeld, F. Corr. Sci. 1981, 21, 647.
35. Fong, D.W. U.S. Patent 4 703 092, 1987.
36. Fong, D.W.; Kowalski, D.J. U.S. Patent 4 678 840, 1987.
37. Fong, D.W.; Kowalski, D.J. U.S. Patent 4 762 894, 1988.
38. Sikes, C.S. U.S. Patent Appl. 07/339 672, 1989.

RECEIVED August 27, 1990

# Chapter 22

# Mechanistic Insights Concerning the Role of Polymers as Scale Inhibitors and Dispersants in Cooling Water Systems

**Kenneth P. Fivizzani, John E. Hoots, and Richard W. Cloud**

**Nalco Chemical Company, Naperville, IL 60563-1198**

Water soluble polymers serve an important function in the treatment of industrial cooling water systems. Classical terms used to describe polymer activity are based on observations of macroscopic properties. A "scale inhibitor" is any material that limits formation of scaling salts, while the term "dispersant" describes the ability of a chemical to prevent particulate settling for an extended time period. Markedly similar performance behavior for scale inhibitors and dispersants has been observed in benchtop activity tests utilizing a variety of scaling ions. In these tests, the precipitation process involves initial formation of microparticles and subsequent agglomeration to produce large particulates. Effective polymers limit the agglomeration of smaller particles. Particle size distribution studies and scanning electron microscopy (SEM) have characterized the particulates formed in benchtop activity tests and provided experimental support for the proposed mechanism which encompasses both inhibition and dispersion.

Most industrial processes generate more heat than can be used. The excessive heat must be dissipated, and water is one of the most effective heat transfer media. Water has a large specific heat ($1 \text{ cal}/\text{g} \cdot {}^\circ\text{C}$) and a large heat of vaporization ($540 \text{ cal}/\text{g} \cdot {}^\circ\text{C}$); a given amount of water can contain (transfer) a large amount of heat relative to other materials. The heat produced by industrial equipment and processes is transferred to water in a heat exchanger. In an open recirculating cooling water system (1-5), the "hot" water returns to a cooling tower which is open to the atmosphere. As the water cascades down the tower, heat is removed from the system due to evaporation of some of the water. Fresh water is added to maintain the desired system volume. During this cooling process, neither dissolved nor suspended solids are removed by evaporation, and the concentration of these materials increases with time.

0097–6156/91/0444–0280$10.00/0

Eventually, dissolved solids will exceed solubility limits and precipitate, and suspended solids will agglomerate and settle out of solution. These deposits can restrict water flow and interfere with heat transfer in the heat exchangers. One way to minimize deposition is by replacing some of the concentrated water with fresh water to dilute the potentially depositing species. A second approach is to add treatment chemicals to prevent precipitation of dissolved materials and keep suspended solids dispersed. Treatment chemicals which minimize precipitation of calcium and magnesium (hardness) salts are known as scale or threshold inhibitors. Dispersants are used to keep particles from settling. The evaluation of current and proposed treatment chemicals has provided insight regarding how such chemicals function in cooling water environments.

Historically, benchtop activity tests have been used extensively and have served an important role in demonstrating the performance of chemical agents as particulate dispersants and threshold inhibitors of scaling salts (6-12). The conclusions reported in numerous technical papers and patents in the field of cooling water treatment have been based upon benchtop activity test results. In some cases, benchtop activity test results have been verified by more sophisticated testing in recirculating water systems which model operating conditions of the applications (13-15). Favorable results in benchtop activity tests have led to successful commercial water treatment programs (16-21). Considering the significance of the benchtop activity tests, relatively little research has been done on the underlying processes of solid formation and growth which determine whether a chemical passes or fails these initial evaluation tests.

Classical water treatment terminology used to describe chemical activity is based upon macroscopic descriptions of a system. For example, a "threshold inhibitor" usually refers to a chemical which limits scale formation when used at a sub-stoichiometric concentration relative to the scaling species (22-24). While that definition describes relative initial concentrations of substances in the system, it does not address the mechanisms whereby the threshold inhibitor affects scale formation, growth, and removal.

A similar situation exists for chemicals which function as particulate dispersants. While the term "dispersant" may describe the ability of a chemical to prevent the settling of particulates as a function of time, it does not necessarily address the mechanism of particle dispersion. Even when an overall mechanism (particulate agglomeration control) has been associated with a process (solids dispersion), additional mechanistic details generally have not been elucidated (e.g., steric versus electrostatic stabilization). There is a fourfold purpose of this research:

1. Evaluate possible mechanisms which explain the performance of threshold scale inhibitors and particulate dispersants in bench-top activity tests.

2. Determine if a dominant mechanism can explain scale inhibition and dispersion activity of chemicals for different metal salts (e.g., calcium phosphate versus zinc hydroxides).

3. Determine whether seemingly different types of polymer performance (e.g., threshold inhibition and dispersion) are really expressions of a single underlying mechanism.

4. Utilize insights gained in the previous areas to improve evaluation methods for new chemicals and formulations to be used in cooling water applications.

## Previously Reported Mechanisms

Most of the basic research on the formation and growth of crystals and particulates which are commonly encountered in cooling water systems (e.g., calcium salts of phosphates, carbonates and phosphonates, iron hydroxides, metal silicates, etc.) has been conducted in academic facilities (25-32). Those studies have exemplified the complexity of crystallization and particle agglomeration processes (Figure 1; 29). Many crystal phases, as well as amorphous solids, can form from just a single type of inorganic salt (e.g., five well-defined crystal phases for calcium phosphates) (33). Fortunately, certain mechanistic pathways and crystal phases tend to be excluded under representative cooling water conditions. For example, tricalcium phosphate, $Ca_3(PO_4)_2$, is not commonly observed in the conversion of soluble phosphate to calcium phosphate particulates (26, 34). Also, amorphous salts (e.g., calcium phosphate) may be the initially formed species in some applications, even though well-defined crystal phases of the mineral are known to exist (35).

The following list provides a general summary of the processes which have been proposed as being significant in crystallization and particulate formation and growth:

**I. Step Edge/Kink Site Model (or Active-Site Model).**  This model has been used to describe the crystallization of a variety of low solubility salts by addition of small ions and molecules to a crystalline surface (29, 36). The crystal growth rate for a general $X_mY_oZ_p$ salt is usually described by the following equation (Equation 1) or one derived from it (26):

$$\text{Rate} = -ks[(X)^{m/(m+o+p)} (Y)^{o/(m+o+p)} (Z)^{p/(m+o+p)} - K_{so}]^n \qquad (1)$$

where k = rate constant, s = number of active growth sites,
n = reaction order (usually $1 < n < 2$), and
$K_{so}$ is the solubility product of the mineral salt under the conditions
being tested

This model is used to describe a low-solubility salt which is kinetically, but not thermodynamically, stabilized. The chemical inhibitor acts as a steric barrier and limits the crystal's growth rate by blocking an active growth site as defined by Steps a-d in Figure 2. The ions comprising this salt are still supersaturated in the liquid phase and resumption of crystal growth can occur if breaches in the

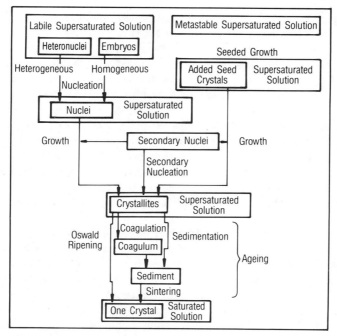

Figure 1. Proposed steps in crystallization processes. Reprinted with permission from ref. 29. Copyright 1976 Ann Arbor Science Publishers.

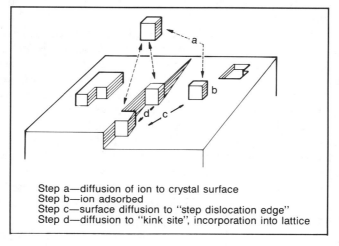

Step a—diffusion of ion to crystal surface
Step b—ion adsorbed
Step c—surface diffusion to "step dislocation edge"
Step d—diffusion to "kink site", incorporation into lattice

Figure 2. "Step Edge/Kink Site" model for crystal growth. Reprinted with permission from ref. 29. Copyright 1976 Ann Arbor Science Publishers.

outer protective layer are not repaired (e.g., excess inhibitor is consumed). This model is most appropriately applied to systems with relatively low levels of supersaturation, where spontaneous homogeneous nucleation does not rapidly occur. This description has been used for approximately 20 years to explain the crystal growth inhibiting effect of various organic and inorganic ions and molecules (e.g., magnesium, carboxylic acids, phosphonates, etc.).

**II. Massive Nucleation Model.** This second model is an example whereby the supersaturation of ions in solution is removed by the formation of large numbers of microparticulates (37). The chemical inhibitor encourages particulate formation and also acts in some unspecified manner to prevent the conversion of small particulates into larger particulates. This mechanism has not received the attention and corroboration that the "active-growth site" model has.

**III. Agglomeration Control/Solid-State Crystallization Model.** This model has been described recently and shares some aspects of the "massive nucleation model" (38). In this model, the supersaturation of ions in solution is largely relieved by formation of large numbers of microparticulates which may be amorphous in nature (Figure 3). A major role of the chemical inhibitor is to limit agglomeration of the microparticulates. If specific crystalline phases appear in the final solids, they are likely formed by a solid-state conversion of the amorphous microparticulates. Overall, the system approaches thermodynamic equilibrium with respect to further loss of ions from solution (due to particle formation and growth). Furthermore, the particles may be kinetically-stabilized against particle agglomeration; however, additional particulate growth and precipitation may occur if that kinetic barrier is breached. This mechanism is more likely to occur in systems which are highly supersaturated.

In the "massive nucleation" and "agglomeration control/solid-state crystallization" models, the concept of particulate dispersion is included. The dispersion of particulates is likely to be a very important factor whether the particulates introduced into the system are pre-existing (clays, silt, iron hydroxides) or formed within the system from the precipitation of supersaturated salts. The concepts and mechanisms of particulate dispersion have been extensively described in the literature (39-41). The major mechanisms invoked are electrostatic repulsion, steric repulsion, and electrosteric repulsion. The dominant mechanism depends upon the charge of particle surface, the type of dispersing agent used, and numerous system operating conditions (e.g., pH, ions present, temperature, etc.). Unfortunately, little research has been done on the impact that each of these mechanisms has on the performance of threshold inhibitors and dispersants in benchtop activity tests for cooling water applications.

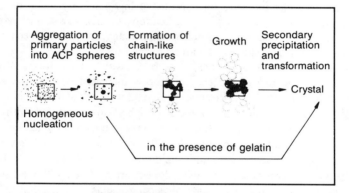

Figure 3. Agglomeration control/solid-state crystallization model .
Reprinted with permission from ref. 38. Copyright 1987 Elsevier
Science Publishers.

## Performance Characteristics of Screening Tests

Benchtop activity tests have been conducted using a broad range of scaling salts:

- Calcium/magnesium phosphate
- Calcium/magnesium phosphate + iron (II)
- Calcium/magnesium phosphate + phosphonates
- Calcium phosphonates
- Zinc hydroxide
- Iron (hydr)oxides
- Iron/manganese (hydr)oxides
- Manganese (hydr)oxides

The term (hydr)oxides indicates the mixture of low solubility metal oxides and hydroxides which can form when metal cations such as $Fe^{+2}$ and $Mn^{+2}$ are oxidized in solution. That term also encompasses any mixture of specific crystal phases and amorphous solids which can be formed.

In general, these benchtop activity tests begin with soluble forms of cations and anions, and the pH of the test mixture is increased so that the solution becomes highly supersaturated. In some cases, oxidation of the metal cation occurs [e.g., $Fe(II) \rightarrow Fe(III)$] and the solution is highly supersaturated with respect to the corresponding Fe(III) salts. By comparing the initial and final test conditions, calculated supersaturation ratios of $>10,000$ are attained in some of the tests (assuming no precipitation occurs until the desired pH level is reached). In order to limit the precipitation of scaling salts, polymers are included in the test solutions as threshold inhibitors and particulate dispersing agents. The significant feature of the benchtop activity tests is that consistent performance patterns are observed. This suggests that a predominant mechanism for particle formation and growth is occurring in all of these benchtop activity tests. The following performance patterns are common to the aforementioned activity tests:

Inhibitor dosage: Below a certain minimum concentration, no benefit is observed and performance increases sharply (to a maximum level) as concentration increases.

Types and ratios of organic functionalities present in the polymer: An intermediate compositional range usually exists where performance is maximized.

Temperature limit: Performance declines sharply as the upper acceptable limit is surpassed.

pH limit: Performance commonly diminishes rapidly as the upper acceptable limit is exceeded.

Test duration: For poor inhibitors, the formation of large particulates occurs rapidly. The process occurs much more slowly as the performance of the inhibitor increases.

Particulate size distribution: In both dispersion and threshold inhibition activity tests, performance decreases as average particulate size and/or distribution range increases.

Presence of a chemical inhibitor: This affects the size of particulates, but rarely affects the mass of particulates formed [except specific cases where oxidation of metal cations is limited (e.g., $Mn^{+2}$ to $Mn^{+4}$)].

Examples of each performance feature of the benchtop activity tests will be discussed in subsequent sections.

In addition to the benchtop activity tests for threshold inhibition, particulate dispersion tests are also conducted. The powdered solids initially added in these tests are clay or iron oxide (hematite). The effects of test duration, water chemistry, and polymer dosage and composition on performance of the dispersing agent will be described in later sections.

## Experimental Procedures

**Analytical Techniques.** Metal ions ($Ca^{+2}$, $Mg^{+2}$, $Zn^{+2}$, $Fe^{+n}$, and $Mn^{+n}$) were quantitatively determined by atomic absorption spectroscopy. Polymer concentrations were quantified spectrophotometrically by a turbidimetric procedure. Orthophosphate content was determined spectrophotometrically (at 700 nm) by the formation of a blue phosphomolybdate complex. Total and organic phosphorus content were determined by complete oxidation of organophosphorus species to orthophosphate and subsequent spectrophotometric analysis. Filtration was conducted using MF-Millipore mixed cellulose ester filters, unless otherwise specified. Initial supersaturation ratios (SR) were estimated using a computer program to calculate mineral solubilities (42).

Performance of chemical agents is commonly measured as percent inhibition (Equation 2) and percent dispersion (Equation 3). The percent inhibition measures the additional amount of particulates and soluble ions passing through a filter when an inhibitor agent is present, as compared to a solution without an inhibitor. The percent dispersion measures the additional amount of solids which did not settle (i.e., precipitate) within a specified time period when a dispersing agent was present, relative to a solution without a dispersant.

$$\% \text{ Inhibition}_Y = \frac{\text{(filtered sample)-(unfiltered blank)}}{\text{(initial sample)-(unfiltered blank)}} \times 100 \qquad (2)$$

$$\% \text{ Dispersion} = \frac{\text{(unfiltered sample)-(unfiltered blank)}}{\text{(initial sample)-(unfiltered blank)}} \times 100 \qquad (3)$$

The "blank" sample represents a test solution which did not contain a scale inhibiting or particulate dispersing agent. "y" represents the pore size of the filter used to determine percent inhibition results.

**Scanning Electron Microscopy (SEM).** Particulates produced in benchtop activity tests were collected on a Nucleopore polycarbonate filter membrane (0.2

$\mu$m pore size), followed by air-drying at ambient temperature. Samples for standard SEM analysis (Cambridge Model StereoScan 250 Mk3) were coated with vacuum-deposited gold/palladium alloy ($\sim$150Å thickness). The elemental composition of particle agglomerates was examined by an energy-dispersive x-ray analyzer (EDAX™, Model PV 9100), and those samples were coated with vacuum-deposited graphite.

**Particle Size Distribution.** The test solutions were passed through a series of Millipore filters with different pore sizes. Alternatively, quasielastic laser light scattering units (NICOMP (Model 200) and Brookhaven Instruments Multi-Angle Light Scattering Goniometer and Correlator) were used to analyze the unfiltered test solutions. A scattering angle of 90° was used in the light scattering measurements.

**Performance Tests.** Calcium Phosphate—A chemical inhibitor (5-20 mg/l actives) is mixed into a test solution containing 250 mg/l $Ca^{+2}$ and 125 mg/l $Mg^{+2}$ (both as $CaCO_3$), then 10 mg/l orthophosphate ($PO_4^{-3}$) is added. The solution is heated, stirred, and the pH is increased to the specified value. The temperature, stirring and pH are maintained for four hours. Finally, the hot test solution is filtered, and percent inhibition (Equation 2) is determined by analyzing the phosphate content of the filtrate.

The calcium phosphate plus $Fe^{+2}$ (3 mg/l as Fe) and calcium phosphate plus organophosphate versions of the previous test involve the respective additions of those species prior to heating and raising the pH of the test solutions. Otherwise, the test procedures are comparable to the basic calcium phosphate inhibition test.

Calcium Phosphonates—This test procedure is comparable to that used for calcium phosphate, except that 360 mg/l $Ca^{+2}$ (as $CaCO_3$) and no magnesium are used. The phosphate is replaced by organophosphonate compounds. Upon completion of the test, the hot test solution is filtered and percent inhibition (Equation 2) is determined by analyzing the phosphate content of the filtrate.

Iron (Hydr)oxides—A chemical inhibitor (5-10 mg/l actives) is mixed into a test solution containing 360 mg/l $Ca^{+2}$ and 200 mg/l $Mg^{+2}$ (both as $CaCO_3$), then 10 mg/l soluble iron ($Fe^{+2}$) is added. The solution is heated, stirred, and the pH is increased to the specified value. The temperature, stirring and pH are maintained for two hours. Next, the test solution is allowed to reach ambient temperature and to remain undisturbed for 24 hours. Finally, an unfiltered sample is taken from the top part of the solution, and percent dispersion (Equation 3) is determined by analyzing the sample for iron content.

Zinc Hydroxide Inhibition—A chemical inhibitor (5-15 mg/l actives) is mixed into a test solution containing 250 mg/l $Ca^{+2}$ and 125 mg/l $Mg^{+2}$ (both as $CaCO_3$), then 5 mg/l $Zn^{+2}$ (as Zn) is added. The pH of the test solution is raised to the specified alkaline value, and the solution is heated for 24 hours. Finally, the hot test solution is filtered, and percent inhibition (Equation 2) is determined by analyzing the zinc content of the filtrate. The term "percent zinc stabilization" has been used previously in the literature, but that term is actually equivalent to the percent inhibition value described in this report.

Iron Oxide (Hematite) Dispersion—The dispersion of hematite ($Fe_2O_3$) particulates by chemical agents was determined according to a previously reported method (6). In this procedure, the iron oxide particulates were initially added to the system rather than being formed *in situ* from a supersaturated solution.

Clay Dispersion—A chemical dispersant (2.5-10 mg/l actives) is mixed into a test solution containing 360 mg/l $Ca^{+2}$ and 200 mg/l $Mg^{+2}$ (both as $CaCO_3$), then the pH is adjusted to the specified value. Next, 100 mg/l of clay (e.g., Engelhard ASP170) is added, and the mixture is agitated for 5 minutes using an impeller-type pump ($\sim$2 l/min flow rate). The percent dispersion (Equation 3) is determined by using a nephelometer to measure the turbidity of the test solution as soon as the pump is turned off and again 24 hours later.

## Performance Characteristics of Screening Tests

**Effect of Inhibitor/Dispersant Dosage.** A strong performance dependence on dosage of inhibitor and/or dispersing agent is a distinctive feature of all the benchtop activity tests. Lower dosages of chemical agents are required for dispersion of particulates compared to good inhibition performance (where filter pore sizes typically range from 0.22-0.45 $\mu$m). In Figure 4, data for calcium phosphate inhibition (0.45 $\mu$m filter) versus polymer dosage are representative of the results typically obtained in benchtop activity tests. Below a specific polymer dosage ($\sim$5 mg/l), very little inhibition of calcium phosphate salts occurs. As the polymer dosage increases (5 to 7.5 mg/l), a sharp increase in inhibition performance occurs. Polymer dosages of 5-7.5 mg/l are approximately stoichiometric with respect to the orthophosphate in the test solution. One might presume that the phosphate which passed through the filter (0.45 $\mu$m pore size) is "soluble" and that few if any particulates of calcium phosphate were present. Benchtop activity tests, however, have clearly shown that presumption to be incorrect (Table I). In addition, Figure 5 demonstrates the effect of a maleic anhydride/sulfonated styrene (MaA/SS) copolymer on performance in the benchtop activity tests for iron (hydr)oxide dispersion. Dispersion of iron (hydr)oxides is almost nil at polymer dosages < 6 mg/l, rises significantly as polymer dosage increases from 6 to 8 mg/l, and almost 100% dispersion is observed with 10 mg/l of polymer. Passing the iron (hydr)oxide test solution through Millipore filters of differing pore sizes revealed that most of the particulates were 0.2-0.8 $\mu$m (Figure 6, Terpolymer A is an acrylate-based terpolymer containing sulfonate functional groups). The polymer dosage and particulate size results in the calcium phosphate and iron (hydr)oxide tests are very similar, except that the particulates produced in the iron (hydr)oxide tests are a little larger.

In the benchtop activity tests, the previous results suggest that a polymeric inhibitor does not change the solubility of the calcium phosphate or iron (hydr)oxide

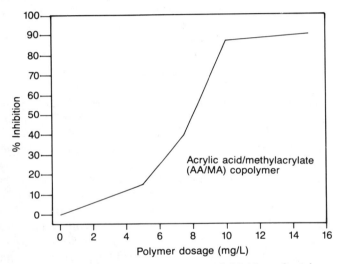

Figure 4. Effect of polymer dosage on threshold inhibition of calcium phosphate.

salts, but affects the overall size of the particulates which are formed. Only small quantities (~5% of the original amount) of phosphate or iron cations remain soluble, even when very effective threshold inhibiting polymers are used. In addition, analyses for the residual polymer and phosphate in the inhibition test solution filtrates indicate that particulate formation causes approximately equal uptake of the polymeric inhibitor and phosphate. This type of behavior has been consistently observed in benchtop activity tests for inhibition of a variety of scaling salts [e.g., iron (hydr)oxides, calcium phosphonate, and zinc hydroxide]. In those tests, good performance can be obtained with the inhibitor present at substoichiometric levels (~1/4 to 1/2) as compared to the inhibited ions. Nevertheless, previous reports have indicated that good threshold inhibition of calcium carbonate can be obtained with inhibitors (e.g., pyrophosphate) at concentrations which are about 1/100 that of the scaling ions (23). The distinct differences in inhibitor concentration required in the previous systems (e.g., calcium carbonate vs. calcium phosphate) suggest that different inhibition mechanisms may be involved.

Table I. Calcium Phosphate Inhibition Versus Polymer Dosage

| Polymer dosage (mg/l) | mg/l phosphate (in filtrate)* | | | |
| | No inhibitor | | AA/Am** | |
| | 0.10 $\mu$m | 0.45 $\mu$m | 0.10 $\mu$m | 0.45 $\mu$m |
|---|---|---|---|---|
| 0.0 | 0.5 | 0.5 | - | - |
| 5.0 | - | - | 0.5 | 0.5 |
| 7.5 | - | - | 0.5 | 5.0 |
| 10.0 | - | - | 0.5 | 9.5 |

* 10 ppm is maximum possible amount of phosphate in filtrate
** AA/Am monomer unit ratio = 23:77 mole %

In the benchtop tests for inhibition of scaling ions, the significant amount of polymer incorporated into the particulates is consistent with precedents in colloidal systems where particle agglomeration is limited by a polymer coating virtually all of the particles' surfaces (39-41). Those reports also indicated that decreases in polymer dosage lead to incomplete coverage of the particle surfaces by the polymer, followed by agglomeration and precipitation of the solids from the system. That performance versus polymer dosage pattern is closely followed in the benchtop activity tests for threshold inhibitors which were previously described.

Dispersancy tests have also shown the strong dependence of performance on polymer dosage, although the particulate solids: polymer weight ratio (~20:1) is much larger in these tests than the scaling ion:polymer ratio (~2:1) in the threshold inhibition evaluations. The dispersancy tests can be divided into two groups: initially homogeneous and initially heterogeneous. When 100 mg/l of

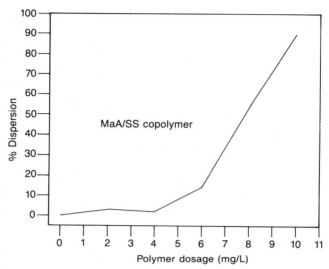

Figure 5.  Effect of polymer dosage on iron (hydr)oxide dispersion.

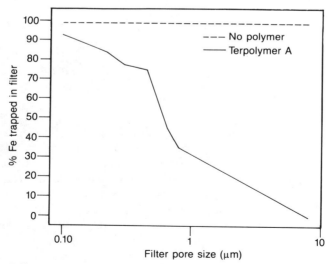

Figure 6.  Effect of polymer on particle size distribution of iron (hydr)oxides.

clay particulates and <1 mg/l of acrylic acid/acrylamide (AA/Am) copolymer (23:77 mole %) are present, very little dispersion occurs after a settling period of 24 hours. As the concentration of AA/Am copolymer increases (1 to 5 mg/ l), a sharp increase in dispersion of clay occurs (Figure 7). The increased dosage of polymer does not completely prevent the settling of the clay particulates, but serves to slow down the settling rate (Figure 8). Scanning electron micrographs have indicated that the original clay powder is comprised of clusters of individual thin platelets with the long axis $> 2\,\mu$m. Mechanical agitation from the recirculating pump breaks up the loosely bound clusters into the individual platelets. Adsorption of the polymer onto the surface of the clay platelets serves to limit reagglomeration of the platelets. Settling of the clay occurs when the platelets reagglomerate into clusters. The relatively large size of the clay platelets predisposes them to settling (43), even if the polymeric dispersant drastically lowers the rate of platelet reagglomeration.

**Effect of Polymer Composition.** Polymer composition, as defined by types/ structures of functional groups and monomer unit ratio, has a very significant impact on the performance of inhibitors and dispersants. In cooling water treatment programs, the performance of a polymeric scale inhibitor/dispersant depends on the ratio of monomer units as well as molecular weight. For the acrylic acid/ acrylamide (AA/Am) copolymer system, calcium phosphate inhibition is strongly dependent on the AA/Am mole ratio; the best "inhibition" of particulates was observed at low levels of polyacrylic acid (AA) monomer units. Figure 9 shows the benchtop activity test results for calcium phosphate inhibition using several AA/Am polymers, as well as acrylic acid and acrylamide homopolymers. The AA/Am copolymers and polyacrylic acid were produced by hydrolyzing polyacrylamide, thus performance differences are associated with changes in composition and not differences in molecular weight or method of polymer preparation. Ideally, a high level of polymer performance will occur over a broad but well-defined compositional range.

Particle size distribution studies, as a function of time, were conducted on two AA/Am copolymers (50/50 and 30/70 mole %, respectively). Those results are depicted in Figures 10-12 . With each AA/Am copolymer, a large number of very small calcium phosphate particles ($< 0.1$-$0.22\,\mu$m) was rapidly formed, as indicated by the small amount of phosphate and polymer present in the filtrate (0.1 $\mu$m pore size filter). Comparison of the upper curves (at elapsed time = 1 and 20 minutes) demonstrates the lack of particle growth with the AA/Am copolymer (30/70 mole %). With the AA/Am copolymer (30/70 mole %), no further significant growth was observed for time periods up to four hours (compare upper curves on Figures 10-12). In the presence of different AA/Am copolymer (50/50 mole %), a second stage of particle growth is apparent with larger particles slowly appearing (0.22-8 $\mu$m). The lower curves in Figures 10 and 11 clearly demonstrate the initial formation of small particles and subsequent development

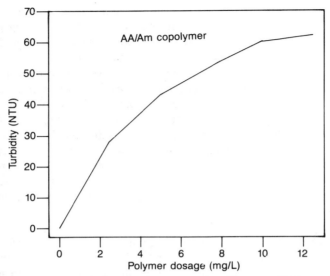

Figure 7. Effects of polymer dosage on clay disperson (24 hr settling).

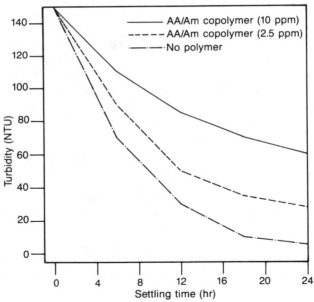

Figure 8. Effect of polymer dosage on settling rate of clay particulates (24 hr settling).

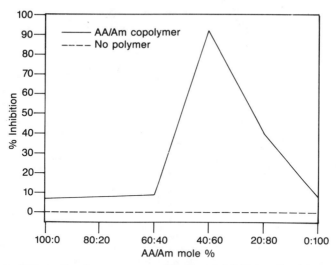

Figure 9. Effect of polymer composition on inhibition of calcium phosphate.

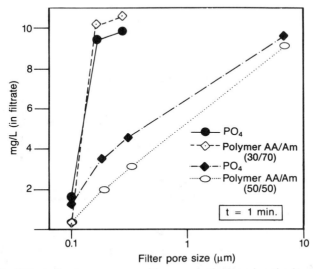

Figure 10. Effect of polymer composition on particulate size distribution of calcium phosphate -1 min. Reprinted with permission from ref. 45. Copyright 1989 National Association of Corrosion Engineers.

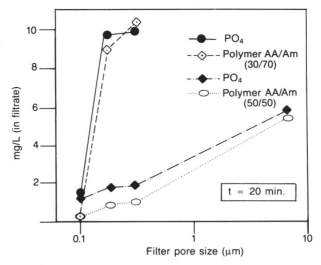

Figure 11. Effect of polymer composition on particulate size distribution of calcium phosphate - 20 min. Reprinted with permission from ref. 45. Copyright 1989 National Association of Corrosion Engineers.

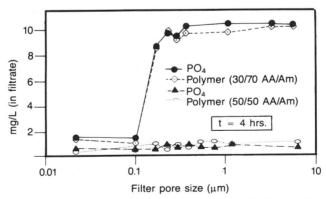

Figure 12. Effect of polymer composition on particulate size distribution of calcium phosphate - 4 hours).

into larger particulates. This particulate growth proceeded even though very little orthophosphate remained in solution.

In addition, comparison of polymer and phosphate concentration shows the intimate relationship between those two species in all the particulates which were formed (compare Figures 10 and 11). Whenever orthophosphate was incorporated into particles, no matter what the size, the polymer was also incorporated at essentially equal levels. It should be noted that virtually all the orthophosphate and polymer (10 ppm of each) was converted into particulates in the presence of either AA/Am copolymer (filter pore size < 0.1 $\mu$m). The main difference in performance between those two AA/Am copolymers lies in the subsequent growth rate of large particulates from the very small particles which were initially formed.

Polymer composition also significantly affects performance in the dispersion tests where particulates are added to the system, rather than being generated *in situ* from supersaturated solutions. In the dispersion of clay particles (Table II), three copolymers and one terpolymer exhibited comparable performance, but the acrylic acid homopolymer had virtually no dispersive effect. Figure 13 shows clay dispersion tests after 24 hours settling; polymers which exhibit poor, fair, and good dispersion of clay particles are shown in this figure. The importance of polymer composition and structure in limiting the growth of particulates has been previously reported for colloidal systems and relates to the ability of a polymer to limit particle agglomeration by modifying the chemical, structural and/or electrostatic character of particle surfaces (39-41).

Table II. Clay Disperson Versus Polymer Composition

| Polymer* | % Dispersion |
|---|---|
| MaA/SS | 35 |
| AA/Am | 34 |
| AA/HPA** | 33 |
| Terpolymer A | 32 |
| AA | 1 |
| Blank (no polymer) | 0 |

  * 10 mg/l polymer actives
** Acrylic acid/hydroxypropylacrylate copolymer

**Effect of Temperature.**  In typical cooling water applications, water temperatures can range from 60-100°F (15-38°C) in the basin to temperatures near 200°F (93°C) in the critical heat exchangers (44). Common mineral scaling salts often are described as being "inversely soluble", that is, less soluble at higher temperatures. Increased temperature puts stress on cooling water systems in two ways: 1) the rate of corrosion reactions (and all other reactions) is accelerated; and 2) the solubility of mineral salts decreases.

Figure 13. Clay dispersion by polymers; from left: poor, moderate, and good
    dispersion.

The effect of temperature on the performance of AA/Am copolymers (50/50 and 30/70 mole %) in benchtop activity tests for calcium phosphate inhibition is demonstrated in Figure 14. At ambient temperature, a significant amount of the phosphate passes through a 0.45 $\mu$m filter, even in the absence of a polymeric inhibitor. As the temperature increases above 22°C, significant amounts of phosphate are incorporated into particulates which are $> 0.45$ $\mu$m. One of the AA/Am copolymers (50/50 mole %) was only able to limit the formation of those particulates until a temperature of about 40°C was reached, and inhibition performance decreased rapidly beyond that temperature. In contrast, the other AA/Am copolymer (30/70 mole %) inhibited formation of the calcium phosphate particulates ($> 0.45$ $\mu$m) until a temperature greater than 80°C was attained.

SEM photomicrographs were obtained on the following filtered samples (0.2 $\mu$m Nuclepore polycarbonate filter membrane):

- No inhibitor present
- AA/Am copolymer (30/70 mole %) at 70°C
- AA/Am copolymer (30/70 mole %) at 90°C, after loss of performance

The SEM photomicrographs indicated that large numbers of microparticles (diameter about 50 nm) were produced in all three samples and that agglomeration of those microparticles led to formation of large particulates (diameter greater than 0.45 $\mu$m). The microparticles appear to be amorphous as their surfaces are relatively smooth and no distinct crystal faces were observed. The microparticles in the test sample heated to 90°C had the same appearance as microparticles in the test solution heated to 70°C. The main difference between those two samples was that a much greater degree of microparticle agglomeration had occurred in the 90°C sample. Similar losses in performance (i.e., growth of particulates) have been observed in the benchtop activity tests for the other scaling salts listed in previous sections. SEM analyses of those samples have shown that large numbers of microparticulates are produced and performance is determined by a polymer's ability to limit aggregation of those microparticulates. As the temperature of the test sample increases, the frequency of collision between particles also increases and raises the probability that particle agglomeration will occur. This increases the likelihood that steric barriers to particle agglomeration, arising from adsorption of polymers to particle surfaces, will be overcome.

**Effect of pH.** In cooling water applications, increasing the pH tends to decrease metal corrosion rates and increase scaling tendencies. The solubility of calcium salts will decrease at a different rate than the solubility of the analogous magnesium salts. Increasing the pH can produce supersaturated solutions of scaling salts by increasing the charge on anionic ions (deprotonation of multiprotic salt ions) and encouraging the oxidation of metal ions (e.g., $Fe^{+2}$ to $Fe^{+3}$).

Figure 15 demonstrates the effect of pH on AA/Am copolymers' performance. In the absence of polymer, virtually all the phosphate passes through a 0.45 $\mu$m filter at pH values less than 7.5, but about 95% of the phosphate is converted

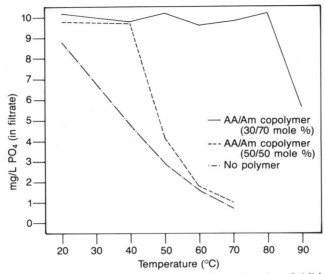

Figure 14. Effect of temperature on calcium phosphate inhibition.
Reprinted with permission from ref. 45. Copyright 1989 National
Association of Corrosion Engineers.

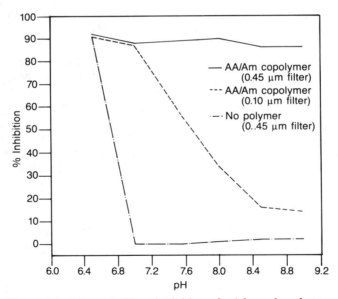

Figure 15. Effect of pH on inhibition of calcium phosphate.
Reprinted with permission from ref. 45. Copyright 1989 National
Association of Corrosion Engineers.

into particulates as the pH increases from 7.5 to 8.5. This formation of particulates mirrors the large increase in the calculated supersaturation ratios (SR $<$ 1 at pH 6.5, SR = 10 at pH 7, and SR $>>$ 100 at pH values between 7.5 and 10) based on the initial concentrations of ions in the test solution (42). The size of those particulates is very large as almost all of the phosphate was trapped by a filter with an 8 $\mu$m pore size. It is commonly observed in benchtop activity tests that polymer performance is strongly dependent upon the pH of the test solution and performance tends to decrease sharply when a certain pH value is exceeded.

In contrast to the test solution without polymer, the presence of an AA/Am copolymer (30/70 mole %) effectively inhibited the formation of calcium phosphate particulates ($>$ 0.45 $\mu$m) over a pH range of 6.5-9.0 (Figure 15). The filtration results (0.10 versus 0.45 $\mu$m) indicate that significant amounts of particles formed between pH 7-9, even in the presence of the AA/Am copolymer. In the benchtop activity tests, that copolymer does not significantly affect the solubility of calcium phosphates but strongly affects the size of the particulates which are formed. SEM photomicrographs of the particulates formed at different pH values indicate that formation of microparticles ($\sim$50 nm) occurred whether a polymer was present or not. SEM photomicrographs of the microparticles and particulate agglomerates formed at pH 8.5 in the absence of a polymer are shown in Figures 16A and 16B, respectively. The microparticles formed in the presence of the AA/Am copolymer (30/70 mole %) are shown in Figure 16D, while any evidence of large particulate agglomerates is absent in Figure 16C. The microparticles and particulate agglomerates do not exhibit distinct crystal faces and appear to be amorphous. The SEM and filtration results indicate that the formation of large particulates occurs by agglomeration of the microparticles, and the AA/Am copolymer forestalls that agglomeration until a pH $>$ 9 is attained.

In some systems, pH can influence formation of supersaturated solutions by affecting the chemical nature of cations, anions, and even the mechanism whereby the ions in solution are transformed into the final precipitated solid. For example, increased pH produces supersaturated solutions by promoting oxidation of metal ions (e.g., $Fe^{+2}$ to $Fe^{+3}$ and $Mn^{+2}$ to $Mn^{+4}$). Figure 17 demonstrates the formation of iron and manganese (hydr)oxide particulates as a function of pH. Between pH 5.5 and 6.5 the formation of iron (hydr)oxide particles occurs, while manganese (hydr)oxide particles form mostly between pH 7.5 and 8.5. When the test solution is maintained at pH 8.5, uncontrolled agglomeration of particles occurs rapidly in the absence of a polymeric agent (Table III). The large particulates of iron (hydr)oxide salts settled quickly with only 5% of the initial amount of iron remaining suspended after 24 hours (unfiltered sample). In the presence of the AA/Am copolymer (30/70 mole %), 100% of the iron (hydr)oxides remained suspended after 24 hours. Nevertheless, the mass of particulates formed was comparable in both test solutions with only 5% of the iron remaining in the filtrate (0.1 $\mu$m filter). The major difference in the two systems is that the AA/Am copolymer

A

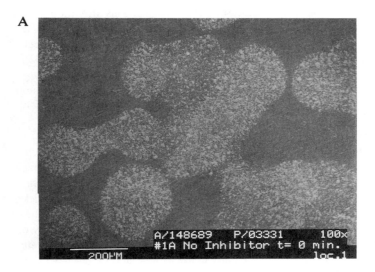

B

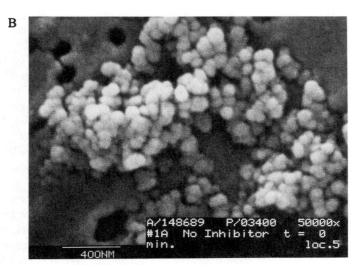

Figure 16. Scanning electron photomicrographs of particle agglomerates produced in calcium phosphate inhibition test. No inhibitor at 1 min. magnified 100 x (A) and 50,000 x (B). AA/Am copolymer at 1 min. magnified 100 x (C) and 50,000 x (D).

C

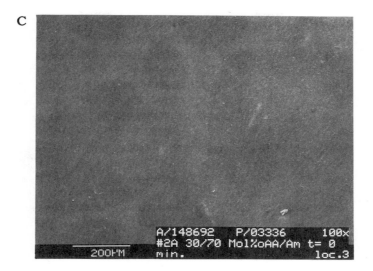

D

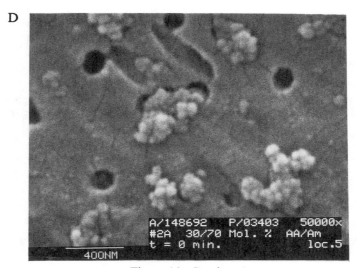

Figure 16.  Continued.

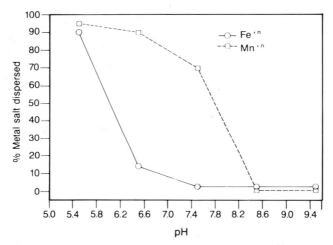

Figure 17. Effect of pH on oxidation and dispersion of iron/manganese salts
(air-oxidation). Reprinted with permission from ref. 45. Copyright
1989 National Association of Corrosion Engineers.

maintains smaller particulate size. The AA/Am does not appear to significantly solubilize the iron (hydr)oxides or prevent the oxidation of $Fe^{+2}$ to $Fe^{+3}$.

The effect of polymers as iron (hydr)oxide dispersants can be demonstrated visually. Figure 18 shows a series of screening tests for Terpolymer A used as an iron (hydr)oxide dispersant. The yellow color of the test solutions indicates

Table III.  Effect of Polymer on Size of Iron (Hydr)oxide Particles

| Polymer | % Fe remaining | | |
| --- | --- | --- | --- |
| | unfiltered | 0.45μm | 0.10 μm |
| AA/Am (30/70 mole %) | 100 | 50 | 5 |
| Blank (no polymer) | 5 | 5 | 5 |

that $Fe^{+2}$, which was added to each test, has been oxidized to $Fe^{+3}$. When the polymer dosage is insufficient to disperse the iron (hydr)oxides, a reddish precipitate settles at the bottom of the solution (far left solution in Figure 18). When sufficient polymer is present to disperse iron (hydr)oxides, the test solution remains yellow and little or no precipitate is observed. It is important to note that the yellow solutions indicate effective dispersion of iron (hydr)oxide particles. These particles can be collected on filters, and their size can be determined by several methods.

In systems containing iron and manganese cations, insights into the mechanism of particle formation and growth have been obtained through the use of scanning electron microscopy and particle size distribution studies. Particulates produced in the standard benchtop activity test using air-oxidation (10 mg/l $Fe^{+2}$, 2 mg/l $Mn^{+2}$, 10 mg/l AA/Am actives at 60 °C) were analyzed. SEM photomicrographs of those particles indicate that the particulate agglomerates are composed of numerous amorphous microparticles with a diameter of approximately 0.05 μm. The uniformity in the microparticle size suggests that they are formed initially. Subsequent "particle growth" occurs mainly through agglomeration of microparticles, rather than continued addition of soluble cations to an existing particle (18). Energy-dispersive x-ray analysis indicated the particulate agglomerates contained both iron and manganese with an Fe:Mn weight ratio of 6:1 (theoretical Fe:Mn ratio = 5:1). Unfortunately, the individual microparticles were too small to have their elemental composition determined by EDAX™. At least four possible compositions for the individual microparticles are conceivable:

1. Individual microparticles composed of mixtures of smaller, separate domains of iron and manganese salts
2. A mixed salt containing both iron and manganese cations
3. Individual microparticles composed of iron or manganese salts and co-agglomeration produces large particulates with both cations present and the microparticles randomly distributed
4. Layered or coated structure with iron salts in one layer and manganese salts in the other

The particulate size distribution and chemical composition studies are summarized in Figure 19. Greater than 90% of the iron and manganese cations have been incorporated into particulates in all cases studied. The AA/Am copolymer limits the particulate size distribution to a narrow range, and about 90% of the particulates were trapped by filters with pore sizes of 0.3- 0.8 $\mu$m. In the absence of an effective polymer, the particulates increase significantly in size ($>> 8$ $\mu$m) and readily precipitate.

The proposed particulate formation mechanism is represented in Figure 20. The ratio of the iron/manganese incorporated into the particulates is virtually the same throughout the entire size distribution. This suggests that the iron and manganese salts form microparticles concurrently. The agglomerated particulates which form are intimate mixtures of the polymeric agent with iron and manganese salts. As previously indicated (Figure 17), the iron and manganese cations are oxidized over different pH ranges and particulates from iron salts appear first. When microparticles containing manganese salts are subsequently formed, they are added to the iron-containing microparticle agglomerates (refer to Figure 20). It is not known whether the manganese-containing microparticles simply coat the initially formed iron-containing agglomerates or whether the microparticles containing each metal become randomly distributed within each agglomerated particulate. A random distribution of microparticulates could occur if the agglomeration process is reversible or the agglomerates are structurally non-rigid. Overall, polymeric agents limit agglomeration processes, thereby minimizing precipitation.

In Figure 6 and Table IV, particle size distribution results (using a series of filter pore sizes and quasielastic light scattering) indicate that certain polymers are capable of limiting the growth of iron (hydr)oxide particulates. Terpolymer A and the AA/Am (30/70 mole %) copolymer are particularly effective in that regard. SEM photomicrographs of the particulates produced in benchtop activity tests without polymer and with the AA/Am copolymer (30/70 mole %) indicate that numerous microparticles ($\sim$50 nm) are formed in both cases. Massive agglomeration of those microparticles occurred in the sample without a polymeric agent present. Only very limited agglomeration of the microparticles occurred in the test with the AA/Am copolymer (30/70 mole %). The results and particulate growth processes appear to be very similar to previously described systems (e.g., calcium phosphates, Figure 16) where no reduction/oxidation reactions are occurring.

Figure 21 shows a qualitative demonstration of iron (hydr)oxide particle size via laser light scattering. The laser beam is first passed through the left solution which contains a well dispersed solution of iron (hydr)oxide. The laser beam is scattered by the small microparticles, but the beam looks continuous to the eye. Next, the beam passes through the right iron (hydr)oxide solution which does not contain an effective dispersant; the solution was agitated to get the

Figure 18.  Fe(Hydr)oxide Dispersion Test, from left: 4, 6, 8, and 10 mg/l polymer.

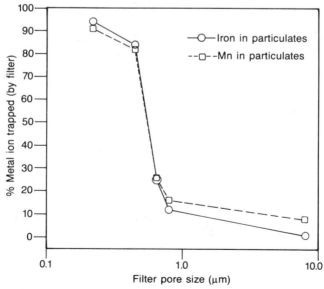

Figure 19.  Particulate size versus chemical composition of iron and manganese salts (air-oxidation).

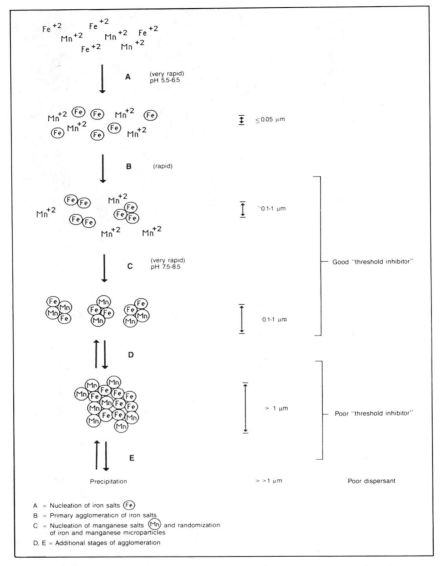

Figure 20. Agglomeration model for particulate and scale formation in system containing oxidizable ions. Reprinted with permission from ref. 45. Copyright 1989 National Association of Corrosion Engineers.

particles into the laser beam. The particles in this solution are much larger, and discontinuity in the scattered laser beam is observed.

Table IV.   Use of Quasielastic Light Scattering to Determine Particle Size Distribution in Iron (hydr)oxide Inhibition Tests

| Dispersant | Effective diameter (microns) |
|---|---|
| Terpolymer A | 0.265 |
| AA/Am Copolymer | 0.270 |
| AA Homopolymer | 1-10 |
| Blank (no polymer) | 1-10 |

## Proposed Mechanisms of Inhibition and Dispersion of Particulates

**Systems Containing Non-Oxidizable Ions.**  In benchtop activity tests containing soluble ions (e.g., calcium and phosphates) which are not readily oxidized, Figure 22 indicates the proposed mechanism for particle formation and agglomeration (45). This mechanism is supported by the SEM photomicrographs and particle size distribution studies of several scaling systems (calcium/magnesium phosphates, calcium/magnesium phosphate plus phosphonates, calcium phosphonates, and zinc hydroxide). In each case, the experimental evidence has indicated that a majority of the soluble ions are converted into many small particles as the solution becomes highly supersaturated (Step A in Figure 22). Filtration results have indicated that the size of the particles was in the 100-200 nm range when an effective threshold inhibitor was present. The SEM results further demonstrated that microparticles (~50 nm diameter) were actually present and the particles trapped by filtration were actually agglomerates of microparticles (Step B of Figure 22). Effective threshold inhibitors strongly limit, but do not completely prevent, the initial (primary) agglomeration of the microparticles. Secondary agglomeration of particles and clustering of agglomerates can occur when a poor inhibitor is present (Step C of Figure 22). If a poor threshold inhibitor/poor dispersant is present, the particulates will continue to increase in size and precipitation will eventually occur (Step D of Figure 22). In the absence of a chemical inhibitor or dispersing agent, microparticle formation and primary/secondary agglomeration processes occur very rapidly, leading to uncontrolled precipitation of the scaling salts. An additional feature of this model is that the agglomerated particles are not tightly bound to each other, and it is possible to reverse the agglomeration process by mechanical agitation. Once that agitation has ceased, the agglomeration process will again cause an increase in the size of particulates.

Of course, some systems contain mixtures of ions from which several scaling salts can form (e.g., calcium phosphate plus phosphonates). In those situations, the initial formation of microparticles (Step A of Figure 22) and agglomeration processes (Steps B through D of Figure 22) may occur sequentially and at different

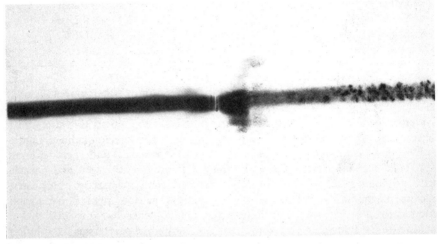

Figure 21. Laser beam passing through well-dispersed solution of small particles (left) and solution with larger particles (right).

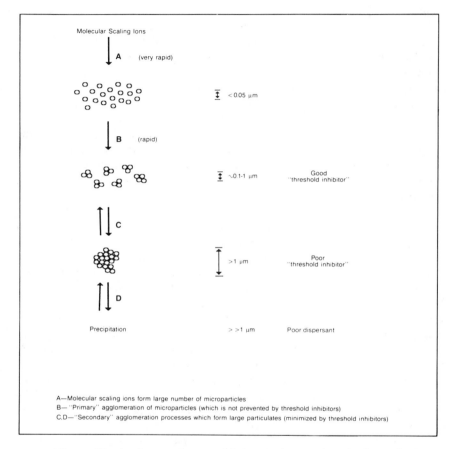

Figure 22. Agglomeration model (particulate and scale formation). Reprinted with permission from ref. 45. Copyright 1989 National Association of Corrosion Engineers.

pH values for each scaling salt. Nevertheless, the overall sequence of steps in these more complex systems would be similar to those in a system containing a single type of scaling salt.

The dispersion of pre-existing particulates is easily accommodated as an extension of the mechanism proposed in Figure 20 (Steps D and E) and Figure 22 (Steps C and D). In the benchtop dispersion tests, the solids added to the system are composed of loosely-bound agglomerated particles. The size of the individual particles tends to be $> 1$ $\mu$m. When solutions containing those agglomerated particulates are mechanically agitated (e.g., impeller pump), individual particles or particulates agglomerated to a lesser extent are produced. The chemical dispersing agent adsorbs onto the surfaces of these solids and limits the reagglomeration and precipitation processes. In essence, dispersion and threshold inhibition of particulates in the benchtop activity tests represent extensions of the same physical phenomenon: particulate size control through limiting microparticle agglomeration.

**Systems Containing Oxidizable Ions.** In benchtop activity tests which contain two types of oxidizable ions, Figure 20 indicates the proposed mechanism for particle formation and agglomeration. This second mechanism is an extension of the mechanism depicted in Figure 22 and is representative of the processes which occur in the benchtop activity tests which initially contain $Fe^{+2}$ and $Mn^{+2}$. The initial formation of microparticles is caused by the oxidation of the soluble ions, which then form salts with low solubility. When mixtures of oxidizable ions are present, oxidation can occur at a different pH value for each type of ion (Steps A and C of Figure 20). After that, the steps of primary and secondary agglomeration and aggregation of agglomerates are very similar to those in systems which do not contain oxidizable ions (Figure 22). Test results have also indicated that even when microparticles of different chemical compounds are sequentially produced, each final agglomerated particulate may still be intimate mixtures of both types of microparticles. Partial reversibility of the agglomeration process (due to mechanical agitation, for example) can lead to randomization of the micro-particles of each chemical salt throughout all the agglomerated particulates.

**Conclusions**

In benchtop activity tests, striking similarities have been observed in microparticle formation and particulate growth processes for a wide variety of scaling salts. The general response of those tests to changes in type and dosage of chemical threshold inhibitor/dispersant, pH, and temperature was also very similar. Initial formation of large amounts of microparticles ($\sim$50 nm diameter), followed by agglomeration of those microparticles, are common features of the benchtop activity tests for "threshold inhibition." The agglomeration processes play a pivotal role in determining whether a chemical agent will act as a threshold inhibitor, dispersant, or be ineffective. Agglomeration of the microparticles appears to be a multistep process. The primary agglomeration step occurs rapidly, and effective threshold inhibitors limit the growth of particulates to between 0.1 and 0.2 $\mu$m in diameter. When an ineffective threshold inhibitor is present, a slow secondary agglomeration step occurs which allows further growth in particulate size. Dispersion of particulates occurs if the chemical agent can limit the extent of the secondary agglomeration process. Differences in the activity of various polymers is related to differences in their ability to limit the agglomeration processes. The mechanistic studies have been corroborated by several techniques including particle size distribution studies (filtration and quasielastic laser light scattering), scanning electron microscopy, and particle growth rate evaluations.

"Threshold inhibition" and dispersion of particulates are closely related and can represent extensions of the same physical phenomenon: limiting the agglomeration of particles. In the threshold inhibition tests, the particles are formed *in situ* from solutions which are supersaturated with respect to certain scaling ions. The dispersion tests involve the addition of particles from an external source. Nevertheless, the key feature to preventing the precipitation of scaling salts in both types of test is the control of agglomeration of individual particles.

### Acknowledgments

We thank the National Association of Corrosion Engineers for permission to republish some of the data contained in this chapter.

### Literature Cited

1. Kemmer, F. N. The NALCO Water Handbook (2nd Edition); McGraw-Hill: New York, 1988; pp. 38.1-38.30.
2. Katzel, J. Plant Engineering April 27, 1989, pp 32-8.
3. Montemarano, J. Water Technology August 1987, pp 40-3.
4. Elliott, T. C. Power December 1985, pp S-1-16.
5. Cheremisinoff, N. P., Cheremisinoff, P. N. Cooling Towers: Selection, Design and Practice; SciTech Publishers: Matawan, New Jersey, 1989.
6. Dubin, L. U.S. Patent 4 361 492, November 30, 1982.
7. Persinski, L. J., Walker, J. L., Boffardi, B. P. U.S. Patent 4 640 793, February 3, 1987.
8. Godlewski, I. T., Schuck, J. J., Libutti, B. L. U.S. Patent 4 029 577, June 14, 1977.
9. Woerner, I. E., Boyer, D. R. CORROSION/84, Paper No. 314, National Association of Corrosion Engineers, Houston, Texas, 1984.
10. Logan, D. P., Rey, S. P. Materials Performance 1986, 25, 38.
11. Nancollas, G. H., Reddy, M. M., Tsai, F. Journal of Physics 1972, 5, 1186.
12. Mickus, J. C., Fordyce, D. B. U.S. Patent 3 580 855, May 25, 1971.
13. Kubo, S., Takahashi, T., Morinaga, H., Ueki, H. CORROSION/79, Paper No. 220, National Association of Corrosion Engineers, Houston, Texas, 1979.

14. Smyk, E. B., Hoots, J. E., Fivizzani, K. P., Fulks, K. E. CORROSION/ 88, Paper No. 14, National Association of Corrosion Engineers, Houston, Texas, 1988.
15. Hoots, J. E., Crucil, G. A. Materials Performance 1987, 26, 17.
16. Yeoman, A. M. U.S. Patent 4 547 540, October 15, 1985.
17. Hann, W. M., Natoli, J. U.S. Patent 4 530 766, July 23, 1985.
18. Hoots, J. E., Johnson, D. A., Fong, D. W., Kneller, J. F. U.S. Patent 4 752 443, June 21, 1988.
19. Persinski, L. J., Ralston, P. H., Gordon, Jr., R. C. U.S. Patent 3 928 196, December 23, 1975.
20. May, R. C., Geiger, G. E. U.S. Patent 4 303 568, December 1, 1981.
21. Hoots, J. E., Fivizzani, K. P., Kaplan, R. I. U.S. Patent 4 869 828, September 26, 1989.
22. Freedman, A. J. Materials Performance 1984, 23, 9.
23. Cowan, J. C., Weintritt, D. J. Water-Formed Scale Deposits; Gulf Publishing Co.: Houston, TX, 1976; pp. 257, 258, 262.
24. Ralston, P. H. U.S. Patent 3 336 221, August 15, 1967.
25. Kallay, N., Tomic, M., Biskup, B., Kunjasic, I., Matijevic, E. Colloids and Surfaces 1987, 28, 185.
26. Koutsoukos, P., Amjad, Z., Tomson, M. B, Nancollas, G. H. J. Am. Chem. Soc. 1980, 102, 1553.
27. Tomson, M. B. J. Cryst. Growth 1983, 62, 106.
28. Weijnen, M. P. C., van Rosmalen, G. M. Desalination 1985, 54, 239.
29. Rubin, A. J. Aqueous-Environmental Chemistry of Metals; Ann Arbor Science Publishers: Ann Arbor, MI, 1976;pp. 219-253.
30. Gill, J. S., Nancollas, G. H. J. Crystal Growth 1980, 48, 34.
31. House, W. A., Donaldson, L. J. Colloid and Interface Science 1986, 112, 309.
32. Melikhov, I. V., Kozlovskaya, E. D., Berliner, L. B., Prokofiev, M. A. J. Colloid and Interface Science 1987, 117, 1.
33. van Kemenade, M. J. J. M., de Bruyn, P. L. J. Colloid and Interface Science 1987, 118, 564.
34. Amjad, Z. J. Colloid and Interface Science 1987, 117, 98.
35. Feenstra, T. P., de Bruyn, P. L. J. Colloid and Interface Science 1981, 84, 66.
36. Nancollas, G. H., Koutsoukos, P. G. Prog. Crystal Growth Charact. 1980, 3, 77-102.
37. Sarig, S., Shiffrin, F. Seawater Desalination 1976, 8-76, 150.
38. Brecevic, Lj., Hlady, V., Furedi-Milhofer, H. Colloids and Surfaces 1987, 28, 301.
39. Napper, D. H. Polymeric Stabilization of Colloidal Dispersions; Academic Press: New York, New York, 1983.
40. Tadros, Th. F. The Effect of Polymers on Dispersion Properties; Academic Press: New York, New York, 1982.

41. Tadros, Th. F. <u>Solid/Liquid Dispersions</u>; Academic Press: New York, New York, 1987.

42. Westall, J. C., Zachary, J. L., Morel, F. M. M. Massachusetts Institute of Technology (Department of Civil Engineering), Technical Note No. 18, 1976.

43. Barth, H. G. <u>Modern Methods of Particle Size Analysis</u>; John Wiley & Sons, New York, New York, 1984, pp. 1-42.

44. Reed, D. T., Nass R. <u>Proceedings of the International Water Conference</u>, Pittsburgh, 1975.

45. Hoots, J.E., Fivizzani, K.P., Cloud R.W., <u>CORROSION/89</u>, Paper No. 175, National Association of Corrosion Engineers, Houston, Texas, 1989.

RECEIVED August 27, 1990

# Chapter 23

# Macromolecular Assemblages in Controlled Biomineralization

C. C. Perry[1], M. A. Fraser[1], and N. P. Hughes[2]

[1]Chemistry Department, Brunel University, Uxbridge, Middlesex
UB8 3PH, United Kingdom
[2]Inorganic Chemistry Laboratory, South Parks Road, Oxford OX1 3PH,
United Kingdom

This article addresses the relevance of membrane structures and organic matrices on the path of controlled mineralisation of amorphous and crystalline materials in biological systems. Our approach to this area of research will be highlighted by experimental data obtained from biological studies of silicified plant hairs from *Phalaris canariensis* and strontium sulphate crystals from the radiolarian *Sphaerozoum punctatum* and by the study of model systems of vesicular precipitation and organic substrate induced crystallisation.

Biominerals result from the regulation and organisation of the inorganic solid state by biological organisms. The materials which result are usually composite inorganic/organic phases with properties characteristic of the composite as a whole (1). The formation of minerals can occur intracellularly, epicellularly and extracellularly and a considerable amount of research effort has been concentrated on defining the 'degree of control' exerted by organisms over both deposition of specific minerals and the development of the mineral phase. Biominerals are usually morphologically and structurally very different from their non-biological analogues and the problem of understanding their deposition is compounded by the close association of organic macromolecules with the inorganic phase during precipitation.

The formation of an inorganic solid from aqueous solution is achieved by a combination of three fundamental physico-chemical steps: supersaturation; nucleation and crystal growth or maturation. A further essential element for controlled *biological* mineralisation is spatial localisation, which may occur either through membrane-bounded compartments or specific cell wall regions. This compartmentalisation allows local regulation and control over physico-chemical factors through selectivity in biochemical processes such as ion and molecular transport. This represents control at a higher level of organisation (ultimately under genetic control) and will not be touched upon further in this article. It must be noted that spatial controls in the production of organised amorphous materials may have special significance.

Each of the steps of mineral deposition will now be discussed in turn with illustrations from appropriate *in vivo* studies. The importance of macromolecular assemblages will also be highlighted.

### Degree of Supersaturation

The activity product, AP, for a salt $M_nX_m$ is given by: $AP = [M^+]^n [X^-]^m$ and precipitation is favoured if this value exceeds $K_s$, the solubility product. There are substantial problems in measuring solubility products in biological systems. For instance, each biological environment is unique and small changes in crystal size and the incorporation of extraneous ions leads to changes in solubility product. Further, the solubility product concept takes no account of kinetic factors, which in certain cases control biological mineral deposition (2).

It has long been known that variations in ion supersaturation have a profound effect on *in vitro* mineral deposition, leading to marked changes in crystal habit. Experiments performed on strontium and barium sulphate deposition in desmids have pointed to similar *in vivo* effects (3), although since it is not possible to measure local ion concentrations *in vivo* this remains somewhat uncertain.

### Nucleation

In order to understand mineral deposition, the processes of crystal nucleation and growth must be tackled separately (4). As a general rule, nucleation in controlled biomineralisation will involve low supersaturation in conjunction with active interfaces. The former can be regulated by transport and reaction-mediated processes (inhibitors, etc.), and the latter generated by organic substrates in the mineralisation zone. Various modes of specific molecular recognition are open to any biomineralisation system, in which interfacial matching of the chemical and electrostatic fields surrounding both the inorganic and organic surface determine crystal type and morphology. Such interactions may be further subdivided into more distinct modes of recognition.

**Spatial Matching of Charge Distribution.** This represents perhaps the most fundamental manner through which an inorganic solid may specifically interact with an organic matrix. Simple topographic variation of the organic matrix surface may occur so as to produce localised pockets and grooves of high charge density which accumulate ions and stabilise nucleation sites. Such a scheme has been proposed for the iron storage protein ferritin (5) and may well be applicable in many biomineralisation systems. Rearrangement of the accumulated ionic cluster into a more ordered three-dimensional structure (crystalline or amorphous) is dependent upon the freedom allowed by the matrix interactions. Hence, strong binding at the matrix surface may favour amorphous nuclei, whereas lower binding affinities will allow movement of ions into periodic lattice sites.

The ultimate level of nucleation control arises through absolute structural correspondence in two dimensions at the inorganic:organic interface. This phenomenon, termed epitaxis, involves lattice matching at the interface resulting in the oriented overgrowth of particular crystal faces. The mechanism has been implicated in a wide range of biomineralisation processes, but little direct evidence for geometrical matching of matrix functional groups and lattice dimensions has been obtained. The only complete study which supports this mechanism involves deposition of the calcium carbonate nacreous layer of the mollusc shell (6). The results obtained showed that in many species the a and b axes of the antiparallel β-pleated sheet of the matrix and the overlying aragonite crystal are matched in orientation at the interface. This suggests epitaxial control. It is worth noting that in contrast to the low affinity binding required for nucleation within matrix cavities, epitaxial interactions require site specific, strongly bound ligation at the matrix surface.

There are several major problems associated with an epitaxial mechanism of deposition control. Firstly, in many cases the disposition of ions in different crystallographic faces is almost identical, making it impossible to select between two structures simply through structural correspondence at the matrix surface (e.g. crystallographic c axis of calcite and aragonite). Secondly, deposition as described requires a rigid two-dimensional template, whereas proteins and polysaccharides in biology commonly display considerable dynamics. Thirdly, it is unlikely that organic matrices can exhibit the molecularly smooth surface which is a prerequisite of epitaxial interactions.

**Stereochemical Matching**. Stereochemistry has been shown to be vitally important in determining the *in vitro* nucleation of specific crystal faces at interfaces (7). Although some consideration of cation coordination geometry is implied in epitaxial theories, this factor is secondary because of the flexible stereochemistry of cation-ligand interactions. No recognition is made, however, of the stereochemical requirement for optimal completion of the anionic coordination geometry in the developing crystal face. This is an important aspect of molecular recognition. Such requirements explain the selectivity between crystal faces with similar inter-cation distances but different anion stereochemistry (e.g. calcite and aragonite above). Consideration of anionic coordination geometry also may explain why faces of optimal charge density do not always interact with the matrix surface.

**Cooperative Effects**. It is probable that in most cases of biomineralisation, nucleation control is exerted through a combination of the mechanisms discussed above. Examples of the cooperative interaction of electrostatic and structural correspondence at the interface arise from the study of oriented calcite nucleation on polystyrene films (8) and the nucleation of calcium carbonate on stearic acid monolayers (9-11). In this way, although a crystal face may not match the ionic arrangement of the underlying organic substrate, a specific nucleation face may be preferred through favourable orientation of the anions within the structure. This is an exemplified by Mann's work on vaterite deposition on stearic acid monolayers, in which spatial matching of charges should lead to preferential nucleation of calcite over vaterite. It is proposed that this does not occur since there is a close stereochemical similarity between the growth face of vaterite observed and the orientation of carboxylate groups at the monolayer surface.

## Crystal Growth

Crystal growth depends, at the most basic level, on the supply of material to the newly formed interface. For morphologically controlled precipitation to occur conditions of low supersaturation must prevail, since under such conditions only one nucleus is favoured. At high supersaturation crystals grow from many surface nucleation sites giving rise to substantial disorder at the surface and resulting in irregularly shaped crystals. The growth rate for large crystals at high supersaturation is likely to be determined by the rate of diffusion of ions to the crystal surface whereas under conditions of low supersaturation and for small crystals attachment of ions to the crystal surface is likely to be the rate limiting step. Extraneous ions can also influence surface-controlled reactions, their incorporation often resulting in striking modifications in crystal morphology.

**Growth Processes**. In the direct growth of crystals from solution through the aggregation of ions and primary particles, structure is controlled by nucleation events. However, crystal chemistry and morphology may be grossly affected by processes occurring in the post-nucleation, growth phase. In many instances the morphology of biominerals bears no relation to the underlying matrix structure but the structures deposited are nonetheless highly reproducible and in many cases quite novel. The ways in which the growth phase may be controlled break down into two distinct types: growth modification, generally considered to have a chemical basis, which has a timed effect on crystal growth (e.g. ion concentrations, small organic molecules, pH variations); and mechanical stress and spatial constraints exerted through cellular membranes and filaments.

**Growth Modification**. Many *in vitro* examples have been documented in which the interaction of biological molecules with growing minerals has produced unique morphological changes. For instance, the influence of a range of closely related carboxylic acids on *in vitro* calcite morphology has been extensively investigated (10), showing that subtle changes in the stereochemistry and flexibility of additives may have a marked effect on habit. Similarly, studies on *in vitro* habit modification of strontium sulphate crystals using simple phosphates (12) have not only shown that very low concentrations of modifiers are necessary, but also that there may be a simple pathway to some of the seemingly complex crystal morphologies which have been observed in biology.

**Mechanical Stress and Spatial Constraints.** Physical forces exerted upon a growing mineral by external membranes are capable of moulding mineral morphology. The biological cases quoted are numerous, however they are best highlighted by examination of an elegant crystallographic example.

In the marine protozoan Acantharian, mineralisation of the strontium sulphate skeleton occurs intracellularly within membrane bound vesicles which are subject to external stresses from cell membranes and filaments. The detailed arguments as to how the external stresses determine crystal morphology are presented elsewhere (13), but in brief the driving forces in the determination of cell morphology are the chemical production of membrane components and on pumping which leads to crystal growth in membrane bound vesicles. The membrane restrains growth of the crystals physically (volume restraints) and chemically (composition), further, molecular rearrangements in the membrane can lead to lateral anisotropy within the crystal. The constant interaction of the stress fields of the crystal and the membrane produce a unique but repeatable morphology. Such ideas in a less elaborate form, have been proposed for the unique crystal morphologies observed in other biological systems, e.g. magnetite crystals in bacteria (14).

**Amorphous Biominerals.** In general, amorphous phases do not prevail in biology as the amorphous states are so much more soluble than the crystalline states. This can be overcome only in systems where unusual barriers to crystallisation exist. One such example is hydrous silica.

Although silica exhibits no long range crystallographic order, microscopic morphological order is often observed. Such order may arise either during the nucleation or the growth processes, since the silica network has similar surface properties to a crystallographic system. Energy considerations support the formation of a silica aggregate with a densely packed, covalently bonded core and a highly hydrated surface. The surface is likely to interact with organic substrates in a biological system in an analogous manner to crystal interactions, lowering the free energy of aggregate formation and controlling aggregate morphology on the microscopic scale. During the growth stage the chemical nature of the mineralisation zone or the physical forces under which the mineral is deposited may control the mineral morphology (15).

### Amorphous Processes

**Nucleation.** Although nucleation processes are not generally considered to play a major role in deposition of amorphous minerals, the ordered aggregation of primary silica particles to form structural motifs in plant emergences may reflect the topographical and stereochemical nature (i.e. the arrangement of charged residues and hydrogen bonding centres) of the polysaccharidic matrix system (16-18). A similar stereochemical role has been inferred in the silicification of diatoms, the template being a protein rich in serine, threonine, and glycine (19).

**Growth or Maturation.** (a) Growth modification. In the case of amorphous silica the role of the ionic environment in regulating morphology has been well documented (20). More recently, *in vivo* studies of silica deposition in higher plants (15,16) have indicated the considerable quantities of inorganic cellular contents present at the silicification site. The specific spatial location of ionic elements suggests that silicification is directly connected to underlying cellular processes which may control the aggregation of silica from solution through the neutralisation of surface charges on initially formed silica particles. It should be noted that such surface charge reduction effects are not capable of regulating morphological features on the micron level, suggesting that further changes in the local environment play a decisive role in morphological control.

(b) Mechanical stress. The packing of amorphous materials under biochemical constraints may have little to do with the associated organic environment, but may be controlled by the principles of flow ordering. On studying small particles (e.g. sand) in a flowing stream, it is observed that where flow is unidirectional in the centre of the stream, the particles appear close-packed. Towards the edges of such a stream, where flow is perturbed, particles are arranged in a much less ordered fashion, including swirls, due to vortices in the flowing liquid. In various plant hair systems (15,16) structural motifs have been observed which can be explained using this approach. The specimens studied all have in common a central axial channel containing cytoplasmic material around which silica is packed in an highly organised manner. Such a cytoplasm provides a pathway for liquid flow.

It should be noted that proposals such as flow ordering are only of relevance in extracellular silicification, i.e. in plants and sponge fibres. The mode of deposition must be quite different in a stationary vesicle, for example the radiolarian in which no substructural motifs are observed.

Silicification in the radiolarian, although complicated by the many different skeletal shapes observed, does appear to obey some simple rules. The basic picture is of vesicles growing, with the growth of the organism, along particular paths (21). Silica is then added to the vesicle as strengthening. The polymer network (tubulins) which ultimately controls vesicle growth can act both radially and circumferentially and thus control skeletal form in all directions. This suggests that the organic polymer deposited in the vesicle is the equivalent of the strontium sulphate crystals in the acantharian. The silica acts simply as a filler and not a form maker (Williams, R.J.P., personal communication).

It is abundantly clear that macromolecular assemblages, whether membranes or cell wall matrices, are extremely important in the regulation of the nucleation and growth of solid state inorganic structures in biological systems. In many instances, the precise nature of the interaction between the inorganic and organic phases is not known and our studies on biological samples are directed towards this end. In this article we present data obtained from two different biomineralising systems. For the plant hair system, the aim of our study of matrix : silica interactions by solid state nmr spectroscopy and cell wall enzymic degradation studies has been to obtain information on the extent and nature of chemical and/or physical interactions between the two phases. In contrast, studies of strontium sulphate precipitation in radiolaria have concentrated on defining the morphology and crystallography of the crystalline phase in relation to surrounding membrane structures and the growth process.

A direct result of our biological studies has been the development of model chemical systems for investigation of crystal or aggregate formation under carefully controlled experimental conditions. Examples presented in this article are taken from our studies of membrane bounded (vesicular) precipitation and studies of the effect of ordered substrate (Langmuir Blodgett films of behenic acid) on the nucleation and growth of crystalline strontium sulphate. For these studies in particular, full experimental details and the results obtained are to be published separately. We anticipate that a combination of biological and model chemical studies will lead to an improved understanding of the principles used by living systems in the production of composite designer phases.

## Materials and Methods

**Sample Preparation**. (a) Biological samples (silica and strontium sulphates). Full details of sample preparation for the biological samples may be found in papers by Perry et al., (16,18) and Hughes et al., (22).

(b) Synthetic samples. (i) Vesicular precipitation: Silica containing vesicles (30-60nm) were prepared using phosphatidylcholine (Lipid Products) with lasalocid A (Sigma Chemicals) in the membrane phase to facilitate rapid proton transfer and an internal aqueous phase of silicate ions or silicon catecholate ions (0.1-0.2M), with or without additional phosphate (pH ~11.5). The external pH of the vesicle containing solution was lowered to ~6 to allow precipitation of the silica. Precipitates of barium phosphate at nominal 100X the solubility for $BaHPO_4$. Synthetic experiments were performed at similar supersaturation. Internal pH changes were monitored using $^{31}P$ nmr and the precipitate by electron microscopy and x-ray microanalysis. Full experimental details will be published at a later date.

(ii) Strontium sulphate precipitation in the presence of Langmuir Blodgett films of behenic acid. Hydrophobic (8 layers) and hydrophilic (9 layers) coated 3 mm diameter tabbed electron microscopy grids were prepared by dipping the grids into a previously prepared compressed monolayer of behenic acid at a rate of $1 \, mm \, min^{-1}$. Crystal growth experiments were performed by placing the microscope grids in 10 ml glass vials and then adding $Sr(NO_3)_2$ and $Na_2SO_4$ solution (1:1) dropwise at various levels of supersaturation. Resultant solution pH was 6.0-6.3. Control experiments were run under identical conditions using similar carbon coated electron microscope grids without the film coating. Relative supersaturations between $10K_{sp}$ and $100K_{sp}$ were used. Grids were removed from the solution at known time intervals and the deposited crystals studied *in situ* using electron microscopy. Full details of the experimental procedures and results obtained may be found in Hughes et al. (22).

**Electron microscopy.** A JEOL 100 CX and JEOL 2000FX both with attached x-ray analytical systems were used for structural studies. High resolution transmission electron microscopy (HRTEM), transmission electron microscopy (TEM) and scanning electron microscopy (SEM) studies yielded structural information at the molecular, microscopic and macroscopic structural levels. Energy dispersive X-ray analysis (EDXA) was used where appropriate to provide information on inorganic elemental composition for all elements with atomic number $> 11$ but no information on the chemical identity, coordination chemistry or oxidation state of a particular element could be directly obtained.

**Solid State $^{29}$Si nmr.** A Varian VXR 300 solid state nmr spectrometer operating at 59.584 MHz for $^{29}$Si was used to investigate the molecular nature of the silica phase and to investigate whether the presence of organic material directly affects the relaxation properties of the silicon nuclei through physical or chemical interactions between the two phases. Magic angle spinning (MAS) spectra were obtained using a 600s recycle time and cross polarisation MAS (CPMAS) spectra were obtained at a range of Hartmann Hahn contact times, 1-10 ms.

**Biochemical Studies.** Enzymic hydrolyses and associated acid hydrolyses of HF treated (polysaccharide only) and intact (silica and polysaccharide) hairs from *Phalaris canariensis* were used to assess the extent of polymer matrix/mineral interactions. Plant hairs (10 mg) were suspended in 1.5 ml 60% w/v HF and sonicated for 30 minutes. The HF was removed by rotary evaporation in vacuo using an NaOH trap. Samples were resuspended in $H_2O$, stirred for 30 minutes, and redried by rotary evaporation 3X to remove HF. Samples were then freeze-dried to yield a white powder. The preferential release of certain monosaccharides in the absence of silica was taken as an indication of polymer/silica spatial interactions. Samples were hydrolysed in 2M TFA, 0.1M TFA and 58% v/v $H_2SO_4$, and also hydrolysed enzymatically using a mixture of purified endo- and exo-xylanases, and Driselase (a commercially available mixture of fungal hydrolases (Sigma), at a concentration of 1 mg ml$^{-1}$ in 0.1M acetate buffer at pH 7.0 at 23°C. Samples were removed for analysis at timed intervals up to 120 hours. The hydrolysates were analysed for monosaccharides by ion chromatography with pulsed amperometric detection using a DIONEX 4500i ion chromatograph and carbopac PAI column. A NaOH gradient operating at 0.9-200mM and flow rate of 1 ml min$^{-1}$ was used for separation of the monosaccharides.

**Matrix-mineral Interactions in Silicified Plant Hairs.** At the molecular level all biogenic silicas currently studied by electron microscopy appear amorphous and no local order above 10Å (3 Si-O repeat units) is visible. Although no long range crystallographic order is exhibited, morphological order exists at the microscopic level. In plant hairs such as *Phalaris canariensis* and the common stinging nettle *Urtica dioica* a range of structural motifs are found within a single cell. It is important to note that the formation of different structural motifs can be correlated with changes in cell wall biosynthesis (18). It has been shown for *Phalaris canariensis* that at the early stages of development arabinoxylans and cellulose are principally synthesised at the same time as sheetlike silica is deposited. At later stages of development the relative rates of synthesis of cellulose and arabinoxylans decreases in importance and the rate of synthesis of non-cellulosic mixed linkage $\beta$(1-3), $\beta$(1-4) glucans increases in importance. This phase is associated with globular silica deposition. Fibrillar silica is only deposited during the latter stages of cell development and is incorporated into/associated with a mature organic matrix where synthesis of cell wall material has ceased. The importance of matrix structure on silica structural components is not understood although it is feasible that the polymers could provide different confining spatial volumes and hydrogen bonding centres for interaction with the developing silica phase.

Direct bonds between the organic and inorganic phases of various silicified systems including plant components, sponges, diatoms and limpet teeth have not been detected. However, a degree of interaction between the two phases certainly exists and is more likely to be of a Van der Waals or hydrogen bonded type interaction. In principle, nmr can be used to investigate the extent of interaction between organic and inorganic phase as it is conceivable that the presence of a proton sink (the organic phase) may play a significant role in altering the spin lattice relaxation properties of the mineral phase. Si-O-C bonds have not been identified in siliceous minerals studied using this technique. Magic angle spinning $^{29}$Si nmr spectra of plant hair cell walls and plant hair silica yield virtually identical spectra at a recycle time of 600s (Figure 1). The properties of $Q_4$, $Q_3$, $Q_2$ species as measured by peak deconvolution techniques are $Q_4$:68%, $Q_3$:22%, $Q_2$:10%. Corresponding data obtained from crosspolarisation experiments performed at a range of Hartmann-Hahn contact times (1-10 ms) show that silicon species associated with nearby protons, e.g. $Q_3$ exhibit large increase in intensity at small contact times. The crosspolarization data for the sample containing silica only and for the intact hair containing silica and organic material are markedly different. The difference in behaviour can be attributed to the organic matrix being sufficiently close in space to act as a pathway for relaxation of the silicon nuclei. In the silica only sample, water is the only other proton source for relaxation of magnetisation. Plots of peak intensity versus contact time show very clearly the difference in behaviour for the $Q_3$ signal, Figure 2. Dipolar interactions clearly exist between the organic and inorganic phases although at the present time

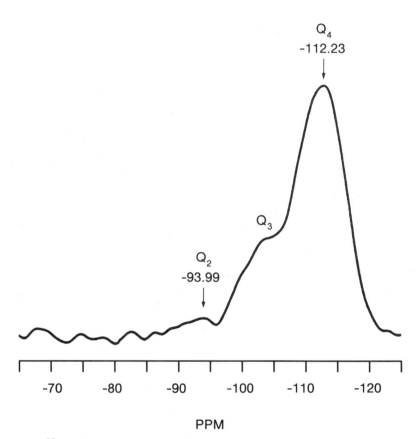

Figure 1. $^{29}$Si mas nmr of silica from *Phalaris canariensis* hairs at a recycle time of 600s. $Q_4$: $Si(OSi=)_4$, $Q_3$: $Si(OSi=)_3OH$, $Q_2$: $Si(OSi=)_2(OH)_2$ are labelled.

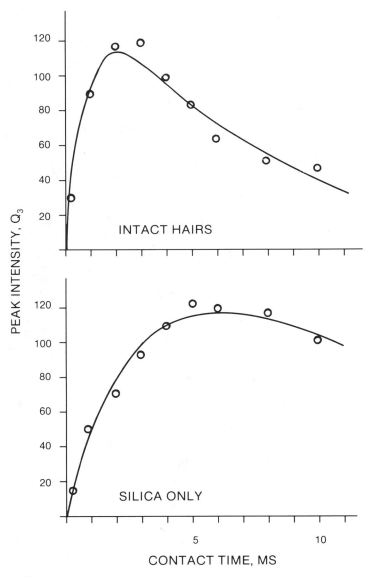

Figure 2. $^{29}$Si cross polarisation data for silica only and intact hair samples. Peak intensity vs contact time is plotted for the $Q_3$ signal showing clearly the difference in relaxation behaviour in the presence of the organic matrix.

further quantitative information is not available. An interesting comparison can be made with gel-like silicas as found in diatoms where the effect of the organic matrix (a significantly smaller weight % than in the plant samples) on relaxation behaviour is even more marked. The implication is that a range of physical interactions (H bonding and Van de Waals type) between organic and inorganic phases exists for biosilicas. This phenomenon is currently being investigated for a wide range of siliceous biominerals in order to obtain a clear idea of the rules governing the extent of matrix: mineral interactions.

Further evidence to substantiate the view that direct interactions exist between organic polymers and silica in plant hairs comes from our studies of selective enzymic degradation studies on silicified and desilicified samples. For certain polymers, degradation is affected by the presence of silica within the sample. Silicified and desilicified hairs from *Phalaris canariensis* (initially contains 40% silica) and *Urtica dioica* (contains a silicified tip only) were treated with a range of enzymes including xylanases, cellulases and driselase which contains a mixture of fungal hydrolases. For both plant systems the action of the cellulase was barely affected by the presence of silica. For the other enzyme systems investigated, the presence of silica within the cell wall matrix was found to confer a marked reduction in the susceptibility of other cell wall polymers to attack by wall degrading enzymes typically produced by plant pathogens. Treatment with xylanase I and II led to a preferential release of arabinose and xylose from desilicified cell walls (Figure 3), and treatment with driselase led to a preferential release of glucose with a smaller increase in arabinose liberation. Comparative studies with largely non silicified plant hair cells showed no difference in saccharide release even after HF treatment (used to remove $SiO_2$ from the silicified samples). The data appear to indicate that silica acts to protect cell wall polymer from attack by enzymes released by pathogenic organisms. The preferential release of certain monosaccharides after silica removal suggests that certain polymers, possibly side chain components of arabinoxylans or glucans are more intimately associated with the siliceous phase than other polymers. Current experimentation on this system involving double labelling of the polysaccharide phase ([14]C) and silica phase ([68]Ge) is aiming to define more precisely which biopolymers are synthesised in association with particular structural motifs - information which is of vital importance in the fuller definition of matrix mineral interactions in such a grossly heterogeneous and structurally complicated system.

**$SrSO_4$ deposition in radiolaria.** *Sphaerozoum punctatum* is a silica-skeletoned, colonial radiolarium found near the ocean surface. During reproduction the species releases celestite ($SrSO_4$) containing flagellated swarmers. Figure 4 shows an SEM of an intact radiolarian just prior to swarmer release. The crystals have been studied by electron microscopy and electron diffraction and have been shown to be single crystals (22). The crystal morphology was found to be based upon an elongated square prism of {011} and possible {023} faces capped with {210} triangular end faces. Crystal surfaces were slightly curved suggesting

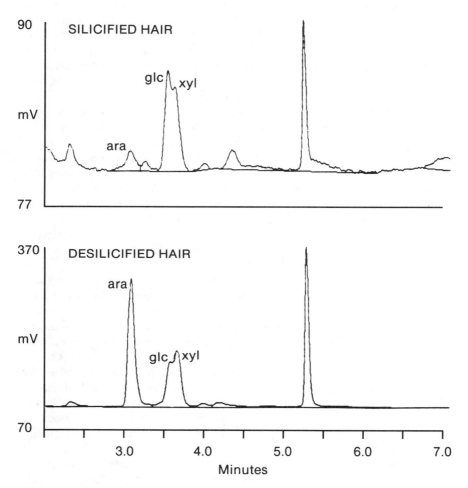

Figure 3. H.P.L.C. traces of cell wall extracts released on enzymic degradation (xylanase I and II) of plant hair samples and desilicified hair samples. The principal monosaccharides are labelled.

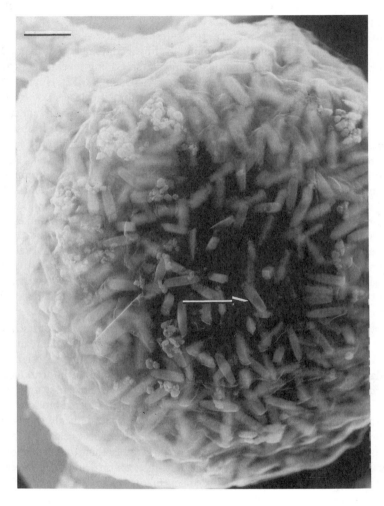

Figure 4. S.E.M. of intact radiolarian just prior to swarmer release. Many swarmer crystals are present, an individual crystal is arrowed. Scale bar represents 10 μm.

membranous involvement/restrictions during growth, Figure 5. A study of the crystals at different stages of development showed that the length of the crystal increases at a far greater rate than the width suggesting that growth along the principal crystal axis is less constrained.

Information on any membrane structures thought to be involved in the development of such crystals has been obtained by a combination of freeze drying and cleavage of intact samples and by staining and sectioning developing swarmers. These results are to be published elsewhere (Anderson, O. R.; Hughes, N. P.; Perry, C. C.; unpublished data) but in essence, the ultrastructural study has shown that each crystal is surrounded by an organic membrane which undergoes a series of changes throughout the crystal deposition cycle. Based on observation of crystals of varying sizes it appears that growth proceeds primarily from one end, resulting in a final crystal which is slightly larger at one end (the base) than the other, Figure 5. From the techniques utilised, at no stage during growth was the observable membrane surface in contact with the crystal surface although this could arise due to crystal dissolution or dehydration during fixation of the sections. It is likely that the membrane structures regulate ion fluxes to the developing crystal.

The SrSO$_4$ of the radiolarian swarmer not only exhibits an unusual growth pattern but also a unique and precise crystallography. Comparison with theoretical, inorganic derived models (24) shows that the morphology of the swarmer crystal is not controlled simply by the energetics of the crystal faces but deposition is controlled by external influences.

The non-equilibrium growth, unique morphology and crystallography of the swarmer crystal may be rationalized in terms of a developmental process which involves the spatial and chemical control of biological celestite deposition. Such regulatory factors have been described for celestite deposition in acantharia (25) and magnetite deposition in magnetotactic bacteria (26). Control may arise through a series of physico-chemical interactions leading to supersaturation control, nucleation and crystal growth controls. For this system it is likely that although the precise mechanisms of morphological regulation remains unclear, the principal effects probably occur during the crystal growth rather than the nucleation stage although it is appreciated that both controlling mechanisms and ion concentration control may act in concert to produce such unique biocrystals. In order to gain a fuller understanding of the possible morphological regulators which may act during crystal development, model studies on crystal growth in synthetic inorganic systems, using organic interfaces and growth modifiers have been carried out.

## Model Inorganic Studies

(a) Vesicular silica deposition: Precipitation of biogenic silicas occurs in confined volumes either membrane or matrix bounded. Studies of silica deposition within phospholipid bilayer vesicles are probing the effects of membrane surfaces, ion composition, pH and small organic molecules and polymers on the nature of the product formed. A generalized reaction scheme is given in Figure 6. $^{31}$P solution nmr has been used to monitor pH changes across the lipid bilayer with protons being transported into the vesicles using lasalocid A as the ionophore. Shortly after pH reduction to 7 (~15 minutes) for the internal phase, small (6-7nm) deposits of silica can be detected at the outer surface of the approximately 100nm diameter vesicle. As time progresses, more particles of similar dimensions are produced until approximately 2 hours after the pH change when a dense vesicular precipitate is observed. Individual silica particles are no longer visible. Representative electron micrographs taken at specific time intervals after the pH change are shown in Figure 7.

(b) Vesicular GpIIa phosphate deposition: Precipitation of strontium and barium sulphates in biology is thought to always take place within membrane-bounded regions as high levels of these ions located intracellularly are toxic. In order to study the precipitation of an insoluble Group IIa salt in a model vesicle system the phosphate salt has been chosen as solubility products are comparable and the extent of precipitation can be monitored by loss of solution phosphate signal. Initially the internal vesicular compartment contains phosphate and sodium ions with barium ions being subsequently transferred across the lipid membrane by complexation with a membrane bound ionophore such as 18-crown-6. Precipitation is monitored by the loss of internal solution phosphate signal. Figure 8 shows a representative series of spectra obtained under conditions of ~100$K_{sp}$ barium phosphate internal. In Figure 8a, the internal phosphate signal is arrowed. After barium ions are added to the system and transfered across the membrane, the internal phosphate signal decreased in size and broadened but did not disappear. It is likely that nucleation of barium phosphate on internal membrane surfaces blocks further barium ion influx into the internal vesicular space resulting in residual solution phosphate species. Figure 9 shows a vesicular precipitate of barium phosphate obtained 30 minutes after addition of barium ions to the external medium. Some discrete vesicles (arrowed) are observed but others aggregate to form larger entities.

(c) SrSO$_4$ Precipitation in the presence of behenic acid Langmuir-Blodgett films. In order to understand the morphological and crystallographic variety observed in biological minerals it is necessary to perform *in vitro* experiments in which supersaturation, nucleation and crystal growth are closely controlled. This has been attempted previously and recent examples are cited in the introduction to this paper.

The most obvious nucleation promoters are ordered substrates and in our experiments crystals of strontium sulphate have been grown in the presence of behenic acid Langmuir Blodgett (LB) films (27). This type of film offers several advantages over Langmuir films deposited at an air/water interface.

A clear advantage of the LB technique is that it may be used to produce stable hydrophobic (methyl tailgroups on the surface) as well as hydrophilic (carboxylic acid headgroups on the surface) interface. This not only allows the influence of a charged surface on crystal growth to be assessed but also the

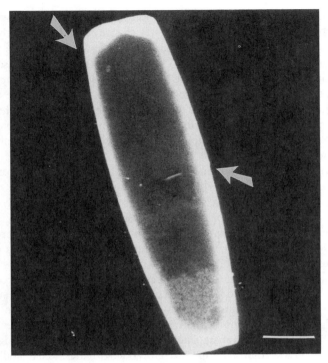

Figure 5. S.E.M. of extracted SrSO₄ crystal showing curved crystal surfaces (arrowed) and one of the basal ends larger than the other. Scale bar represents 1 $\mu$m.

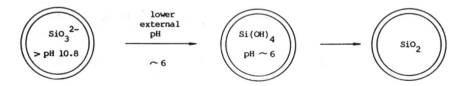

Figure 6. General scheme for vesicular precipitation experiments. Silicate (phosphate) is enclosed in a lipid bilayer at high pH. Internal pH reduction occurs when the external pH is lowered and proton transport takes place across the membrane via an ionophore. Polymerisation then takes place. For Gp IIa phosphate precipitation, $M^{2+}$ is added to the external medium (no pH change necessary) and transported into the vesicle by an ionophore. Precipitation then takes place.

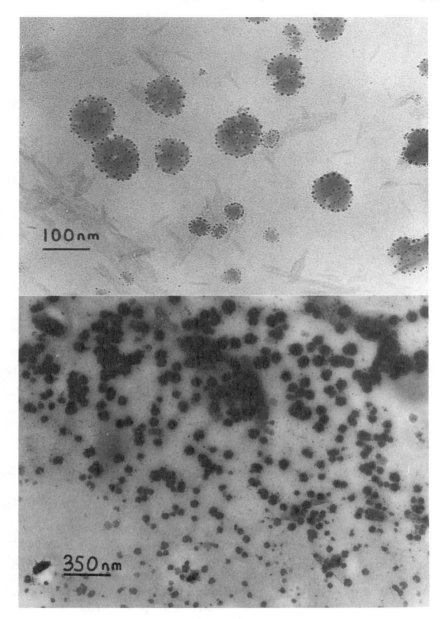

Figure 7. T.E.M. of vesicular silica deposits (top) 30 minutes after external pH change: (bottom) 2 hour after the external pH change. Small (6- 7 nm) particles form initially and eventually coalesce to produce particles 70 nm in diameter.

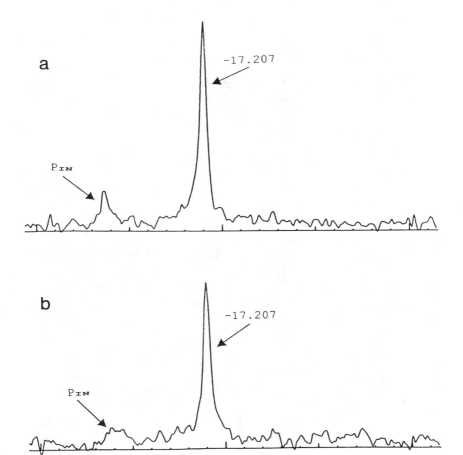

Figure 8. $^{31}P$ nmr spectra for $Ba^{2+}$ phosphate precipitation using 18-crown-6 as ionophore. Peaks arising from the lipid headgroup at -17.207 ppm and internal phosphate are labeled. a) initial spectrum, internal phosphate only. b) 30 minutes after addition of barium ions to the external phase. The signal arising from the internal solution phosphate has broadened and significantly decreased in intensity.

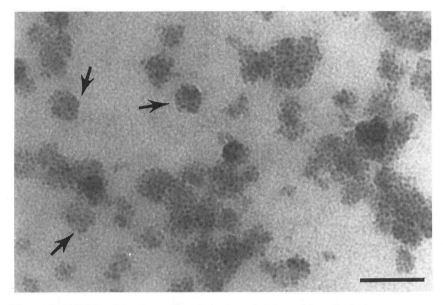

Figure 9. T.E.M. of barium phosphate containing vesicles. Scale bar represents 100 nm.

influence of an ordered 'non-charged' surface. Ongoing work on the structure of LB film surface (27,28) is important in definition of the LB film surface for subsequent crystallisation studies.

The deposition of LB films has been developed so that carbon-coated electron microscope grids may be used as a substrate (29). This means that crystals grown on such films may be studied *in situ* allowing much earlier stages of development to be studied than using a Langmuir monolayer, where specimen transfer is required.

In biological mineralisation it is generally assumed that crystal deposition occurs from solutions at low supersaturation since this allows close control over the crystal deposition process. Experiments were performed under a range of supersaturation conditions (10-100 $K_{sp}$) with and without growth modifiers such as citrate at concentration of $\sim 10^{-6}$M. The highest level of control was observed at the lowest supersaturation levels investigated.

Examples of crystal morphologies obtained for experiments without films and in the presence of hydrophobic and hydrophilic films are given in Figure 10. Rhombic crystals were observed for the blank (no film) samples which lay perpendicular to the [001] zone axis (Figure 10a,b). The addition of trace citrate levels produced poorly formed octagonal crystals often showing evidence of secondary nucleation sites (Figure 10c). Precipitation of $SrSO_4$ at 10 $K_{sp}$ in the presence of hydrophobic LB films produced disc-like crystals which lay perpendicular to the [010] zone axis (Figure 10d). Crystals deposited on the hydrophilic behenic acid film at 10 $K_{sp}$ showed a floret morphology with outgrowth of needles away from the film surface (Figure 10e). The tips of the crystallites extending from the film surface were highly beam sensitive. Electron diffraction of either the floret core or needle-like crystallites did not yield easily recognisable celestite diffraction patterns. It is presently thought that the underlying substrate causes distortion of the crystal lattice but further experimentation to identify the crystal faces expressed is underway. Experiments of this type show that small perturbations in an inorganic system may directly affect crystal morphology and crystallography. Clearly these studies may have considerable biological importance since in any compartimentalised cellular system it is possible to regulate the local supersaturation, the levels of extraneous ions and the structure of nearby organic interfaces. Small variations only are required to induce fundamental changes in crystal morphology. A full account of experimental procedures, result obtained and interpretation of the data can be found in a separate publication (23).

## Conclusions

For the two biological minerals discussed in this article it is likely that deposition may be directly affected at more than one developmental stage. In each case one particular deposition control mechanism may predominate, but undoubtedly more than are control mechanism will be acting. The deposition of minerals can therefore be considered as competitive, with ion concentrations, nucleation and growth all acting to define the final morphology and crystallography. It

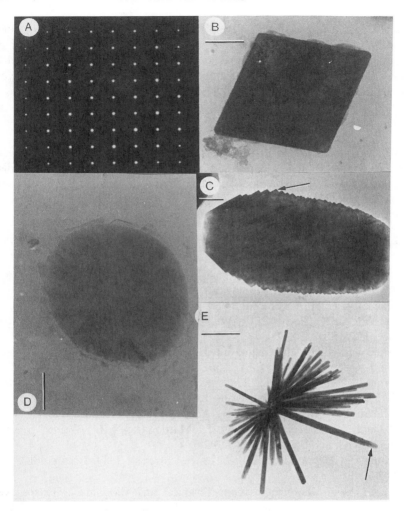

Figure 10. T.E.M. of synthetic SrSO₄: a) electron diffraction pattern for (b). b) well formed rhombic crystal lying perpendicular to the {001} zone axis. Scale bar represents 0.3 $\mu$m. c) poorly formed octagonal crystal taken from a solution at 10 $K_{sp}$ with trace levels of citric acid after 2 weeks. Note sites of secondary nucleation (arrowed). Scale bar represents 0.2$\mu$m. d) disc-like crystal deposited on a hydrophobic behenic acid film from a solution at 10 $K_{sp}$ removed after 3 weeks. The roughened end of a crystallite is arrowed. Bar represents 0.2 $\mu$m. The crystal is lying perpendicular to the {010} zone axis. e) floret-like crystal deposited on a hydrophilic behenic acid film from a solution at $K_{sp}$ removed after 3 weeks. Bar represents 2 $\mu$m. The corresponding electron diffraction pattern cannot readily be defined.

is of paramount importance that simple model systems are developed to look at these factors both in isolation and in conjunction with one another if we are going to understand how inorganic and organic phases are coproduced as nature's designer composites. Clearly much remains to be accomplished before a simple mechanism for biological mineralisation processes can be proposed!

## Acknowledgements

The authors would like to thank the SERC and AFRC Funding Councils for financial support, Professor O.R. Anderson for interest in the radiolarian studies, D. Heard for assistance in preparation of L.B. films and Professor R.J.P. Williams, F.R.S. for enjoyable stimulating discussion of the subject matter presented in this article. N.P.H. is in receipt of an SERC studentship.

## Literature Cited

1. Williams, R. J. P. Phil. Trans. B Soc. Lond. 1984, B304, 411-24.
2. Robertson, W. G. In Biological Mineralisation and Demineralisation, Nancollas, G. H., Ed. Springer-Verlag: New York, 1982, pp 5-21.
3. Wilcock, J. R.; Perry, C. C.; Williams, R. J. P.; Brook, A. J. Proc. Roy. Soc. 1989, B238, 203-21.
4. Nancollas, G. H. Adv. Colloid Interface Sci. 1979, 10, 215-52.
5. Ford, G. C.; Harrison, P.M.; Rice, D. W.; Smith, J. M. A.; Treffry, A.; White, J. L.; Yariv, J. Phil. Trans. Roy. Soc. 1984, B304, 551-65.
6. Weiner, S.; Traub, W. Phil. Trans. Roy. Soc. 1984, B304, 425-434.
7. Addadi, L.; Weiner, S. Proc. Natl. Acad. Sci. USA 1985, 82, 4110-4.
8. Addadi, L.; Moradian, J.; Shay, E.; Maroundas, N. G.; Weiner, S. Proc. Natl. Acad. Sci. USA 1987, 84, 2732-6.
9. Mann, S.; Heywood, B. R.; Rajam, S.; Birchall, J. D. Nature 1988, 334, 692-5.
10. Mann, S. In Biomineralisation: Chemical and Biochemical Perspectives; Mann, S.; Webb, J.; Williams, R. J. P., Eds. VCH: Weinheim, 1989, p 35-62.
11. Mann, S.; Heywood, B. R.; Rajam, S.; Birchall, J. D. Proc. Roy. Soc. 1989, A423, 457-71.
12. Otani, S. Bull. Jap. Chem. Soc. 1960, 30, 1543-54.
13. Perry, C. C.; Wilcock, J. R.; Williams, R. J. P. Experimentia 1988, 44, 638-50.
14. Mann, S.; Moensch, T. T.; Williams, R. J. P. Proc. Roy. Soc. 1984, B221, 385-93.
15. Perry, C. C. In Biomineralisation: Chemical and Biochemical Perspectives; Mann, S.; Webb, J.; Williams, R. J. P., Eds., VCH: Weinheim, 1989, p 223-56.

16. Perry, C. C.; Mann, S.; Williams, R. J. P. Proc. Roy. Soc., 1984, B222, 427-38.

17. Mann, S.; Perry, C. C. In Silicon Biochemistry, Evered, D.; O'Connor, M., Eds. Ciba Foundation Symposium 121, 1986, J. Wiley: New York, p 40-58.

18. Perry, C. C.; Williams, R. J. P.; Fry, S. C. J. Plant. Physiol. 1987, 126, 436-448.

19. Hecky, R. E.; Mopper, K.; Kilham, P.; Degens, E. T. Mar. Biol. 1973, 19, 323-31.

20. Iler,R. K. The Chemistry of Silica. Wiley-Interscience: New York, 1979; p 97.

21. Anderson, O. R. In Biomineralisation in Lower Plants and Animals. The Systematics Association Special 30, Leadbeater, B. S. C.; Riding, R., Eds., 1986; Clarendon Press: Oxford, p 375-91.

22. Hughes, N. P.; Perry, C. C.; Anderson. O. R.; Williams, R. J. P. Proc. Roy. Soc. Lond. 1989, B238, 223-33.

23. Hughes, N. P.; Perry, C. C.; Williams, R. J. P. J. Appl. Phys. 1990, in press.

24. Hartman, P.; Perdock, W. G. Acta Cryst. 1955, 8, 525-9.

25. Wilcock, J. R.; Perry, C. C.; Williams, R. J. P. Proc. Roy. Soc. Lond. 1988, B233, 393-405.

26. Mann, S.; Sparks, N. H. C.; Blakemore, P. P. Proc. Roy. Soc. Lond. 1987, B231, 477-87.

27. Peterson, I. R.; Russell, G. J. Thin Solid Films 1985, 134, 143-52.

28. Robinson, I.; Sambles, J. R.; Peterson, I. R. Thin Solid Films 1989, 172, 149-58.

29. Heard, D.; Roberts, G. G.; Holcroft, B.; Goringe, M. J. Thin Solid Films 1988, 160, 491-9.

RECEIVED August 7, 1990

# Chapter 24

# Some Structural and Functional Properties of a Possible Organic Matrix from the Frustules of the Freshwater Diatom *Cyclotella meneghiniana*

**D. M. Swift and A. P. Wheeler**

**Department of Biological Sciences, Clemson University, Clemson, SC 29634–1903**

Organic materials extracted from frustules of the diatom *Cyclotella meneghiniana* exhibit some characteristics common to calcium-biomineral organic matrices. Isolated siliceous frustules cleaned with 5% NaOCl, pH 8, were dissolved in an $NH_4/HF$ solution. The remaining organic materials contained large portions of protein, some carbohydrate, and lesser quantities of phosphate and sulfate. The protein(s) in the aqueous soluble fraction contained 20% serine, 24% glycine and 10% aspartic acid. The extracted materials at concentrations up to 1 $\mu$M had no effect on *in vitro* $SiO_2$ polymerization in solution, but did exhibit strong inhibitory capacities against *in vitro* $CaCO_3$ crystallization at approximately $10^{-2}$ $\mu$M. This level of activity was roughly the same as for proteins obtained from carbonate biomineral indicating that the siliceous proteins, like the carbonate proteins, may contain highly anionic domains of polyaspartic acid or polyphosphoserine.

Biosilicification has been studied far more intensely in diatoms than in any other group of organisms that utilize biosilica. Morphological studies of the elaborate diatom shells or frustules date from the 1800's. More recently, investigations have focused on the silicification process as reviewed by Sullivan (1). From these studies it is evident that the frustule is produced within a silica deposition vesicle (SDV) which is delimited by a membrane, the silicalemma. The SDV is formed adjacent to the newly formed plasmalemma after cytokinesis. There is evidence that in some species the SDV is formed by coalescence of small vesicles of unknown origin. The Golgi has been found in close proximity to the SDV and is considered a likely source of the vesicles. Silica appears to be deposited centripetally within the SDV shortly after its appearance (2).

Temporal patterns of uptake and incorporation of silica by diatoms have been characterized (1) and silica ionophores have been identified and localized (3).

0097–6156/91/0444–0340$06.00/0

However, the SDV itself, its properties, and the processes that take place within it, remain uncharacterized. Heckey et al. (4) characterized the organic casing of the mature frustule which consists of old plasmalemma, part of the silicalemma, and various carbohydrates secreted after frustule formation is completed. The amino acid composition of the casing was found to differ from that of the cell contents and this difference was suggested to reflect the occurrence of a unique protein or proteins directly involved in silica deposition. In particular, the casing proteins contained a relatively high serine content and a hypothesis was presented in which the -OH groups of the serine distributed along the backbone of the protein would interact with the silanol groups of silica forming a template for silica polymerization. To date no additional analyses of the proteins or experimental evidence to support this hypothesis have been presented.

Ideally, to isolate the organic materials involved in the process of frustule formation, nacent valves still surrounded only by the silicalemma should be obtained. From these the silica would then be removed and the remaining organic component, free of post-silicification components, could be analyzed. However, the isolation of frustules exclusively at this stage would be technically difficult to achieve. One alternative is to isolate frustules from cultures of growing diatoms when the division rate of the cultures is at its greatest, that is, during exponential growth. During the exponential growth phase, division, and therefore frustule formation, occurs at its most rapid rate. Consequently, organic molecules involved in frustule formation should be synthesized at a maximal rate during this phase of growth.

To this end, cultures of *Cyclotella meneghiniana,* a centric freshwater diatom, were grown until mid- to late exponential growth phase and harvested. Frustules were isolated free of cellular debris and the silica dissolved leaving only organic materials which were then characterized. In addition, the organic materials thus isolated were further tested for any functional activity using *in vitro* mineralization assays.

## Methods

Cultures of *Cyclotella meneghiniana* (clone CyOH) were maintained at 15 °C with constant stirring and were aerated through 0.45 $\mu$m teflon filters. Light was provided by two banks of four fluorescent 40 watt cool white bulbs on one side of the carboys and four 300 watt quartz halogen bulbs on the opposite side. Batch cultures consisting of four carboys at ten liters each were inoculated to a concentration which allowed only 1-2 days before the onset of exponential growth. Cell concentrations were determined by a Coulter Counter (Model TA II). Cells were harvested at mid to late exponential growth ($1.75$-$2.5 \times 10^5$ cells/ ml) by flow-through centrifugation.

Concentrated cells were sonicated and the frustules were then collected separately from cell contents by differential centrifugation five times in distilled deionized water ($dH_2O$) at 1800 x g for 2 min (4). Further purification of the frustule fraction was achieved by washing it 2-3 times for 15 minutes in 100% dimethylformamide (DMF). This step removed pigments still detectable in the frustule fraction after the water washes. Frustules were collected and washed once more with $dH_2O$. This final wash was followed by dialysis to remove any remaining DMF. Frustules were then frozen and lyophilized and the weight of the silica plus organic material determined. In addition, after the lyophilization, some material was treated for 3-5 minutes with 5% NaOCl, 10 mM Tris, pH 8.0 followed by centrifugation and washing with $dH_2O$. Frustules were then dialyzed, lyophilized and weighed. The removal of the organic casing from the mature frustules was determined by exposing dried samples to ruthenium red stain and examining them using light microscopy. The casing contains mucopolysaccharides known to stain specifically with ruthenium red (2,5). Lack of staining was taken to indicate that the organic casing had been removed. When the non-NaOCl cleaned (untreated) frustules were exposed to ruthenium red they stained intensely, suggesting there was still some contamination from the organic casing or cell contents. However, the NaOCl cleaned (treated) frustules exhibited no staining.

To dissolve the silica from the frustules, both the untreated and treated samples were then suspended in a solution of 1% HF and 3% $NH_4F$, pH 5 for 48 hours at 4 °C with constant shaking. The remaining organic material was then dialyzed exhaustively, lyophilized and weighed. The dried material was suspended in $dH_2O$ for 48 hours with constant stirring followed by centrifugation at 24,000 x g for 20 minutes. The supernatant and pellet were collected separately, lyophilized and weighed. Fractions from treated frustules will be designated as supernatant (T) and pellet (T) whereas untreated fractions will be followed by (NT). A variety of compounds were exposed to the same extraction procedure to determine the effect of the dissolution medium on molecular size and structure. The test molecules included cytochrome C, bovine serum albumin, blue dextran, pepsin, and myoglobin. The dissolution method was deemed not especially deleterious to structure as both cytochrome C and myoglobin retained their heme groups and all retained their visible and ultraviolet absorbance profiles. All appeared to retain their approximate molecular sizes as well.

Chemical analyses were performed on the supernatant fraction which was always resuspended in $dH_2O$ at 1 mg/ml by dry weight. The pellet material, when analyzed, was suspended and sonicated to achieve a uniform suspension. Protein content was determined by the Miller (6) modification of the Lowry assay and carbohydrate by the Dubois et al. (7) method. Phosphate and sulfate analyses were performed on dried aliquots of both soluble and insoluble materials. Phosphate was determined by the Wheeler and Harrison (8) modification of the Marsh (9) method and sulfate by a modified method of Wainer and Koch (10). For this latter analysis, samples were oxidized in 30% $H_2O_2$ at 70 °C until all liquid was evaporated. Oxidized samples and standards resuspended in 0.4 ml of $dH_2O$ were vortexed with 0.1 ml of 0.1 M sodium acetate-acetic acid buffer. An ethanol (95%) suspension containing 3 mg of barium chloranilate in 0.5 ml was added and the samples vortexed about once per minute for 10 minutes. After centrifugation, 0.5 ml of the supernatant was mixed with 0.2 ml of a 1:1 solution of glacial acetic acid:water and the absorbance of the resulting solution was determined at 530 nm .

Extracts were also tested for an effect on *in vitro* silica polymerization rates using a modification of the method of Weres et al. (11). In this assay, the rate of polymerization of silicic acid is determined by following the rate of disappearance of free silica monomers from a supersaturated solution. Free silica concentrations were determined using the beta silicomolybdate reaction (12). In this assay, silica monomers, short chain linear oligomers and perhaps even some small cyclic species will react with molybdic acid in less than 5 minutes and are termed molybdate active silica (MAS) (13). The pH of the polymerizing solution was maintained at 7 with BES buffer (163 mM). The initial silica concentration was 750 ppm as $SiO_2$ (12.5 mM). The reaction was initiated by the addition of $SiO_2$ and the pH was adjusted rapidly ($<$1 min.) to 7.00 $\pm$0.05 pH units. Aliquots were periodically removed from the stirred reaction vessel to determine the concentration of MAS in solution. Test materials were introduced to the reaction mixture before the addition of $SiO_2$.

The effect of the extracts on $CaCO_3$ crystallization rates was determined using the pH-stat system as described in Wheeler and Sikes (14). Molecular weights of the soluble materials were determined by gel filtration chromatography using HPLC and a TSK-30 column with a mobile phase of 0.05 M tris-acetate, pH 7.4. Amino acid analyses were performed using the method of Waite and Benedict (15) on a Beckman 6300 Autoanalyser. All dialyses were done using SpectraPor 3 membranes (3500 MWCO) at 4°C against $dH_2O$ with constant stirring.

## Results

The results of the chemical analyses are shown in Table I. The relative amount of protein detectable following NaOCl treatment decreases in both the supernatant and pellet fractions. The supernatant (NT) has a higher proportion of protein when compared to carbohydrate than does the supernatant (T). The pellet has more detectable protein than carbohydrate from both the untreated and treated samples. Both supernatant extracts have roughly the same relative amount of phosphate, whereas both pellet fractions contain little detectable phosphate. Both pellet fractions contained small amounts of sulfate, with about twice as much appearing in the pellet (NT) as the pellet (T).

Amino acid analyses of soluble and insoluble fractions from both untreated and treated frustules are given in Table II. Also included is the amino acid composition of cell contents collected from the supernatant of the first centrifugation after cells were sonicated. The most prominent amino acids in the supernatant fractions were glycine and serine, which for the treated supernatant were present at approximately two times the concentrations found in the cell contents fractions. Additionally, there were larger relative quantities of hydrophobic amino acids found in the pellet fractions than the supernatant fractions.

Table I. Chemical Composition[1] of Fractions Extracted from Frustules

| Fraction[2] | Protein | Carbohydrate | PO$_4$ | SO$_4$ | n |
|---|---|---|---|---|---|
| Supernatant (NT) | 509 ± 178 | 227 ± 77 | 66 ± 22 | n.d.[3] | 6 |
| Pellet (NT) | 307 ± 171 | 63 ± 17 | 16 ± 12 | 25 ± 8 | 3 |
| Supernatant (T) | 274 ± 11 | 325 ± 102 | 73 ± 4 | n.d.[3] | 3 |
| Pellet (T) | 190 ± 30 | 60 ± 10 | 4 ± 2 | 14 ± 2 | 3 |

[1] ($\mu$g/mg dried wt.)
[2] Fractions designated T were obtained from 5% NaOCl cleaned frustules and fractions designated NT were from untreated frustules.
[3] n.d. not determined

Table II. Amino Acid Composition of Various Extract Fractions

| Amino Acid | T supernatant | T pellet | Cell Contents | NT supernatant | NT pellet |
|---|---|---|---|---|---|
| asp | 10.03 | 10.17 | 10.45 | 9.72 | 9.86 |
| thr | 5.30 | 3.93 | 5.30 | 5.25 | 5.03 |
| ser | 20.31 | 9.44 | 10.41 | 15.64 | 12.83 |
| glu | 9.16 | 11.33 | 10.32 | 10.14 | 10.50 |
| pro | 4.43 | 6.08 | 5.83 | 4.72 | 5.66 |
| gly | 24.46 | 19.03 | 13.87 | 17.86 | 18.46 |
| ala | 8.22 | 9.66 | 10.85 | 9.01 | 8.77 |
| cys | 0.02 | 0.02 | 0.27 | 0.29 | 0.25 |
| val | 4.35 | 6.79 | 5.68 | 4.20 | 5.39 |
| met | 0.00 | 0.00 | 0.00 | 0.00 | 0.00 |
| ile | 2.53 | 2.93 | 3.83 | 3.25 | 2.70 |
| leu | 4.31 | 9.52 | 6.98 | 5.31 | 7.90 |
| tyr | 0.79 | 0.93 | 2.54 | 2.38 | 1.83 |
| phe | 2.23 | 4.16 | 3.49 | 2.80 | 3.02 |
| his | 0.86 | 1.70 | 1.56 | 2.49 | 1.57 |
| lys | 1.81 | 1.78 | 4.55 | 3.34 | 3.23 |
| arg | 0.26 | 0.88 | 2.46 | 2.10 | 2.08 |

Analyses for degree of similarity of amino acid compositions were made using the method of Cornish-Bowden (16). In this method, a difference index between two compositions is determined according to the equation

$$S\Delta n = 1/2\Sigma(n_{iA} - n_{iB})^2$$

where $n_{iA}$ or $n_{iB}$ is the quantity of the $i^{th}$ type of amino acid contained in protein A or B. If $S\Delta n < 0.42N$ (N=100 residues when using mole percent) there is a 95% certainty that proteins A and B are related. When $0.42N < S\Delta n < 0.93N$, there is only a weak relationship, and when $S\Delta n > 0.93N$ the proteins are unrelated. $S\Delta n$ values comparing the supernatant fractions to the cell contents indicates that the supernatant (T) by these criteria are different from the cell contents ( $S\Delta n = 1.24$). Much of the difference lies in the higher serine and glycine content of the treated supernatant and somewhat lower percentage of hydrophobic amino acids. While the Cornish-Bowden index is most useful when comparing purified proteins, it is used here to illustrate the increasing purity of the proteins being separated in the supernatant of extracts from the NaOCl treated frustules. No other fractions test significantly different from the cell contents using this index.

The elution profiles of both the supernatant fractions as a result of gel permeation chromatography contain two peaks (Figure 1). The first peak represents material eluting in the void volume of the column thus having molecular weights greater than 200 kD. The second peak falls in the molecular weight range of 45-60 kD. The supernatant (T) contains a reduced quantity of material eluting at the void volume of the column compared to the supernatant (NT) and significantly more material in the second peak. In addition, this second peak is shifted to a slightly later elution time in the treated supernatant. The supernatant (NT) has multiple small peaks which represent sizes below 1000 daltons.

The rate of polymerization of $SiO_2$ at 750 ppm initial concentration is not affected by the presence of the soluble material from untreated frustules at the concentrations tested (Figure 2). Based on the effects of calcium biomineral matrices on mineral growth, it was anticipated that these substoichiometric concentrations of materials (relative to silica in the assays) might have an inhibitory effect on the rate of polymerization. In addition, no putative analogs of the organic materials (including serine, polyserine, glycine) exhibited any effect on the rate of silica polymerization measured in this assay. After an initial increase in polymerization rate, borate, a known silica inhibitor, did reduce the rate of polymerization when present in superstoichiometric quantities. It should be pointed out that supernatant (NT) extract does, however, bind to both fumed (Sigma) and pellicular (Whatman) silica (data not shown).

The effects of both the untreated and treated frustule supernatant extracts on the rate of $CaCO_3$ crystallization are shown in Figure 3. Both fractions are able to decrease, and at higher concentrations stop, $CaCO_3$ growth. That the supernatant (T) fraction was more active on a weight basis than the supernatant (NT) can be determined by comparing the dose response curves in Figure 3. To more readily compare additives the quantity required to produce 50% inhibition of crystal growth ($I_{50}$) is determined by graphical extrapolation of the data set for each additive (14). At 0.477 $\mu$g/ml the treated supernatant $I_{50}$ was roughly

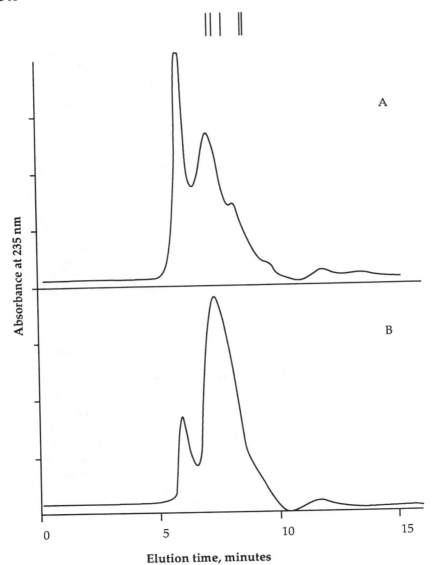

Figure 1. Typical elution profiles of 100 μg by dry weight of the soluble fractions of material extracted from: A) non-treated frustules and B) treated frustules. Elution peaks of standards are marked by vertical lines and in order of increasing elution time are BSA (66 kD), ovalbumin (45kD), B-lactoglobulin 36.8 (kD), myoglobin (14 kD), and cytochrome C (12 kD).

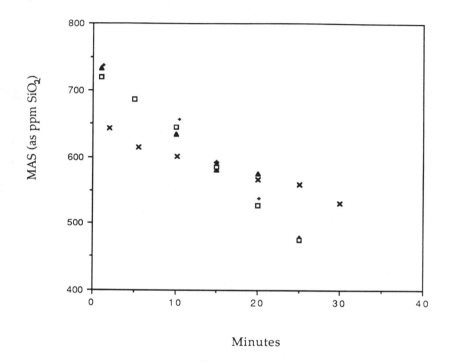

Figure 2.  Effect on polymerization of SiO₂ by supernatant (NT) at 20 μg/ml(+), 50 μg/ml(△), and borate at 200 mM(x). Control curve (□).

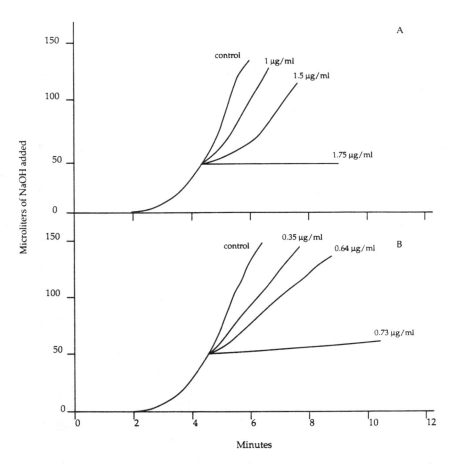

Figure 3. Effect of untreated and treated frustule supernatant extracts on the rate of CaCO$_3$ crystallization from solution. The rate of crystal growth is taken as the rate of base titrated to maintain pH during calcium carbonate precipitation. A) untreated frustule extract, I$_{50}$ = 1.02 $\mu$g/ml, B) treated frustule extract, I$_{50}$ = 0.47 $\mu$g/ml.

half that of the untreated supernatant (1.06 $\mu$g/ml), which means that only half as much of the former fraction is required to produce a 50% decrease in crystal growth rate. The untreated pellet fraction has no effect on CaCO$_3$ growth in this assay at concentrations up to 235 $\mu$g/ml (data not shown).

## Discussion

The existence of an organic matrix which interacts with the silica during mineralization in diatoms was suggested by Heckey et al. (4) based on their discovery of an amino acid composition unique to the cell wall which was high in serine. The basic premise of the Heckey model was expanded on by Sullivan and coworkers. They were able to show that *de novo* protein synthesis was necessary for biomineralization in the diatom *Navicula saprophilia*. Further, they determined from Q$_{10}$ values of both *in vivo* and *in vitro* polymerization that autopolycondensation alone could not account for *in vivo* polymerization of silica in *N. saprophilia* (17). These findings suggest the possibility of an organic matrix associated with the biosilica of diatoms.

The materials remaining after NaOCl treatment in this study were assumed to exist in an intimate association with the mineral, possibly even locked within the silica itself, and thus would qualify as an organic matrix. However, the results of a study by Perry (18), in which adsorption techniques were used to examine acid cleaned diatom frustules for pore structures, suggest that if an intra-mineral location of organic material exists in frustules, it must consist of extremely fine strands, as no obvious pore structures were found. In addition, Perry found no evidence of Si-O-C bonds using NMR, suggesting that interactions between silica and any organic materials do not utilize this type of covalent bond.

The evidence from the current study suggests that an organic material unique to silica exists. For example, cleaning the frustules with NaOCl affects the material subsequently extracted in several ways. The proportions of protein to carbohydrate are reversed in the supernatant fractions as compare to materials extracted from non-NaOCl cleaned frustules. The increase in the proportion of carbohydrate to protein after NaOCl cleaning is somewhat counterintuitive in that it has been suggested that just prior to and during SiO$_2$ deposition the majority of materials added to "new" walls are proteinaceous, while most of the material added to "old" walls is carbohydrate (19). If NaOCl treatment removes the organic casing found on mature valves, one might expect to see a decrease in the proportion of carbohydrate present. However, after cleaning with NaOCl, the proportion of material testing as carbohydrate increases in the soluble fraction, and the proteinaceous material decreases by one third to one half. One possible explanation for this seeming incongruity may be that the observed changes in material deposited onto maturing frustules is reflected in the composition of the insoluble material. It must be remembered that the values for protein content of the insoluble material may not accurately reflect the actual quantities of those materials present as the values were obtained using sonicated suspensions. Lowry techniques are most useful when determining quantities of protein in solution. In addition, NaOCl treatment may change the reactivity of the type of carbohydrate associated with the frustules.

NaOCl treatment of the frustules changes the proportions of the size classes produced using gel chromatography, probably resulting from fragmentation or elimination of certain groups. The peak which elutes at the void volume of the column is reduced by two thirds, the second peak is doubled in size and the small peaks are eliminated. The reduced amounts of large molecular weight materials may explain the increase in activity of the supernatant (T) fraction in the $CaCO_3$ crystallization assay as it has been shown that the activity of carbonate matrix proteins against $CaCO_3$ crystallization increases with decreasing molecular weight (20). In fact, the material in the second peak may be responsible for most of this activity. In this light, it would be interesting to determine if this material is the principal source of the phosphate found in the soluble fraction, as phosphorylated materials have been shown to be active $CaCO_3$ inhibitors in this particular assay (21,22).

While all the frustule fractions differ to some extent from the cell contents, the amino acid composition of the soluble fraction from NaOCl cleaned frustules is the only fraction clearly different from the cell contents. Five amino acids, glycine, serine, aspartic and glutamic acids, and alanine, comprise roughly 72% of the protein compared to 56% for cell contents. Interestingly, Volcani (23) indicates unpublished results that suggest the amino acids found in the highest concentrations in new cell walls are those listed above. The aspect of the amino acid composition that is unique, however, is the relatively high levels of serine and glycine. This may indicate a role for these amino acids in the process of silica deposition. Typically the only other naturally occurring proteins with such high serine levels are matrix proteins (24). Serine groups in oyster matrix are modified with O-linked phosphate groups (22). There is some evidence that the phosphate and/or carbohydrate present may be O-linked to serine (Swift and Wheeler, unpublished observation). At this time we know nothing about the primary structure of the proteins extracted. However, the activity of the extracted proteins in the carbonate mineralization assay rivals that of carbonate matrix. The activity of carbonate matrix and matrix analogs depends on runs of polyaspartic acid and phosphoserine (20). Consequently, it may be conjectured that similar structures are present in frustule matrix.

Heckey et al. (4) suggested that the role of a high serine protein in mineralization is one of a template for ordered silica deposition. However, there are multiple roles an organic matrix for silica biomineral could take, one of which is a mobile nucleation site for relatively unordered binding, capable of being moved by microtubules to sites of silica deposition within the SDV (25). An attempt to further elucidate the role of a high serine protein in Si polymerization illustrated that polyserine does bind to $^{68}Ge(OH)_4$. Roth and Werner (26) incubated cell-free extracts, polyserine, and polymethionine with $^{68}Ge(OH)_4$ for 2 hours and then spotted them on DEAE-cellulose TLC plates. Compared to the migration of $^{68}$Ge(OH)$_4$ alone, polyserine reduced the spread of the label as it migrated in each of two directions. The cell extracts reduced the R$_f$ of the migrating label, but not the spread. Polymethionine showed no association with the germanium label. The authors concluded that the hydroxyl groups of the polyserine afforded a means of interaction with the germanate.

The lack of an effect by the materials tested thus far on the rate of SiO$_2$ polymerization is interesting in light of the fact that at low concentration the polymers are effective inhibitors of calcium mineral formation. It was anticipated that soluble extracts might inhibit polymerization of silica out of solution in a manner similar to inhibition of CaCO$_3$ deposition. Failure to observe this may result from any of several possibilities. For example, the soluble materials tested thus far consisted only of non-NaOCl cleaned frustule extracts at a maximum concentration of 50 $\mu$g/ml. No effect was observed at concentrations of up to 0.5 mg/ml insoluble NT material. However, it is possible that there would be an inhibitory effect at higher doses. If these materials do have an effect on SiO$_2$ polymerization rate, then this concentration may be below the level necessary to detect said effect. Also, it may be that the treated supernatant fraction might exhibit some effect on polymerization where the NT fraction does not. Other compounds tested in this assay that exhibited no effect on the rate of polymerization of silica include polyaspartic acid, copolymers of serine and aspartic acid, bovine serine albumin, polyacrylate, glycerol, and oyster soluble matrix. Borate did exhibit an inhibitory effect after initial silica binding, but at superstoichiometric levels (200mM). Meier and Dubin (27) also found that only stoichiometric levels of borate, of all the industrial compounds tested, produced an effect on the rate of silica polymerization. In this light, perhaps the lack of any measurable effect by the diatom extracts at the concentrations tested is not surprising. Because the concentration of free SiO$_2$ at the site of polymerization is unknown, the concentration of SiO$_2$ in the *in vitro* assays may be unrealistic as a model for the concentrations found *in vivo*. Also, if the materials extracted offer nucleation sites for deposition, then their function may be masked at the high rate of polymerization (hence, large number of nucleation events) occurring in the solution. Obviously, concentrations of extracts above those used thus far must be employed to test this idea. Also, various levels of SiO$_2$ supersaturation should be evaluated. This process will be difficult to execute because of the small quantities of the soluble materials available. It is possible that the compounds extracted may interact with soluble or forming silica polymers only when immobilized on a substrate. This hypothesis is being explored. Additionally, the effect of a siliceous matrix may be to regulate the form or architecture of the silica deposited, with little or no effect on rate of deposition. Silica is known to occupy various morphs in plant hairs, the deposition of which correspond temporally with changes in organic components associated with the deposits (28). Finally, it is important to keep in mind that the mineralization process in diatoms is not one of crystallization but polymerization. There are no discrete surfaces formed, as the silica polymers exist in a substantially random network (18). At any point in time there are an unknown number of free silanol groups available for polymerization, not a fixed surface with discrete lattice sites as is found in growing calcium carbonate crystals.

In conclusion, we have identified materials which are closely associated with the siliceous frustule, as evidenced by their resistance to attack by NaOCl, that have unique properties which suggest that they may have a role in biosilicification. The origin of these materials, their location in relation to the mineral, further physical characterization, and their *in vivo* function are all areas which are being explored.

## Acknowledgments

We gratefully acknowledge Dr. Herbert Waite for performing the amino acid analyses and Dr. Susan Kilham for providing starter cultures of the diatom. This work was funded in part by a grant from the South Carolina Sea Grant Consortium (APW) and a grant-in-aid of research from Sigma Xi (DMS).

## Literature Cited

1. Sullivan, C. W. In Silicon Biochemistry; Evered, D.; O'Conner, M., Eds.; John Wiley and Sons: N.Y., 1986; pp 59-89.
2. Schmidt, A-M. M.; Borowitzka, M. A.; Volcani, B. E. In Cytomorphogenesis in Plants; Werner, D., Ed.; 1981; pp 63-97.
3. Bhattacharyya, P.; Volcani, B. E. Biochem. Biophys. Res. Comm. 1983, 114(1), 365-72.
4. Hecky, R. E.; Mopper, K.; Kilham, P.; Degens, E. T. Marine Biology 1973, 19, 323-31.
5. Duke, E. L.; Reimann, B. E. F. In The Biology of Diatoms; Werner, D., Ed.; Univ. of California Press: Berkeley, 1977; pp 65-109.
6. Miller, G. L. Anal. Chem. 1959, 31, 964.
7. Dubois, M.; Gilles, K.; Hamilton, J. K.; Rebers, P. A.; Smith, F. Anal. Chem. 1956, 28(3), 350-6.
8. Wheeler, A. P.; Harrison, E. W. Comp. Biochem. Physiol. 1982, 71B, 629-36.
9. Marsh, B. B. Biochem. Biophys. Acta 1959, 32, 351-61.
10. Wainer, A.; Koch, A. Anal. Biochem. 1962, 3, 457-61.
11. Weres, O.; Yee, A.; Tsao, L. J. Coll. Inter. Sci. 1981, 84(2), 379-402.
12. Iler, R. K. The Chemistry of Silica. Wiley-Interscience: N.Y., 1979; p 97.
13. Marsh, A. R.; Klein, G.; Vermeulen, T. Report LBL-4415, 1975; Lawrence Berkeley Laboratory.
14. Wheeler, A. P.; Sikes, C. S. Am. Zool. 1984, 24, 933-44.
15. Waite, H.; Benedict, C. V. Meth. Enz. 1984, 107, 397-413.

16. Cornish-Bowden, A. Meth. Enz. 1983, 91, 60-75.
17. Blank, G. S.; Robinson, D. H.; Sullivan, C. W. J. Phycol. 1986, 22, 382-389.
18. Perry, C. C. In Biomineralization: Chemical and Biochemical Perspectives; Mann, S.; Webb, S.; Williams, R. J.P., Eds.; VCH: London, 1989; p 223-56.
19. Coombs, J.; Volcani, B. Planta (Berl.) 1968, 80, 264-79.
20. Wheeler, A. P.; Rusenko, K. W.; Swift, D. M.; Sikes, C. S. Mar. Biol. 1988, 98, 71-89.
21. Volcani, B. E. In Silicon and Siliceous Structures in Biological Systems; Simpson, T. L.; Volcani, B. E., Eds.; Springer-Verlag: N.Y., 1981; p 157-200.
22. Rusenko, K. W.; Donachy, J. E. In Surface Reactive Peptides and Polymers: Discovery and Commercialization. Sikes, C. S.; Wheeler, A. P., Eds.; ACS Books: Washington, D.C., 1990.
23. Sikes, C. S.; Wheeler, A. P. CHEMTECH 1988, 620-6.
24. Wheeler, A. P.; Sikes, C. S. In Biomineralization: Chemical and Biochemical Perspectives; Mann, S.; Webb, S.; Williams, R. J. P., Eds., VCH: London, 1989; p 95-131.
25. Robinson, D. H.; Sullivan, C. W. TIBS 1987, 12151-4.
26. Roth, R.; Werner, D. Z. Pflanzenphysiol. 1977, 83, 363-74.
27. Meier, D. A.; Dubin, L. Corrosion, 1987, Paper 334.
28. Perry, C. C.; Wilcock, J. In Chemical Aspects of Regulation of Mineralization; Sikes, C. S.; Wheeler, A. P., Eds., University of South Alabama Publication Services: Mobile, AL, 1988; p 39-48.

RECEIVED August 27, 1990

# Chapter 25

# Silica Inhibition in Cooling Water Systems

## Leonard Dubin

### Nalco Chemical Company, Naperville, IL 60563–1198

Amorphous silica ($SiO_2$) deposition is a significant problem in recirculating cooling water systems due to the low solubility of silica. Because silica deposits are difficult to remove, avoidance of the problem by minimizing cycles of water concentration is the common practice. In water short areas, this is a significant problem. A discussion of the chemistry of silica as it relates to cooling water systems is presented. The operation of a cooling tower is briefly described. Experimental laboratory work on the evaluation of chemical treatments as amorphous silica inhibitors and the identification of active materials is presented. Process simulation and field data on the efficacy of an active silica inhibitor are presented. A possible mechanism by which the inhibitor operates is discussed.

## General Background

Amorphous silica ($SiO_2$) and magnesium silicate ($MgSiO_3$) deposition in cooling water systems is a serious problem in high silica water. Deposition problems worsen when iron, calcium, and phosphate are present in the cooling water as they also can cause fouling. These and other metal ions can precipitate as salts as well as exacerbate deposition of silica.

In various parts of the world, the efficient use of water in cooling water applications is limited by high silica content of the water. In the western hemisphere, Mexico and portions of the United States suffer from high silica water. For example, in the United States, the following areas have surface and well waters containing silica levels as high as 100 mg/L: Hawaii, the Pacific Northwest, New Mexico, Nevada, Arizona, Texas, Louisiana, Oklahoma and portions of lower California.

Because amorphous silica has a very limited solubility in water (approximately 150 mg/L $SiO_2$, as molybdate reactive silica at neutral pH and normal ambient temperatures), it can readily foul heat transfer equipment and other surfaces in cooling water systems, and in a relatively short time force a plant to shutdown. The composition, quantity, and rate of silica deposition depend on pH, temperature,

0097–6156/91/0444–0354$07.50/0
© 1991 American Chemical Society

the ratio and concentration of calcium to magnesium, and the concentration of other polyvalent ions in the water (1-3). Furthermore, precipitation of mineral salts (e.g. calcium carbonate and iron phosphate) and corrosion products can occur under the same water conditions. These can significantly affect the quantity and the rate of silica deposits.

The end result of reactive silica (silicic acid monomer) polymerization to form the nonmolybdate reactive colloidal silica is the deposition of a high molecular weight silica polymer known as amorphous silica. Fouling by amorphous silica is unique compared to other foulants that are commonly found in cooling water applications. Common mineral scales that cause fouling by precipitation onto a heat transfer surface include $CaCO_3$, $Ca_3(PO_4)_2$, $CaSO_4$, $Fe_2O_3$, and $FePO_4$. Most of these inorganic salts have the property of inverse solubility and, therefore, are less soluble in hot water. In contrast, amorphous silica has normal solubility; it is less soluble in cold water. Consequently, conventional treatments (4-7) used to control common cooling water foulants do not necessarily apply and, in effect, may be opposite to that required. This is especially true with the following parameters such as calcium, pH/alkalinity, and temperature. For example, cool bulk water temperatures and low heat exchanger skin temperatures are desirable for limiting $CaCO_3$ scale, but are undesirable for preventing silica scale.

The amorphous character of silica deposits precludes the use of conventional crystal modification and dispersant techniques. This means phosphonates and polyacrylates, (4-7) that are useful in preventing $CaCO_3$ and $Ca_3(PO_4)_2$ precipitation, provide little benefit. The only exception is a limited use of polyacrylate to disperse low levels of colloidal or particulate silica.

Silica scale is extremely tenacious and very difficult to remove. It is not uncommon that fouled equipment has to be cleaned with a combination of mechanical methods and hydrofluoric acid. Since the cleaning process is both difficult and dangerous, and sometimes not even possible, most plants choose to restrict the number of cycles that cooling water may be concentrated. To ensure no fouling, this limit may be only two to three cycles in a high $SiO_2$ make-up water. In areas where the water supply is limited, this can be a severe problem. Even in areas where water is not in short supply, the consequent use of large volumes of water can be a significant problem for waste treatment facilities. For these reasons, a research project was undertaken to find a method to inhibit silica deposition and scaling in cooling water systems.

This paper will report the work involved in the development of a borate-based silica inhibitor that directly inhibits amorphous silica deposition and thereby helps to control "magnesium silicate" formation in recirculating cooling water systems with up to 400 ppm (mg/L)$SiO_2$ (8-9). Some of the initial laboratory testing, pilot cooling tower studies and field data are discussed. Since this work was slanted to recirculating cooling water applications, a brief and simplified explanation of the operation of a cooling water system is provided. This is necessary as some of the experimental procedure(s) were developed with recirculating cooling system requirements in mind.

## Cooling System Technology

Figure 1 shows a mid-size industrial cooling tower. The plume (cloud) that is above the tower is from some of the "hot" water that evaporated during cooling of the bulk of the water and has begun to condense. Water from heat generating processes is cooled in the cooling tower where some of it evaporates. The cooled water is then recycled through the heat sources. Hence, such towers are called recirculating cooling systems. The heat generating process can be electrical generating turbines in a power plant, oil refinery process equipment such as crude oil cracking units, chemical processes where specific chemicals such as polyethylene are made, or even an air conditioning unit such as found in hospitals, universities or other buildings.

In a typical industrial operation, the cool water is passed over or through the process which requires cooling for temperature control by the use of metal tubes. The metal surface that has hot process components on one side and cooling water on the other is called the heat transfer surface. This metal surface is the primary concern of plant operations because deposition of mineral salts and silt will result in a scaled surface which is less effective as a heat transfer vehicle for cooling. In extreme conditions, when heat transfer equipment becomes plugged the resulting loss of cooling water may result in a plant shutdown.

Figure 2 is a schematic of a heat transfer system. In this diagram, cooling water is flowing on the outside of the tube in a direction opposite that of the hot process liquid flowing on the inside. This means that the coldest cooling water is at the point of initial exposure (inlet) of the cooling water to the process, and the hottest water is at the opposite end (outlet). It is this increasing temperature differential on concentrated cooling water that causes deposition of most types of mineral scales.

After leaving the heat transfer equipment (outlet), the now heated cooling water is returned to the cooling tower where it enters at the top. This water flows downward over a "fill", a surface area increasing medium, where it contacts cool air being drawn upward. The counter flowing mixture of cool air and hot water allows evaporative cooling of the water. Because of this evaporative process, mineral concentration in the water can increase above saturation levels. This can result in precipitation in the fill and/or down stream in places such as the heat transfer equipment unless chemical treatment and mechanical methods are used.

Mechanical methods are used in combination with specific chemical treatments for different minerals. Mechanical methods generally refer to replenishing evaporated water by fresh water (make up) as well as removing the highly concentrated water in the basin at the bottom of the cooling tower (blowdown), and replacing it with make up water. The purpose of this is to maintain recirculating water within the operating capabilities of the specific chemical treatment being used.

Figure 1. Typical medium size industrial cooling tower.

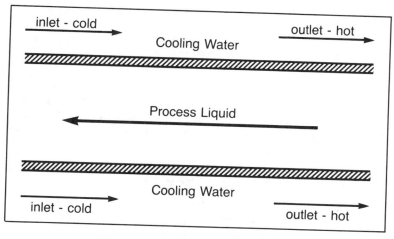

Figure 2. Heat transfer schematic.

In the operation of a cooling system, the degree of concentration of minerals from the make up water is referred to as the cycles of concentration. The "fresh" make up water, whatever its source (rivers, lakes, wells, groundwater), and whatever hardness or other minerals in it, is defined as one cycle. The purpose of the chemical/mechanical control of the cooling water system is to maintain identical ratios of hardness and minerals in the recirculating water as it concentrates by evaporation, leaving behind minerals in solution. If the water in a cooling system is said to be in balance at 10 cycles, this means that mineral components found in the original make up water (one cycle) have been concentrated exactly by 10, and no minerals have been precipitated. If a specifically identified mineral is missing, then the cooling water is described as out of balance or out of cycles, and deposition has occurred. Further details on the operation and problems of cooling systems are available elsewhere (10-14).

## Theoretical and Practical Chemistry of Silica

Silica deposition as amorphous (colloidal) silica can be explained by several equilibria. These can occur in such water treatment situations as boilers, cooling water systems, seawater evaporators, and possibly geothermal applications. In cooling water applications, silica fouling can be caused by three mechanisms:

1. Deposition from silicic acid [$Si(OH)_4$] via condensation to form colloidal amorphous silica.

2. Breakdown of colloidal silica with subsequent coagulation and scale formation.

3. Precipitation of silicic acid or colloidal silica with cations such as Al(III), Mg(II), Fe(III), or Fe(II) and Ca(II) through $CaCO_3$ formation.

All of the above mechanisms involve equilibria which are pH sensitive. they are also accelerated by metal ions, especially those that form metal hydroxides [e.g., $Fe(OH)_2$, $Fe(OH)_3$]. Condensation of silicic acid [$Si(OH)_4$] or hydrolysis reactions involving silica and water are reversible. The following equations are given to illustrate some of the equilibria involved:

$$SiO_2 + 2H_2O \longleftrightarrow Si(OH)_4 \text{ (Silicic Acid)} \tag{1}$$
$$Si(OH)_4 + OH^- \longleftrightarrow HSiO_3^- + 2H_2O \tag{2}$$
$$2HSiO_3^- \longleftrightarrow Si_2O_5^{-2} + H_2O \tag{3}$$
$$HSiO_3^- \ OH^- \longleftrightarrow SiO_3^{-2} + H_2O \tag{4}$$

Polymerization and depolymerization of silicic acid/amorphous silica is catalyzed by $OH^-$ [Equations (2) and (4)] and is the dominant reaction above pH 3. It is first order from pH 3 to 7 with $OH^-$, maximal at pH 8 to 9, and decreases at higher pH (1-3). The decrease is due to formation of the more soluble silicate ion by the following reaction.

$$Si(OH)_4 + OH^- \longleftrightarrow H_3SiO_4^- + H_2O \tag{5}$$

The effect of pH on solubility is shown in Table I ([1, 15, 16]).

Table I.  Solubility of Amorphous Silica 1,15,16

| pH | mg/L(at 25 °C) |
| --- | --- |
| 6.0–8.0 | 120 |
| 9.0 | 138 |
| 9.5 | 180 |
| 10.0 | 310 |
| 10.6 | 876 |

The concept for operating a cooling water system at a high pH is based on the fact that silica is more soluble at higher pH's (Table I). There are two difficulties with this approach:

1. A pH of at least 10 is needed for a significant increase in silica solubility by the mechanism in Equation (5).

2. At a pH of 8 to 9, the rate or driving force for the polymerization is at a maximum ([17,18]).

In principle then, a high pH operation (8.5 to 9.5) should not be feasible, especially because iron oxides (hydroxides) will promote silica deposition through a heterogeneous catalysis process. Furthermore, high alkalinity can aggravate iron and phosphate deposition. In practice, there is evidence that a dispersant program can be useful at an operating pH of 8 to 9, if the $SiO_2$ concentration is kept at a level less than 200 mg/L. This requires a strong dispersant and $CaCO_3$ inhibitor package, as well as an effective corrosion inhibitor to prevent formation of nucleation sites that will lead to silica condensation.

Polymerization of silicic acid is believed to occur through an SN2 steric inversion reaction involving an ionized silanol and a nonionized silanol group ([1,17]). This is illustrated in equation 6 which shows a neutralization reaction between a Lewis acid and base, and is consistent with the concept that the nonionized silicic acid is considered acidic.

$$(HO)_3Si\text{-}O:^- + \overset{\displaystyle OH}{\underset{\displaystyle OH}{\overset{|}{\underset{|}{HO\text{-}Si\text{-}OH}}}} \longrightarrow (OH)_3\text{-}Si\text{-}OSi\text{-}(OH)_3 + OH^- \quad (6)$$

Alkalinity therefore neutralizes the acidic but nonionized silicic acid so that it can react with other silicic acid. Silicic acid then, does not readily condense under acidic conditions pH(<4), and at very high pH (>9), there are insufficient nonionized surface groups to act as a substrate for further condensation. The use of acid to minimize silica condensation is based on mechanistic considerations. The use of acid to prevent phosphate fouling is based on similar reasons.

The proposed method by which amorphous silica precipitates involves six steps. The first three are critical and of practical interest in cooling water (19).

1. Formation of silica minipolymers of less than critical nucleus size.
2. Nucleation of an amorphous silica phase in the form of colloidal particles.
3. Growth of the supercritical amorphous silica particles by further chemical deposition of silicic acid on the surface.

For chemical inhibition and practical chemical analysis, step one encompasses the critical and/or rate determining step where the monosilicic acid condenses to pentamer-size units. This step is critical both analytically and mechanistically for controlling scaling by amorphous silica. The second step "nucleation" is associated with the observed "induction" or lag time and is influenced by other foulants, such as $Ca_3(PO_4)_2$ and iron deposits. At this point colloidal silica can be detected analytically and it is first observed visually.

In detecting silica, it is useful to distinguish between total and reactive silica. In theory, molybdate reactive silica (MAS) is only monosilicic acid as based on the molybdate test for silica. However, with sufficient time, molybdate will react positively with silica up to approximately pentamer size (1,3). This size is acceptable in cooling water, especially if technology can be found that limits condensation of silicic acid to this degree of polymerization. If atomic absorption (AA) is used, total rather than reactive silica is found analytically, but only if colloidal silica is kept to a particle size of $<1$ $\mu$m.

From a mechanistic view, the dimerization reaction of silicic acid [Equation(3)] is the slow step within Step 1. If an inhibitor/stabilizer could be successfully introduced, the dimerization reaction and possibly subsequent steps would be slowed further. In addition, dispersants may prevent deposition of the small silica minipolymers, but their greatest benefit would be in preventing fouling by calcium or iron phosphate and iron oxides.

Once dimerization occurs, the dimer converts almost instantly to a trimer and then to tetramer and pentamer, but at progressively slower rates (17). The nucleation step can thus be thought of as the point where hexamer formation initiates because, at this point, the side reactions, isomer formation and cross linking, begin to occur. These reactions lead to colloidal silica formation and eventual gelation and scaling (17).

Polyvalent metal ions can cause or enhance condensation of silica under common cooling water conditions. The effects of metals that are commonly found in a cooling water system are given in Table II (1,20). The metals of most interest are zinc, iron, and magnesium. However, for this paper, the focus will be strictly on magnesium.

Fouling by what is described as "Mg/silicate" ($MgSiO_3$) precipitation is presumed to occur when a cooling water system is run at a pH above 8 and where high levels of magnesium (100 to 200 ppm) and silica (150 to 250 ppm) are present.

Table II. Effect of Polyvalent Metals on Amorphous Silica Precipitation under Common Cooling Water Conditions

| Metal | Effect |
|---|---|
| Calcium | None at pH 6 to 9 unless $CaCO_3$ precipitates |
| Chromium | None at pH 6 to 9 for chromate (Cr VI) |
| Molybdenum | None at pH 6 to 9 |
| Aluminum | Enhance precipitation at pH $> 4$ and is basis for the insoluble clays (silts) |
| Iron | Enhance precipitation at pH 6 to 9, especially at surfaces where corrosion is taking place |
| Zinc | Can cause precipitation under selected conditions, e.g., high pH, high carbonate and hard water |
| Magnesium | A coprecipitation can occur above pH 8 under conditions of relatively high concentrations of magnesium and alkalinity. The precipitation is commonly thought of as $MgSiO_3$, but this is probably incorrect at pH below about 11. |

This presumption is based on the presence of high levels of Mg and $SiO_2$ in deposits (Table III), but this may not be the case. When a cooling water system is operating at a pH of 9, its pH is still two units lower than the theoretical pH for $MgSiO_3$ precipitation. Furthermore, crystalline $MgSiO_3$ formation at a pH below 11 is not found. The "Mg/silicate" found in cooling water systems is generally reported as amorphous. A better explanation for $MgSiO_3$ precipitation is a coprecipitation or coagulation of a positively charged colloidal metal hydroxide and negatively charged colloidal silica (21,22). This is consistent with the observation that many polyvalent ions (Fe, Zn, Mg) are absorbed by silica when the pH is raised within 1 to 2 units of the pH at which the metal hydroxide will precipitate by itself (21,22). This is illustrated in Table III with a deposit analysis from a cooling water system that was operated at pH 9 and 200 ppm $SiO_2$. The higher percentage of silica and the greater quantity of silica scale buildup on the cold-side condenser are consistent with the lower solubility of silica in cold water (1-3). The $CaCO_3$ and $Ca_3(PO_4)_2$ deposition on the hot-side condenser is consistent with their inverse temperature solubility. The phosphate is from degradation of phosphonate. In calculating the mineral fractions, it is assumed that all magnesium precipitates as $MgSiO_3$; however, this does not account for all the silica, especially on the cold-side condenser. This deposit analysis is very strong evidence that the reported $MgSiO_3$ found in some cooling water applications is really a coprecipitation and coagulation of amorphous silica and a metal hydroxide such as $Mg(OH)_2$. This means that the mineral fractions in Table III should really be calculated and presented as percent $Mg(OH)_2$ and percent amorphous silica. A treatment program, therefore, would require a method of inhibiting amorphous silica formation and/or $Mg(OH)_2$ precipitation but not a $MgSiO_3$ inhibitor.

Table III. Deposit Analysis of pH 9, high silica cooling water operation (silica typical - 160 to 190 mg/L $SiO_2$ at 15 to 20 cycles)

|  | Hot side condenser | Cold side condenser |
|---|---|---|
| Buildup gm/ft$^2$ | 15 | 40 |
| %$SiO_2$ | 36 | 73 |
| %CaO | 21 | 3 |
| %MgO | 15 | 17 |
| %$P_2O_5$([1]) | 16 | 2 |
| %$Fe_2O_3$ | 4 | 2 |
| %$CO_3$ | 2 | 0 |
| %Other | 6 | 2 |
| %Total | 100 | 99 |
| %Mineral fractions (back calculated) |  |  |
| as $Ca_3(PO_4)_2$ | 36 | 5.0 |
| as $MgSiO_3$ | 37 | 42.5 |
| as Amorphous $SiO_2$ | 14 | 47.5 |
| as $CaCO_3$ | 3 | – |
| % as $Fe_2O_3$ | 4 | 2.0 |
| as Clay Silts ([2]) | 6 | 2.0 |
| Total mass balance | 100% | 99% |

[1] Source - degradation of phosphonate by excessive chlorination and high temperature.
[2] Includes Al, and very low levels of Cu and Zn

Phosphate, and specifically orthophosphate, is frequently found in cooling water systems. Its source can be the makeup water or chemical treatments. The $Fe_2O_3$ is either from iron present in the make-up water, or more likely from corrosion.

## Experimental

**Bench Testing.** The screening test has been described ([8,9]). The apparatus is shown in Figure 3. Fifty ml of a 1% w/v sodium metasilicate solution (as $SiO_2$) is passed through a strong cation (HGRW2 acid form resin) exchange column into a three necked round bottom flask containing 300 ml of water under reflux. The flask contains calcium hardness and any desired chemical treatment. The column is eluted with 100 ml distilled water; after the pH is adjusted from 4-5 to 8.2-8.3 with 1% NaOH, a final 50 ml of distilled water is eluted through the column. The result is a reactor containing 500 ml of solution with 500 mg/

Figure 3. Silica screening apparatus.

L silicic acid as $SiO_2$, 300 mg/L $Ca^{2+}$ as $CaCO_3$ and a test treatment at typically 100 mg/L. To avoid unnecessary complications, magnesium is not present as this test was designed specifically for amorphous silica. The value of 500 ppm silicic acid as $SiO_2$ was chosen because it represents the upper limit of $SiO_2$ solubility under these conditions (23).

After refluxing for one hour, the solution is cooled overnight (18 hours). Physical observations are made first, then part of the solution is filtered (0.45 micron membrane filter) and both filtered and unfiltered samples are analyzed for total $SiO_2$ by AA and reactive (soluble) silica (molybdate test).

A blank without added inhibitor, depending on the final pH, will give up to 150 mg/L molybdate reactive silica (MAS) and a total silica of about 200 mg/L. Moderate silica inhibition activity is associated with a total filtered silica value of approximately 250 mg/L $SiO_2$. High activity is associated with silica values of a least 300 mg/L $SiO_2$. Treatments which allow an obvious visible scale film on the reactor walls are severely down graded. This is most likely to occur with treatment of modest activity.

In developing this test protocol, a number of variables were carefully considered to assure that the procedure would conform and predict known chemistry, especially in regard to the blank. Table I shows the expected/theoretical solubility values for silica for a variety of pH conditions. Table IV gives the results with the test apparatus. The data in Table IV compare favorably with the expected values. Further, analysis of the precipitate by x-ray confirmed the presence of amorphous silica.

Table IV. Effect of pH on Silica Blank Screening Results (Standard Conditions with 300 mg/L $CaCO_3$)

| pH initial | pH final | Total $SiO_2$ - mg/L (after filtering through 0.45 micron filter) |
|---|---|---|
| 8.30 | 9.10 | 170 |
| 8.29 | 9.05 | 170 |
| 8.36 | 9.18 | 200 |
| 8.38 | 9.22 | 210 |
| 8.41 | 9.21 | 230 |
| 8.64 | 9.40 | 270 |
| 8.60 | 9.40 | 260 |

The effects of calcium hardness and pH were evaluated and the pertinent data are given in Table V. As predicted theoretically, condensation of the silicic acid to amorphous silica does not occur under acidic (pH<4) conditions. The reaction is driven by alkalinity, and, as predicted by equation 6, the pH does rise. So that results could be obtained within a reasonable length of time, a pH of 8.2-8.3 was chosen for the screening test.

Table V. Effect of $Ca^{2+}$ and pH on Condensation of Silica

| $Ca^2$ (mg/L) as $CaCO_3$ | Temperature °C | pH Initial | pH Final | mg/L $SiO_2$ Soluble (Reactive) | Total (AA) |
|---|---|---|---|---|---|
| None | 100 | 7.8 | 8.6 | 150 | 470 |
| 300 | 100 | 4.4 | 4.3 | 460 | 470 |

**Process Simulation Testing.** Process simulation was used for advanced screening to confirm activity of the most active silica inhibitor. In these evaluations, a Pilot Cooling Tower (PCT) was used for the testing (Figure 4).

The PCT used has been described in more detail elsewhere (24). These units have all the features of a standard cooling tower. Cooling water flows from the basin over eight heat transfer tubes in series, then through two holders for metal coupons and a probe for corrosion measurements (Corrater, Rohrback Corp.) before returning to the tower where it is sprayed over film-type packing. A fan atop the tower is thermostatically controlled, based on the basin water temperature. Conductivity and pH are controlled automatically. Cycles of concentration are maintained by a blowdown pump connected to the conductivity controller. System capacity is maintained by controlling the make-up water flow to the basin using a level controller.

Results are monitored in several ways. Corrosion rate data are obtained by weight loss determinations on coupons and heat transfer tubes, and instantaneously with Corrater readings. The heat transfer tubes are also evaluated for pitting on mild steel and dezincification of admiralty brass. Fouling factors are determined using special heat transfer tubes equipped with thermocouples to measure their surface temperatures. Deposit weight and composition are determined and used to calculate fouling rates.

The initial PCT studies involved the use of a concept known as "limit" testing. In this method, the cycles of concentration are increased at a controlled rate (based on conductivity) until the solubility "limit" is reached and precipitation occurs. The recirculating water is sampled after 18 hours at each concentration level, and analyzed for soluble (MAS) silica and total silica (AA) and compared against theoretical silica levels. Inhibition of silica deposition is demonstrated when analytical results match theoretical. The "silica limit" is defined as the highest silica concentration where the water chemistry is in mass balance with respect to the initial make-up water.

In the initial PCT studies, eight tubes of mixed metallurgy were used with a heat load of 500 watt on each tube. This testing was done using a pH 7 chromate/zinc treatment program with and without the borate inhibitor. The make-up was as follows:

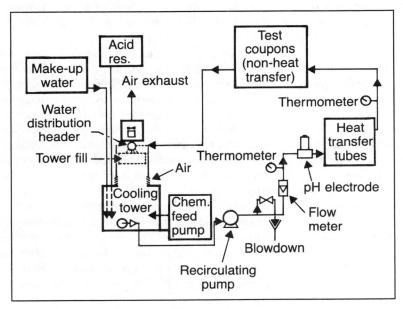

Figure 4.  Pilot cooling tower schematic.

Initial PCT Studies - Make-up Water

| | |
|---|---|
| $Ca^{2+}$ as $CaCO_3$ | 60 mg/L |
| $Mg^{2+}$ as $CaCO_3$ | 26 mg/L |
| Silica as $SiO_2$ | 61 mg/L |
| $HCO_3^-$ as $CaCO_3$ | 130 mg/L |
| pH | ~7.5 |

Advanced PCT studies were done using a high pH (8.0 - 9.5), all-organic program. In these tests, high silica waters were cycled up to the desired concentration and maintained in steady state condition for at least two weeks. The purpose of these tests was to determine if the silica inhibitor would prevent both magnesium silicate and amorphous silica deposition. Test conditions for the advanced PCT studies are given in Table VI.

Table VI.  Advanced Steady State PCT Test Conditions

| | |
|---|---|
| Test Duration | 2 weeks |
| Holding Time Index* | 24-30 hours |
| Basin Temperature | 35 - 38 °C |
| Return Temperature | 40.5 - 43.5 °C |
| System Volume | 50 Liters |
| Flow Rate | 7.6 - 9.5 L/m |
| Flow Velocity | 0.9 - 1.2 m/s |
| pH | 8.8 - 9.3 |
| Cycles of Concentration | 4 - 5 |

| Component | Make-up Water For Test | (mg/L) |
|---|---|---|
| Ca Hardness ($CaCO_3$) | | 60 |
| Mg Hardness ($CaCO_3$) | | 30 |
| Total Alkalinity ($CaCO_3$) | | 80 |
| $SiO_2$ | | 60 |
| pH | | 8.1 |
| Conductivity ($\mu$mhos) | | 600 |

(Both heated and unheated tubes were used consisting of a mixture of stainless, admiralty and mild steel.)

* Time required for one half of water volume to be replaced by new water.

**Field Trial Conditions.**  A field study was done by the addition of a supplemental silica inhibitor to the cooling water program already in use. At this site, the water source was from nine wells ranging in silica from 70 - 100 mg/L. Inhibition of silica deposition was measured by water analysis and visual observation of the on-site deposit monitor which was set with a heat flux of 28.125 $Kw/m^2$ (9,000 $Btu/ft^2/h$) and a flow rate of 0.9 m/s (3 ft/s). Fouling of the chiller condenser by $SiO_2$ deposition was measured by chiller efficiency studies. Table VII gives the operating conditions of the cooling tower used in the field test.

Table VII.  Operating Conditions of Field Test Site Cooling Towers

| | |
|---|---|
| Total Volume | 7570 L |
| Recirculating Rate | 2271 L/m |
| $\Delta T$ | 4.4 °C |
| Make-up Water analysis* | |
| $Ca^{2+}$ ($CaCO_3$) | 48-60 mg/L |
| $Mg^{2+}$ ($CaCO_3$) | 48-80 mg/L |
| $HCO^-_3$ ($CaCO_3$) | 100-132 mg/L |
| $SiO_2$ | 76-100 mg/L |
| Condenser Metallurgy | Copper |
| Flow Velocity | 1.2 - 2.1 m/s |
| Average Load | 0.492 x $10^6$ watts |
| Cycles with silica inhibitor | 4 |
| Cycles without silica inhibitor | 2 - 2.3 |

*Source was variable, as water came from twelve wells which provided different compositions and were not always used simultaneously or in same proportions.

**Results**

**Screening Test Results.** Some of the screening test data demonstrating the ability of a borate compound to inhibit amorphous silica deposition are given in Table VIII and in references 8 and 9. For comparison purposes, data for several other chemical treatments, frequently described as useful for silica control, are also provided. Unless specified, all treatments were evaluated at 100 mg/L.

The data show that boron compounds can be very effective in preventing deposition of amorphous silica. Polyhydroxy compounds, such as glycerine, have a modest level of activity, but are limited by other considerations that are found in cooling water systems, e.g., microorganism growth, chlorination and chemical reactivity.

Table VIII.  Silica Inhibition Screening Data

| Chemical Treatment (100 mg/L Actives) | Total Silica Filtered (mg/L SiO$_2$) |
|---|---|
| 1.  Boron-based inhibitor | 390 |
| 2.  BF$_3$ ethylamine complex | 360 |
| 3.  Glycerine | 260 |
| 4.  N-(Tris hydroxymethyl) methylglycine | 240 |
| 5.  EO/PO block polymer, MW-4400 | 240 |
| 6.  2000 MW polyacrylic acid | 240 |
| 7.  EO/PO block polymer, MW-2200 | 230 |
| 8.  Tris (hydroxymethyl) aminomethane | 230 |
| 9.  Mannitol | 230 |
| 10.  N,N,N',N'-tetrakis (2-hydroxypropyl) ethylene diamine | 210 |
| 11.  Pentaerythritol | 210 |
| 12.  Polyimine of approx. 50,000 MW | 210 |
| 13.  Aluminum citrate | 210 |
| 14.  Triethanolamine | 200 |
| 15.  2-nitro-2-ehtyl-1,3 propanediol | 190 |
| 16.  10,000 MW acrylamide/acrylic acid copolymer | 190 |
| 17.  Linear 1800 MW polyimine | 190 |

Surfactants that were tested had minimal activity. Polymers such as the polyacrylates of the type commonly used in cooling water applications were also ineffective. Further, both aluminum citrate and a commercially available 1800 molecular weight polyimine, which are reported in the literature (1,3,22,25-29) as being useful, were found to be ineffective in this test procedure.

Table IX indicates that the inhibition activity appears to be strictly related to orthoborate. A variety of orthoborate compounds or compounds that generate an orthoborate anion were able to inhibit silica deposition. In contrast, other boron compounds were ineffective. No activity was observed for metaborate, tetraborate and octaborate. Figure 5 is a photograph of reaction vessels, an untreated blank (left) and one treated with orthoborate, after standing overnight. The untreated system is cloudy and also contains precipitated amorphous silica (by X-ray). The orthoborate treated system is clear with only a minimal amount of precipitate.

According to theory, silica condensation should generate alkalinity. This was observed. The pH of both reaction vessels went up 0.8 units (from an initial pH of 8.3 to a pH of 9.1).

**Process Simulation Testing - Limit Testing.** An orthoborate compound was evaluated in a PCT as an amorphous silica scale inhibitor (8,30). In this study, orthoborate was used in addition to a conventional pH 7 chromate/zinc corrosion inhibition program. For comparison, a control study was run with the identical

Table IX.  Activity of Boron Compounds

| Chemical Type | Total Silica - mg/L SiO$_2$ |
|---|---|
| Orthoborate (5 sources) | 330-390 |
| Metaborate | 170 |
| Tetraborate | 210 |
| Octaborate | 180 |
| Blank | 170 |

program but without the silica inhibitor to establish the upper silica limit that could be achieved before water chemistry mass balance was lost. The limit for the control was found to be a total silica value by AA of 275 mg/L with a concurrent soluble (molybdate reactive) silica value of 225 mg/L. The results in Figure 6 show that, for these conditions, as little as 30 mg/L silica inhibitor can stabilize up to a total silica value of 425 mg/L with a concurrent reactive silica value of 315 mg/L before mass balance is lost.

**Advanced Process Simulation Studies - Steady State.**  This Pilot Cooling Tower experiment focused on the ability of orthoborate to inhibit silica deposition under steady state, high pH (8.0-9.5) cooling water conditions (30).

The criteria of success for these tests is slightly different from that of the Limit Test. In the Limit Test, soluble SiO$_2$ in the recirculating water was the determining factor. In the high pH steady state PCT tests, the average deposit rate on heat transfer tubes was the primary factor in determining success or failure. The soluble SiO$_2$ levels were also monitored. The deposit criteria (for heat transfer surfaces) used to evaluate the tests are given below:

| Metal | Deposit/Rate (mg/cm$^2$-year) |
|---|---|
| Mild Steel | 70 |
| Copper | 10 |
| Admiralty | 10 |

The deposit rate must be less than or equal to the stated rates to be successful. These deposit rates have been established based on PCT and field data correlations.

The initial PCT test (No. 1; Table X) was a blank to determine the amount of deposit formed in a high SiO$_2$ make-up (Table VI) water and establish baseline data. The corrosion and scale control program was an all-organic program, fed at 100 mg/L. No silica inhibitor was used in this test. The water chemistry during the two week test averaged 210 mg/L SiO$_2$ by reactive molybdate out of a theoretical of 300 mg/L (5 cycles) at pH 9.0.

Figure 5. Photograph of reactor vessels blank (left) vs orthoborate treated (right), after standing overnight.

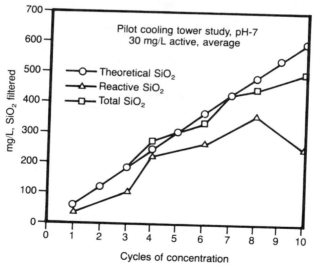

Figure 6. Pilot cooling tower limit test result for silica inhibition Reprinted with permission from ref. 30. Copyright 1987 National Association of Corrosion Engineers.

With no silica inhibitor in the system, heavy scaling resulted. The copper heat transfer surfaces averaged 151.1 mg/cm$^2$-year of deposit. The mild steel tubes averaged 74.8 mg/cm$^2$-year of deposit. Both results were considered unacceptable.

In the next PCT test (No. 2; Table X) the boron based silica inhibitor was used. The inhibitor feed rate was 75 mg/L. There was virtually no deposit or corrosion on any of the metals in the system. The theoretical $SiO_2$ level was again 300 mg/L, but averaged 240 mg/L in the water. The low $SiO_2$ value may be due to the inability of the molybdate reactive wet chemistry method to react with the 5-plus unit soluble silica polymers (oligomers). The deposit results on copper tubes averaged 1.1 mg/cm$^2$-year with mild steel 13.0 mg/cm$^2$-year. These results are considered excellent. For Test No. 2, there was insufficient deposit on the tubes for a deposit analysis, but a small amount of basin silt was found which consisted primarily of $SiO_2$. Figure 7 shows that the deposit obtained from Test No. 1, as computer calculated from X-ray fluorescence analysis, consisted primarily of amorphous silica and 'MgSiO$_3$' complexes.

In summary, not only did the orthoborate silica inhibitor provide acceptable results in comparison to defined deposition criteria, it provided approximately two orders of magnitude less deposition on copper and a 83% reduction in deposition on mild steel.

Table X.  Comparison of Average Water Chemistry and Deposit Rates for PCT Tests No. 1 and No. 2

| Test | pH | Conductivity** | Alkalinity* | Ca* | Mg* | SiO$_2$*** | Inhibitor*** |
|------|-----|------|------|------|------|------|------|
| PCT No. 1 | 9.1 | 3000 | 301 | 301 | 164 | 210 | 0 |
| PCT No. 2 | 9.1 | 2720 | 293 | 308 | 150 | 240 | 75 |

| | Test No. 1 | Test No. 2 |
|------|------|------|
| Metal | Deposit Rate mg/cm$^2$-year | Deposit Rate mg/cm$^2$-year |
| 1. Mild steel | 76.7 | 9.1 |
| 2. Mild steel | 73.0 | 16.8 |
| 3. Copper | 80.3 | 1.1 |
| 4. Copper | 87.6 | 0.4 |
| 5. Admiralty brass | 105.9 | 0.7 |
| 6. Copper | 153.3 | 1.5 |
| 7. Copper | 186.2 | 0.4 |
| 8. Copper | 248.2 | 2.2 |

  * mg/L as CaCO$^3$
 ** $\mu$mhos conductivity
*** mg/L as is

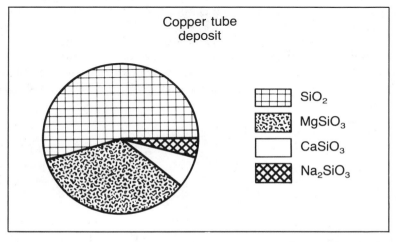

Figure 7. Pilot cooling tower heat transfer tube deposit analysis for test without borate inhibitor. Reprinted with permission from ref. 30. Copyright 1987 National Association of Corrosion Engineers.

**Field Trial Results.** A trial was conducted at a facility which was limited to 2-2.3 cycles of concentration in its air conditioning cooling water systems due to the use of high silica well water. By maintaining 125-140 mg/L of orthoborate based silica inhibitor in a test cooling water system, it was hoped to achieve two goals:

1. Maintain 300-340 mg/L $SiO_2$ in the recirculating water.
2. Maintain chiller efficiency.

Previously, this chiller system would have to be shut down and cleaned of silica deposition after one month if operated at 150-200 mg/L $SiO_2$. The trial lasted about two months. The $SiO_2$ levels averaged 295 mg/L molybdate reactive silica throughout the trial. The initial chiller efficiency studies indicated a clean system operating in excess of 100% efficiency. The high efficiency was achieved because of over design of the chiller condenser by the manufacturer to account for waterside fouling of the tubes. When the tubes are clean, 100+% efficiency could be achieved. This efficiency did not change throughout the trial. A summary of the water chemistry and efficiency data are given in Table XI and Figure 8 (30).

Table XI. Summary of Field Trial Data

| Date | % Efficiency | Ca* | Mg* | Total* Alkalinity | ** Conductivity | $SiO_2$*** | Inhibitor*** |
|------|--------------|-----|-----|-------------------|-----------------|-----------|--------------|
| 10/22 | 101.3 | 80 | 80 | 170 | 450 | 134 | — |
| 10/23 | 102.0 | 120 | 160 | 300 | 900 | 286 | 125 |
| 10/24 | 102.4 | 130 | 180 | 332 | 920 | 300 | 103 |
| 10/28 | 103.0 | 120 | 148 | 318 | 800 | 245 | 45 |
| 10/29 | 101.1 | 160 | 170 | 390 | 980 | 300 | 88 |
| 10/30 | 102.0 | 180 | 160 | 390 | 1000 | 350 | 147 |
| 11/01 | 101.4 | 184 | 200 | 440 | 1000 | 300 | 175 |
| 11/05 | 100.2 | 300 | 284 | 664 | 1450 | 360 | 210 |
| 11/08 | 102.2 | 388 | 346 | 800 | 2050 | 420 | 210 |
| 11/12 | 103.0 | 68 | 80 | 140 | 350 | 110 | 84 |
| 11/15 | 101.4 | 188 | 180 | 360 | 1050 | 240 | 158 |
| 11/18 | 101.0 | 184 | 180 | 396 | 1000 | 280 | 158 |
| 11/22 | 100.8 | 164 | 204 | 392 | 1000 | 375 | 115 |
| 11/25 | 102.0 | 184 | 200 | 428 | 1050 | 300 | 137 |
| 12/03 | 101.5 | 184 | 208 | 420 | 1050 | 281 | 63 |
| 12/06 | 102.0 | 184 | 212 | 420 | 1050 | 300 | 73 |

* mg/L Concentration as $CaCO^3$
** $\mu$mhos Conductivity
*** mg/L as is

The cooling system did not run smoothly during the trial. The system over-cycled several times because of problems with the conductivity controller. The

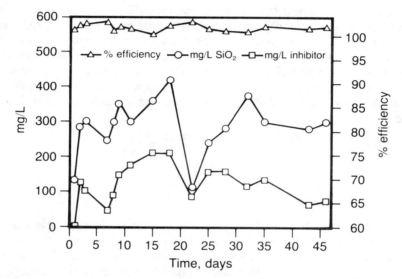

Figure 8. Silica scale inhibition study at field test site relating chiller efficiency with silica concentrations. Reprinted with permission from ref. 30. Copyright 1987 National Association of Corrosion Engineers.

highest silica concentration found in the tower was 420 mg/L (reactive molybdate). No fouling was found with either the chiller efficiency studies or by deposit formation on the deposit monitor tube.

These data support all other experimental work which showed that orthoborate is an effective inhibitor of silica deposition.

## Discussion

**Proposed Inhibitor Mechanism.** A brief review of boron chemistry shows that it has a chemistry similar to silica/silicic acid [$Si(OH)_4$] (31,32). The orthoborate ion in contact with water will convert from a trigonal geometry into a tetrahedron. From a structural viewpoint, this would allow the orthoborate ion to fit into the familiar tetrahedral coordinated silica lattices where it can interfere with silicic acid condensation as represented by equations 3-6. From a simple molecular weight and size perspective, this is quite reasonable. As shown below, the orthoborate anion is only slightly smaller in size and molecular weight than silicic acid.

| Species | Formula Weight |
|---------|----------------|
| $SiO_2$ | 60.1 |
| $B(OH)_3$ | 61.8 |
| $Si(OH)_4$ | 97.1 |
| $B(OH)_4^-$ | 78.8 |

This suggests that silica inhibition by orthoborate ion may be strictly a size-shape-steric phenomenon, but this is probably not entirely accurate. Unlike silica, the borate polymers that are formed in solution are very soluble. This suggests the possibility that the borate inhibitor prevents silica deposition through the formation of a more water soluble borate-silicate copolymer (oligomer). The amount of borate needed would be a function of the pH, amount of silica present, and the concentration of magnesium in the water. The size of the soluble borate-silicate (oligomer) would also be a function of pH and silica concentration. A mechanism that could allow orthoborate to function as both an inhibitor of the silicic acid dimerization step as well as allow formation of borate-silicate copolymers is possible through the use of Lewis acid and base concepts.

Within this context, "magnesium silicate" (at pH $\leq$ 11) is believed to be a co-precipitation of a positively charged colloidal metal hydroxide and negatively charged colloidal silica (21,22). Precipitation is prevented by having sufficient borate incorporated into the borate-silicate oligomer to stabilize the co-oligomer in the presence of magnesium.

This is illustrated by the field test results in Table XI which show that the silica to orthoborate ratio always varied from 1:1 to 5:1 with a typical value of 3:1. The mechanism by which orthoborate operates is probably kinetic and not thermodynamic in nature. The ability of orthoborate to inhibit silica condensation is probably greatest at the dimerization step (equations 3,6) but can probably occur, though less effectively, at other degrees of condensation. Evidence for this kinetic mechanism comes from observations of reactive molybdate silica analysis. Analysis of samples after several weeks storage show much lower reactive molybdate silica values than originally found at the time of sampling.

## Summary and Conclusions

The purpose of this work was to identify chemistry that would be useful for preventing silica deposition in cooling water systems and possibly geothermal applications. In this work, a bench-type screening test was developed that appears to be consistent with the known chemistry of silica condensation and cooling water physical parameters. Advanced process simulation testing and field test experiments confirmed the activity of the most highly ranked chemical agent based on the screening test. The most important conclusions are as follows:

1. Orthoborate has a unique ability to limit/inhibit condensation of amorphorus silica and thereby avoid silica deposition and scaling in cooling water systems.

2. Polymers (polyacrylates, surfactants) and phosphonates are not effective in inhibiting condensation of silica.

3. Process simulation test data and field test data show that orthoborate can stabilize up to 400 mg/L $SiO_2$ in the pH range of 7-9.5.

4. Other borate compounds are not effective in stabilizing $SiO_2$.

5. The proposed mechanism by which orthoborate operates is apparently kinetic in nature and seems to work through the formation of a water soluble borate-silicate oligomer.

## Acknowledgments

I would like to thank Ted Krol and the Nalco Laboratory Water Analysis Group for their analytical work, Daniel A. Meier who did the advanced steady state PCT work and brought laboratory technology to commercial application by carrying out the field testing, Dan Pruss who set up and operated the pilot cooling towers and Roy I. Kaplan who assigned me this project and supported my efforts. I also thank the National Association of Corrosion Engineers and John Wiley and Sons for permission to republish some of the data contained in this chapter.

## Literature Cited

1. Iler, R. K., The Chemistry of Silica; John Wiley and Sons, New York, N.Y., 1979; Chapters 1, 3, and 6.

2. Weres, O., Yee, A., Tsao, L. J. of Colloid and Interface Science. 1981, 84, No. 2, 379-402.
3. Weres, O., Yee, A. Tsao, L. Kinetics of Silica Polymerization (LBL- 7033, UC-4); May 1980, University of California, Earth Science Division, U.S. Dept. of Energy Contract W-7405-Eng. 48.
4. Dubin, L., J. of the Cooling Tower Institute. 1982, 3, No. 1, pps. 17-26.
5. Nancollas, G. H.; Kazmierczak, T. F.; Schuttringer, E. Corrosion. 1981, 37, No. 2, pp. 76.
6. Gill, J. S., Nancollas, G. H. Corrosion. 1981, 37 No. 2, p. 120.
7. Dubin, L., Fulks, K. E., Corrosion 84, Paper 118, 1984, National Association of Corrosion Engineers (NACE), Houston, Texas.
8. Dubin, L., U. S. Patent 4,584,104, 1986.
9. Dubin, L., U. S. Patent 4,532,047, 1985.
10. Kemmer, F. N. The Nalco Water Handbook; Crawford, H. B., Margolis, R. T., Eds.; McGraw-Hill, Inc., 1988.
11. Elliot, T. C., Ed., Power; Sp. Report "Cooling Towers", March 1973, pp S2-S24.
12. Elliot, T. C., Ed., Power; Sp. Report "Cooling Towers", Dec. 1985 pp. S1-S16.
13. Katzel, J. Ed., Plant Engineering; "Fundamental of Cooling Towers," April 27, 1989, pp. 32-38.
14. Montemarano, J., Water Technology; "Chemical Treatment Basics for Cooling Towers and Boilers", August 1987, pp. 40-43.
15. Alexander, G. B.; Heston, W. M.; Iler, R. K. J. Phys. Chem. 1954, 58, pp. 453-455.
16. Goto, K. J. Chem. Soc, Jap., Pure Chem. Sect. 1955, 76, pp 1364.
17. Weres, O., Yee, A. Tsao, L. Kinetics of Silica Polymerization (LBL-7033, UC-4); May 1980, pp. 25-33; University of California, Earth Science Division, U.S. Dept. of Energy Contract W-740-5-Eng. 48.
18. Marsh, A. R. III, Klein, G., Vermeulen, T,. Polymerization Kinetics and Equilibria of Silicic Acid in Aqueous Systems; Report LBL-4415, Lawrence Berkeley Laboratory, Oct. 1975.
19. Were, O., Yee, A., Tsao, L. Kinetics of Silica Polymerization (LBL-7033, UC-4); May 1980, p. 63; University of California, Earth Science Division, U.S. Dept. of Energy Contract W-740-5-Eng. 48.
20. Dubin, L., Dammeier, R. L., Hart, R. A. Materials Performance. Oct. 1985, pp. 27-33.
21. Iler, R. K., The Chemistry of Silica; John Wiley and Sons, New York, N.Y., 1979; pp. 162, 667.
22. Iler, R. K., The Colloid Chemistry of Silica and Silicates; Cornell University Press, Ithaca, N.Y., 1955; p. 182.
23. Iler, R. K., The Chemistry of Silica; John Wiley and Sons, New York, N.Y., 1979; pp. 40-43.

24. Reed, D. T., Nass, R. Proc. - Int. Water Conf. Eng. Soc. Western Penn., 1975, pp. 1-12.

25. Harrar, J. E.; Locke, F. E.; Lorensen, L. F.; Otto, C. H. Jr.; Deutscher, S. B.; Frey, W. P.; Lim, R. On-Line Test of Organic Additives for the Inhibition of the Precipitation of Silica from Hypersaline Geothermal Brine; April 3, 1979, University of California, Lawrence Livermore, U. S. Government Contract Report No. UCID-18091.

26. Harrar, J. E.; Locke, F. E.; Otto, C. H. Jr.; Lorrensen, L. E.; Moneco, S. B.; Frey, W. P. Society of Petroleum Engineers Journal. Feb. 1982, pps. 17-27.

27. Harrar, J. E.; Locke, F. E.; Otto, C. H. Jr.; Lorrensen, L. E.; Frey, W. P.; Snell, E. O. On-Line Test of Organic Additives for the Inhibition of the Precipitation of Silica from Hyersaline Geothermal Brine IV; Feb. 1980, University of California, Lawrence Livermore, U. S. Government Report No. UCID-18536.

28. Harrar, J. E.; Locke, F. E.; Otto, C. H. Jr.; Deutscher, S. B.; Frey, W. P.; Lorensen, L. E.; Snell, E. O.; Lim, R.; Ryon, R. W.; Quong, R. Final Report on Test of Proprietary Chemical Additives as Antiscalents for Hypersaline Geothermal Brine; Jan. 1980, University of California, Lawrence Livermore, U. S. Government Contract Report No. UCID-18521.

29. Harrar, J. E.; Locke, F. E.; Otto, C. H. Jr.; Lorensen, L. E.; Frey, W. P.; Snell, E. O. On-Line Tests of Organic Additives for the Inhibition of the Precipitation of Silica from Hypersaline Geothermal Brine III. Scaling Measurements and Tests of Other Methods of Brine Modification; University of California, Lawrence Livermore, U. S. Government Contract Report No. UCID- 18238.

30. Meier, D. A., Dubin, L., Corrosion 87, Paper 334, 1987, National Association of Corrosion Engineers (NACE), Houston, Texas.

31. Meutterties, E. L., The Chemistry of Boron and Its Compounds; John Wiley and Sons, New York, N.Y., 1967.

32. Cotton, A. F., Wilkinson, G. Advanced Inorganic Chemistry; John Wiley & Sons, New York, N. Y., 1980.

RECEIVED August 27, 1990

# INDEXES

# Author Index

# Affiliation Index

# Subject Index

## Other ACS Books

*Biotechnology and Materials Science: Chemistry for the Future*
Edited by Mary L. Good
160 pp; clothbound, ISBN 0–8412–1472–7, paperback, ISBN 0–8412–1473–5

*Chemical Demonstrations: A Sourcebook for Teachers*
Volume 1, Second Edition by Lee R. Summerlin and James L. Ealy, Jr.
192 pp; spiral bound; ISBN 0–8412–1481–6
Volume 2, Second Edition by Lee R. Summerlin, Christie L. Borgford, and Julie B. Ealy
229 pp; spiral bound; ISBN 0–8412–1535–9

*The Language of Biotechnology: A Dictionary of Terms*
By John M. Walker and Michael Cox
ACS Professional Reference Book; 256 pp;
clothbound, ISBN 0–8412–1489–1; paperback, ISBN 0–8412–1490–5

*Cancer: The Outlaw Cell,* Second Edition
Edited by Richard E. LaFond
274 pp; clothbound, ISBN 0–8412–1419–0; paperback, ISBN 0–8412–1420–4

*Chemical Structure Software for Personal Computers*
Edited by Daniel E. Meyer, Wendy A. Warr, and Richard A. Love
ACS Professional Reference Book; 107 pp;
clothbound, ISBN 0–8412–1538–3; paperback, ISBN 0–8412–1539–1

*Practical Statistics for the Physical Sciences*
By Larry L. Havlicek
ACS Professional Reference Book; 198 pp; clothbound; ISBN 0–8412–1453–0

*The Basics of Technical Communicating*
By B. Edward Cain
ACS Professional Reference Book; 198 pp;
clothbound, ISBN 0–8412–1451–4; paperback, ISBN 0–8412–1452–2

*The ACS Style Guide: A Manual for Authors and Editors*
Edited by Janet S. Dodd
264 pp; clothbound, ISBN 0–8412–0917–0; paperback, ISBN 0–8412–0943–X

*Personal Computers for Scientists: A Byte at a Time*
By Glenn I. Ouchi
276 pp; clothbound, ISBN 0–8412–1000–4; paperback, ISBN 0–8412–1001–2

*Chemistry and Crime: From Sherlock Holmes to Today's Courtroom*
Edited by Samuel M. Gerber
135 pp; clothbound, ISBN 0–8412–0784–4; paperback, ISBN 0–8412–0785–2

For further information and a free catalog of ACS books, contact:
American Chemical Society
Distribution Office, Department 225
1155 16th Street, NW, Washington, DC 20036
Telephone 800–227–5558